# Geophysical Monograph Series

Including

IUGG Volumes
Maurice Ewing Volumes
Mineral Physics Volumes

# GEOPHYSICAL MONOGRAPH SERIES

**Geophysical Monograph Volumes**

41 **Gondwana Six: Stratigraphy, Sedimentology, and Paleontoloty** *Garry D. McKenzie (Ed.)*
42 **Flow and Transport Through Unsaturated Fractured Rock** *Daniel D. Evans and Thomas J. Nicholson (Eds.)*
43 **Seamounts, Islands, and Atolls** *Barbara H. Keating, Patricia Fryer, Rodey Batiza, and George W. Boehlert (Eds.)*
44 **Modeling Magnetospheric Plasma** *T. E. Moore, J. H. Waite, Jr. (Eds.)*
45 **Perovskite: A Structure of Great Interest to Geophysics and Materials Science** *Alexandra Navrotsky and Donald J. Weidner (Eds.)*
46 **Structure and Dynamics of Earth's Deep Interior (IUGG Volume 1)** *D. E. Smylie and Raymond Hide (Eds.)*
47 **Hydrogeological Regimes and Their Subsurface Thermal Effects (IUGG Volume 2)** *Alan E. Beck, Grant Garvin and Lajos Stegena (Eds.)*
48 **Origin and Evolution of Sedimentary Basins and Their Energy and Mineral Resources (IUGG Volume 3)** *Raymond A. Price (Ed.)*
49 Slow Deformation and Transmission of Stress in the Earth **(IUGG Volume 4)** *Steven C. Cohen and Petr Vaníček (Eds.)*
50 **Deep Structure and Past Kinematics of Accreted Terranes (IUGG Volume 5)** *John W. Hillhouse (Ed.)*
51 **Properties and Processes of Earth's Lower Crust (IUGG Volume 6)** *Robert F. Merev, Stephan Mueller and David M. Fountain (Eds.)*

**IUGG Volumes**

1 **Structure and Dynamics of Earth's Deep Interior** *D. E. Smylie and Raymond Hide (Eds.)*
2 **Hydrogeological Regimes and Their Subsurface Thermal Effects** *Alan E. Beck, Grant Garvin and Lajos Stegena (Eds.)*
3 **Origin and Evolution of Sedimentary Basins and Their Energy and Mineral Resources** *Raymond A. Price (Ed.)*
4 **Slow Deformation and Transmission of Stress in the Earth** *Steven C. Cohen and Petr Vaníček (Eds.)*
5 **Deep Structure and Past Kinematics of Accreted Terranes** *John W. Hillhouse (Ed.)*
6 **Properties and Processes of Earth's Lower Crust** *Robert F. Mereu, Stephan Mueller and David M. Fountain (Eds.)*

**Maurice Ewing Volumes**

1 **Island Arcs, Deep Sea Trenches, and Back-Arc Basins** *Manik Talwani and Walter C. Pitman III (Eds.)*
2 **Deep Drilling Results in the Atlantic Ocean: Ocean Crust** *Manik Talwani, Christopher G. Harrison, and Dennis E. Hayes (Eds.)*
3 **Deep Drilling Results in the Atlantic Ocean: Continental Margins and Paleoenvironment** *Manik Talwani, William Hay, and William B. F. Ryan (Eds.)*
4 **Earthquake Prediction—An International Review** *David W. Simpson and Paul G. Richards (Eds.)*
5 **Climate Processes and Climate Sensitivity** *James E. Hansen and Taro Takahashi (Eds.)*
6 **Earthquake Source Mechanics** *Shamita Das, John Boatwright, and Christopher H. Scholz (Eds.)*

**Mineral Physics Volumes**

1 **Point Defects in Minerals** *Robert N. Schock (Ed.)*
2 **High Pressure Research in Mineral Physics** *Murli H. Manghnani and Yasuhiko Syono (Eds.)*

Geophysical Monograph 52
IUGG Volume 7

# Understanding Climate Change

**A. Berger**
**R. E. Dickinson**
**John W. Kidson**
*Editors*



American Geophysical Union
International Union of Geodesy and Geophysics

*Geophysical Monograph/IUGG Series*

**Library of Congress Cataloging-in-Publication Data**

Understanding climate change.

(Geophysical monograph; 52/IUGG series; 7)
1. Climatic changes—Congresses.
I. Berger, A. (André).
II. Dickinson, Robert E. (Robert Earl).
III. Kidson, John W. IV. Series.
QC981.8.C5U48 1989 551.6 89-6746
ISBN 0-87590-457-2

---



Printed in the United States of America.

# CONTENTS

## VI. MODELLING CLIMATE, PAST, PRESENT AND FUTURE

# PREFACE

The principal aim of this symposium was to describe the contributions which are made by each of the disciplines represented in the IUGG to the study of climate change. In order to present a balanced program, the Symposium was composed of invited reviews but other viewpoints were put forward during general discussion. The themes covered reflect the interests of the seven IUGG Associations and include volcanism; biogeochemistry; land hydrology; modeling climate, past and present; cryosphere; paleoclimates; land-surface processes; tropical oceans and the global atmosphere; clouds and atmospheric radiation; aeronomy and planetary atmospheres; and modeling future climate changes.

This symposium attracted enough papers for 3 full days plus an overflow in poster form. G.A. McBean opened the session with an overview of the main objectives of the World Climate Research Programme. This program is conceptually organized on the basis of three streams directed towards monthly and seasonal climate prediction, interannual variability and climate predictability out to decades. Interrelationships between WCRP and other geophysical programs were also noted.

The impact of the time dependent global distribution of aerosols after the Agung (1963) and El Chichon volcanic eruptions was investigated against natural variability and model assumptions. H. Sigurdsson stressed the importance of the type and mass yield of volatiles released during an eruption. Petrologic estimates of volatile degassing show, for example, that some eruptions have yielded of the order of $10^{10}$ to $10^{11}$ kg of sulfur, chlorine and fluorine compounds to the atmosphere, with highest yield from Tambora in 1815, Laki in 1783, Eldgja in 934AD and Katmai in 1912.

The role of the circulation and biogeochemistry of the Atlantic Ocean in the determination of the precise airborne fraction of anthropogenic emission of $CO_2$ to the atmosphere was underlined by B. Bolin. The processes affecting the ocean for the explanation of the ice age $CO_2$ changes, like the biotic effect on upper ocean chemistry, was discussed by J.R. Toggweiler. In this respect, the presence of unutilized nutrients in high latitudes and the production of refractory dissolved organic compounds by marine organisms seem to be two sensitive points in the ocean's biological system.

S.E. Nicholson examined the current state of knowledge of African climate variability and drought. Paleo-, historical and recent data conclude that surface feedback is likely linked to drought in the Sahel, while sea-surface temperature anomalies appear to be more clearly linked with variability in equatorial and southern regions than in the Sahel. The crucial roles of soil moisture and vegetation in the hydrological processes were discussed by Duzheng Ye through studies of climate models sensitivity. Deforestation and large-scale irrigation have profound effects on the climate, the influence of the hydrological disturbances being not necssarily limited to the disturbed region.

P.R. Rowntree reviewed the present capability of 3-D models of the general circulation of the atmosphere and ocean to simulate existing and past climates. These experiments call for consideration of the realism of simulation of air-sea interactions, of cloud optical properties, of land-surface processes, of snow accumulation on ice caps and of sea-ice distributions. The leading candidates for explaining warm global Cretaceous climate (which are the very different continental geometry, the high eustatic sea level, the higher $CO_2$ atmospheric concentration and the greater role of the ocean in poleward heat transport) were tested by E.J. Barron using a variety of climate models. Water isotopes and desert dust particles were included by S. Joussaume in the LMD general circulation model to better simulate the last glacial maximum and to help transfering paleo-data into atmospheric parameters.

The role of land ice in climate was presented by M. Kuhn who showed that one third of this century's sea level rise are caused by melting mountain glaciers whereas polar ice did not give an appreciate contribution. The non linear W.R. Peltier's model, which incorporates the physics of the isostatic sinking of the Earth under the weight of the ice, was shown to explain the dominant $10^5$ year cycle in the Pleistocene ice volume, as a subharmonic resonant relaxation oscillation which characterizes its response to realistic astronomical forcing.

Paleoclimatic data are now recognized to play an important role in climatic research by providing a multiple opportunity to test the sensitivity of our climate models and so to better understand the behavior of the climate system. J.Cl. Duplessy has reviewed 3 of the major recent developments in paleoceanography: (i) the deep water circulation can now be reconstructed, in particular during the last glacial maximum and the peak of the last interglacial; (ii) carbon isotope records of planktonic and benthic foraminifera show that over the last glacial-interglacial cycle the atmospheric $CO_2$ varied in phase with the global climate, in agreement with the ice core data; (iii) accelerator mass spectrometer has provided a detailed chronology of the retreat of the polar front during the last deglaciation. Variations in climate parameters over the last 150,000 years as deduced from the Vostok ice core were given by Cl. Lorius. They concern Antarctic surface temperature, continental and marine aerosols and atmospheric $CO_2$. Spectral analysis of the isotope temperature and $CO_2$ profiles show clearly peaks around the orbital frequencies. Besides a long term trend, rapid $CO_2$ variations are also recorded in the ice cores, and data from Siple (Antarctica) has allowed to reconstruct the atmospheric $CO_2$ increases since 1800, as clearly shown by H. Oeschger.

Understanding the land-surface processes is one of the high priority in climatic research. The first results, reviewed by J.Cl. André, of the HAPEX-MOBILHY program show that variability within the experimental zone of $10^4$ $km^2$ is due to differences in vegetation cover and crop development, and that the water budget locally balances accurately at the monthly time scale. The climate-soil-vegetation interactions were also investigated by P.S. Eagleson using simplified physical formulations.

Variations in tropical sea-surface temperatures, of which the El Niño phenomenon in the Pacific is the best example, and the associated changes in the global atmospheric circulation were discussed by K.E. Trenberth, J.M. Wallace and D. Halpern. The historical record shows that ENSO events have certain features that evolve in a systematic fashion, but that each event has also its individual character. The southern oscillation (worldwide) and the El Niño (more regional) phenomena are undoubtedly related but they do not exhibit a one-to-one relationship. This is why, although the mechanisms responsible for these large fluctuations in the climate system are beginning to be understood in general terms, there remain many fundamental questions that will have to be answered before their amplitude and duration can be forecasted. Given the impact of large scale SST anomalies, reliable estimates of their evolution up to several months in adance is a goal of the TOGA program. But achieving this goal requires increased understanding of the roles of small scale near surface mixing processes, of upper ocean 3-D circulation and tranport field and of the air-sea fluxes in changing SST. All the data related to these processes must thus be recorded (mainly from satellites) at the same time that realistic ocean general circulation models developed.

The Earth's radiation budget plays a crucial role in understanding the Earth's climate and potential changes caused by external and internal forcings. The Earth Radiations Budget Experiment will thus be of primary importance in particular for validating radiative transfer and GCM's, as emphasized by B.R. Barkstrom, but also for cloud studies. These are the primary objective of the International Satellite Cloud Climatology Project described by E. Raschke who reviewed also some other national and international research projects related to clouds.

The uncertain feedback effects of cloud-radiative interactions was also claimed by V. Ramanathan as being one of the largest source of errors in model estimates of the trace gas impacts on climate. Ramanathan showed that the non-$CO_2$ trace gases in the atmosphere are now adding to the greenhouse effect by an amount comparable to the effect of $CO_2$ increase and that the trace gas warming will become large enough to rise above the background climate "noise" before the end of this century. The significance of chemical-radiative-dynamical interactions to climate response has also been underlined.

Solar-terrestrial and aeronomic interactions with the Earth's global climate played a key role throughout the evolution of our planet. J.Cl. Gérard demonstrated that the apparent stronger response of the precambrian atmosphere to solar cycle activity was possibly the consequence of a lower $O_2$ atmospheric content and a less developed stratosphere. G.E. Hunt showed that the fundamental processes which govern the wide range of atmospheric motions in the planetary atmospheres are often analogous to processes controlling the circulation of the Earth's atmosphere and oceans and therefore the climate of our planet. In relation to the impacts of the stratospheric ozone to the surface climate and the recent Antarctic ozone "hole", C. Leovy reviewed the modes of transport of trace gases into the lower stratosphere and upward into the photochemically most active regions near 40 km.

The last session was devoted to modeling our future climate. S. Manabe discussed one interesting feature of the southern ocean circulation in relation with the air-sea-ice interactions and with the interhemispheric asymmetry in the transient response of climate. A hierarchy of numerical climate models of varying complexity are now available to study a problem which may affect global climate: W. Washington has illustrated their potential to simulate changes in the climate system which may result from increase in $CO_2$ and/or in other trace gases. Finally M. Schlesinger analyzed the feedbacks which amplify the direct radiative forcing of the increased $CO_2$ concentration into three different kinds of models: EBMs, RCMs and GCMs. He concluded that such an analysis is a useful method of model intercomparison that provides insight on the causes of the differences in the models' simulated $CO_2$-induced warming.

All the sessions were very well attended except the poster session. The nature of the papers and the discipline of the attendees clearly demonstrate that the study of climatic changes is a multidisciplinary one involving all branches of geophysics and others, like chemistry and biology, in particular. It remains thus one of the most relevant examples for global changes studies within an International Geosphere-Biosphere Program. As is customary, these papers were all reviewed by the authors' peers in order to ensure their strict adherence to the high standards of international research publishing. The IUGG editors express their appreciation to all the scientists who gave time and energy as referees for this expression of the important work done throughout the International Union of Geodesy and Geophysics.

A. Berger is from the Université Catholique de Louvain, the Institute of Astronomy and Geophysics G. Lemaître in Louvain-la-Neuve, Belgium; R. E. Dickinson is from National Center for Atmospheric Research in Boulder, Colorado and John W. Kidson is from New Zealand Meteorological Service, Wellington, New Zealand.

# Section I

# WORLD CLIMATE RESEARCH PROGRAMME

# THE WORLD CLIMATE RESEARCH PROGRAMME

Gordon A. McBean

Canadian Climate Centre, Institute of Ocean Sciences, Sidney, B.C., Canada
and
Vice-Chairman
WMO/ICSU Joint Scientific Committee for the WCRP

## Introduction

In 1979, the Global Atmospheric Research Programme (GARP), jointly sponsored by the International Council of Scientific Unions and the World Meteorological Organization, culminated with the Global Weather Experiment. In the same year, the First World Climate Conference was held in Geneva and focussed the scientific community's attention on the importance of the variability and possible change of the earth's climate. The result was the establishment of the World Climate Programme, which has components for applications, data, impacts and research.

The objectives of the WMO/ICSU World Climate Research Programme (WCRP) are to determinine:
- to what extent climate can be predicted; and
- the extent of man's influence on climate.

To achieve these objectives, it is necessary (WMO, 1984):

(a) to improve our knowledge of global and regional climates, their temporal variations, and our understanding of the responsible mechanisms;

(b) to assess the evidence for significant trends in global and regional climates;

(c) to develop and improve physical-mathematical models capable of simulating and assessing the predictability of the climate system over a range of time and space scales; and

(d) to investigate the sensitivity of climate to possible natural and man-made stimuli and to estimate the changes in climate likely to result from specific disturbing influences.

## Strategy and Structure of the WCRP

Whereas GARP focussed primarily on atmospheric dynamics, the WCRP must consider the fully-coupled, global climate system (atmosphere, oceans, lithosphere and cryosphere) which varies on essentially all time and space scales. At the beginning of GARP, the range of deterministic weather prediction was 1-2 days; by the early 1980's, improvements in observing systems and forecast models had extended this to 5-7 days and forecasts up to two weeks are deemed possible. This provides a shorter time scale limit for the climate program. In terms of impact on national and international policies, the upper limit has been set at several decades. Thus, the WCRP is concerned with time scales of weeks up to one century. To focus the activities more sharply this wide range has been divided into three streams (Houghton and Morel, 1984).

The First Stream is concerned with extended-range (monthly to seasonal) weather forecasting. In certain instances, such as blocking situations, the atmosphere appears to have a longer period of predictability. Further, the influence of surface boundary conditions, such as sea surface temperature and soil moisture anomalies, provide a basis for extended prediction. To take advantage of this situation we need to improve: the internal dynamics and thermodynamics of atmospheric models; our capability of observing surface boundary conditions; and modelling of their interactions with the atmosphere. If we consider the earth as four interacting components (Figure 1), then the First Stream of the WCRP considers the atmosphere and its direct, short-time scale interactions with the other components.

The next major time scale of change, the Second Stream, is interannual variability for which the strongest signal appears to be the interaction of tropical oceans with the global atmosphere. Secondary effects may occur due to variations in sea-ice boundaries. Initial studies have shown that a global atmospheric model interacting with the tropical oceans will provide useful predictions of climate variability on interannual timescales. Thus, Second Stream

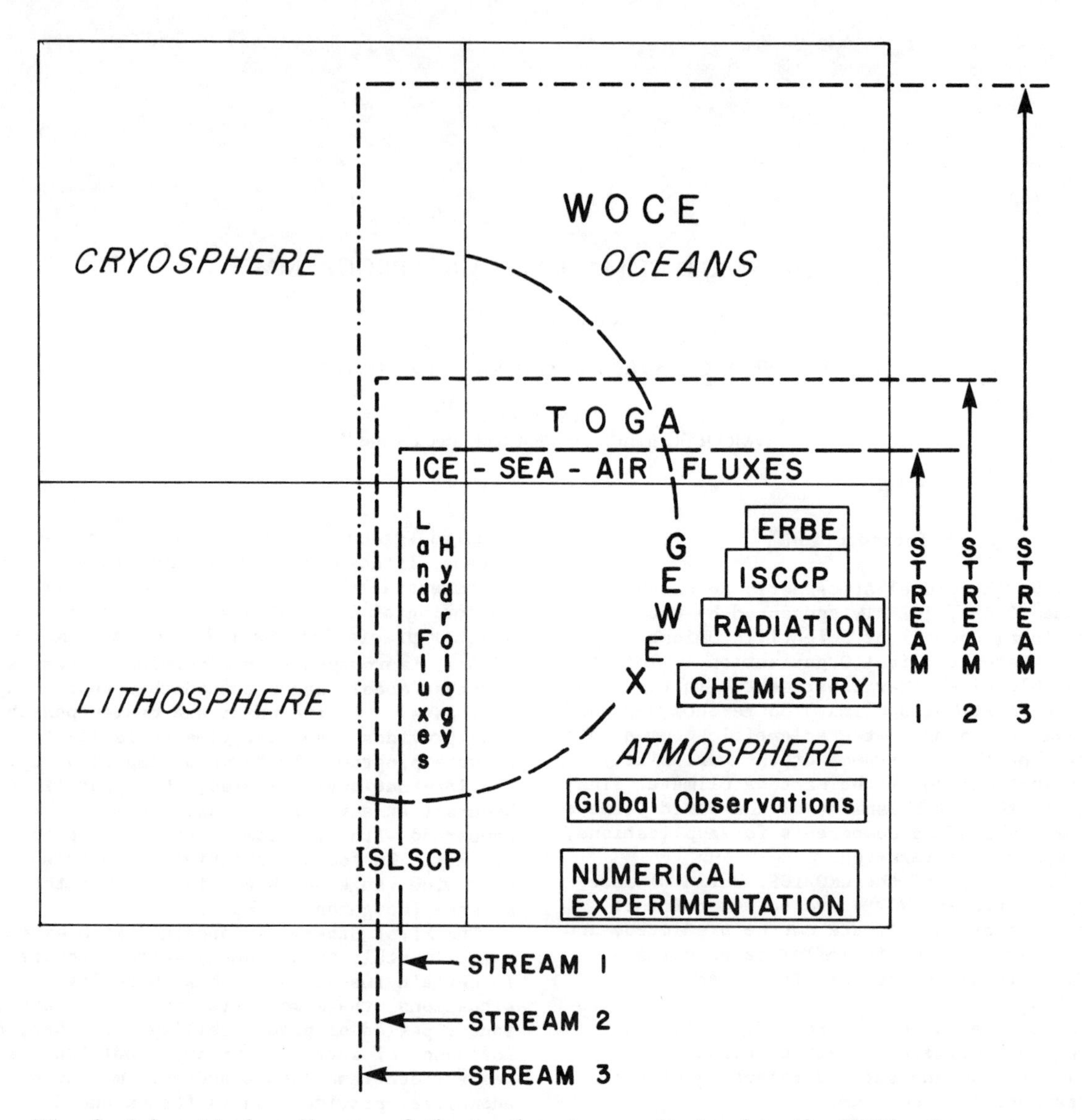


Fig. 1. Schematic box diagram of the earth sciences showing how the WCRP's three stream strategy includes study of increasingly larger fractions of the earth. The main WCRP activities are indicated on the boxes. See the text for further explanation.

research must include the tropical oceans to a depth of a few hundred metres.

In the Third Stream, for periods of decades and longer, we must deal with interactions of the atmosphere with the global ocean and sea ice, and impacts of changing atmospheric concentrations of radiatively-important gases. Some aspects of the oceans and major parts of the cryosphere and the lithosphere will remain beyond the scope of the WCRP.

It is important to stress that the strategy of the WCRP is a building block approach; everything required for the First Stream is required for the Second, and so on.

## WCRP Activities

### Numerical Experimentation

Numerical experimentation, including model development and data assimilation, is being used to test physical concepts and to identify modelling and observing system deficiencies. The numerical experimentation project is closely linked with numerical weather prediction activities. The areas of study include: systematic errors and climate drift in atmospheric circulation simulations; estimation of ocean-atmosphere fluxes; sensitivity of the atmospheric circulation to tropical sea-surface temperature anomalies; and formulation of cloud and radiation feedback processes. The Working Group on Numerical Experimentation is developing a strategy of model development for refining estimates of the climatic impact of radiatively-active gases.

Clouds and Radiation

Diabatic heating becomes more important as the length of the time integration increases. The WCRP has placed high priority on the study of clouds and radiative transfer. The International Satellite Cloud Climatology Project (ISCCP) was started in 1982 to provide consistent global data sets of cloud types, distribution and optical properties. Radiances from all operational meteorological satellites, except INSAT, are now being routinely collected and delivered to a Global Processing Centre. The Project also provides for periodic satellite-to-satellite radiance intercalibration. Another major activity has been the organization of inter-comparisons of radiation codes used in climate models. The Earth Radiation Budget Experiment has provided the first calibrated and global set of moderately high-resolution radiation flux data. One of the ERBE instruments has measured the solar constant as 1364.8 $Wm^{-2}$, with an accuracy of better than 0.2 %. A strategy for on-going measurement of global radiation fluxes is being developed.

Land-Surface Processes and Climate

Several studies (see Mintz, 1984) have shown that changes in soil moisture and evaporation rates can have significant impact on long-range forecasts over periods of one month or more. The interaction of the atmosphere with the land surfaces also has obvious important economic implications for agriculture and forestry. However, the extreme heterogeneity of the land surfaces has made parameterization of their interactions with the atmosphere very difficult to model. To study albedo, evapotranspiration, changes in soil moisture, runoff, etc., which are very complicated processes, the WCRP has initiated a series of Hydrological-Atmospheric Experiments (HAPEX). The first, HAPEX-MOBILHY (Andre et al., 1986), was conducted in a 100-km square area of southwest France in 1986. It is hoped that similar experiments will be conducted in seasonally snow-covered, boreal forest regions, in the tropical rain forests and in other representative land-surface areas. The WCRP is also cooperating with the International Satellite Land-Surface Climatology Project (ISLSCP, co-sponsored by IAMAP, COSPAR and UNEP). A series of satellite pixel-scale experiments are underway. The First ISLSCP Field Experiment (FIFE) was conducted in a 10-km square area of the central United States in 1987.

Global Observing Systems

The global observations needed for the WCRP must come from an enhanced World Weather Watch (WWW) and utilize space-based observing techniques whenever possible. The WCRP is cooperating with the WWW, the World Climate Data Programme, and others to improve the data bases for improving our knowledge of regional and global climates and their variability. In the last section of this report, still another initiative, a Global Energy and Water Cycle Experiment (GEWEX), built upon the next-generation of earth observing satellite system, will be discussed.

Tropical Oceans - Global Atmosphere - TOGA

The Tropical Ocean - Global Atmosphere Programme (WMO, 1985) has process and modelling studies and an enhanced observation programme including numerous tropical sea level stations and XBT lines, plus seven data management centres (level III atmospheric; sea surface temperature; precipitation; marine climatology; tropical upper-air winds; sea level; ocean sub-surface). TOGA started in 1985 and will continue until 1995. There are still some important data gaps, particularly over the Indian Ocean. Promising results are now being obtained from uncoupled tropical ocean models with realistic specified interface conditions and there is considerable optimism in the TOGA community of developing fully-coupled models. The TOGA Scientific Steering Group has recently agreed that the main thrusts of TOGA should be:

Thrust I: Development of an operational capability for dynamical prediction of the coupled tropical ocean/global atmosphere system beginning with the current state, that is, prediction of time-averaged anomalies up to several months in advance.

Thrust II: Exploration of the predictability of the longer time-scale climate variations of the tropical ocean/global atmosphere system on time scales of one to several years and understanding the mechanisms and processes underlying the predictability.

In view of the global impact of El Nino and other tropical ocean phenomena, there would be great benefits in being able to predict the development and evolution of an El Nino.

World Ocean Circulation Experiment - WOCE

It is clear that decadal and longer variations in climate depend critically on the behaviour of the global oceans. We can now see new space-based (altimeters and scatterometers) and in situ observing techniques to study the global oceans as was never feasible before. At the same time, advances in computing power and analytical techniques are making it possible to consider comprehensive global-ocean models. The World Ocean Circulation Experiment (WMO, 1986) has two basic goals:

I To develop models useful for predicting climate change and to collect the data necessary to test them; and

II To determine the representativeness of specific WOCE data sets for the long-term behaviour of the ocean, and to find methods for determining long-term changes in the ocean circulation.

The WOCE scientific plan identifies three main thrusts or "Core Projects":

(1) The Global Description - essentially to raise the level of description of the global oceans to that presently available for the North Atlantic, with the addition of the measurement of sea level and wind stress by satellite-borne systems.

(2) The Southern Ocean Experiment - concerned with the Antarctic Circumpolar Current and its interaction with the oceans to the north.

(3) The Gyre Dynamics Experiment - designed to support the development of eddy-resolving oceanic circulation models, with priority given to studying small-scale dynamical processes that may need to be parameterized in global ocean models.

The intensive WOCE observations will be concentrated within a five-year period in the early 1990s to coincide with the time- frame of the altimetry and scatterometer satellites. Some ship-borne surveys are already underway.

Polar Regions Programme

While the tropical regions are the main recipients of solar radiation, the polar regions are the main sink of radiative energy. To fully understand the global climate we must examine both. Variations in the extent and thickness of sea ice in the polar regions (Untersteiner, 1984) have an impact on both interannual and longer time-scale variability. The uptake of atmospheric carbon dioxide into the oceans occurs principally at high latitudes and so does the formation of deep ocean water. The Sea Ice and Climate Programme is the main WCRP activity in the polar regions but other studies are also relevant and hence included here. Towards the objective of developing fully interactive atmosphere-ice-ocean models, sea ice models will be tested with time-dependent atmospheric forcing fields (from observations) and specified oceanic heat fluxes for their ability to reproduce the time-dependent distribution of sea ice over periods of several years. The performance of atmospheric models in polar regions is being evaluated. The WCRP cooperates with the Arctic Ocean Sciences Board in its study of the Greenland Sea. The determination of the transports of mass, heat and ice through Fram Strait will provide a constraint on Arctic Ocean models. The air-ice-sea component of the Greenland Sea Programme will provide data for development and validation of sea ice models. Ways of measuring the vertical oceanic heat flux are being explored.

Although Arctic sea ice variations are reasonably well documented by existing observations, the situation in the Antarctic is much more difficult. In cooperation with SCOR and SCAR, an Antarctic sea ice project is being developed to coordinate ship-based hydrographic surveys, deploy suitably instrumented data buoys within the Antarctic sea ice zone and examine the formation of deep ocean and bottom water. These activities will constitute a part of or will be closely co-ordinated with the WOCE Southern Ocean Project.

The optical properties of clouds and their effect on radiative fluxes are very important to the heat balance of the polar regions. The ISCCP (see (ii) above) has a special project on developing algorithms for detection of polar clouds from space. Improvements in modelling the albedo of sea ice and snow are also essential.

The important hydrological and snow-albedo feedback processes at high latitudes are to be included in a boreal forest HAPEX of the Land-Surface Processes and Climate Programme (iii).

Global Energy and Water Cycle Experiment - GEWEX

The distribution, duration and amounts of precipitation constitute the most significant manifestations of climate for human activities and are sensitive indicators of climate change. Alterations of precipitation by a $CO_2$-induced climate change may have more socioeconomic impact than a temperature rise. However, at present, the representation of the water cycle in global weather and climate models is still too crude to allow useful projections of future conditions.

Looking ahead to the next decade, it is expected that there will be increased computer capability to deal with the finer-mesh models that will be required to deal adequately with clouds and precipitation. The WCRP's and others' programmes on clouds and radiation, numerical experimentation, land-surface processes and other areas will have greatly improved our understanding of the processes of the global water cycle.

There is now growing concern about major global changes. Water and energy cycles are critical and are the basis on which all other global cycles must rely.

In the period 1995 and beyond, it is expected that new space platforms will offer greatly increased opportunities for new remote sensing instruments and earth observation. They will make possible the deployment of much larger and higher-power instruments, such as active remote-sounding radars and lidars.

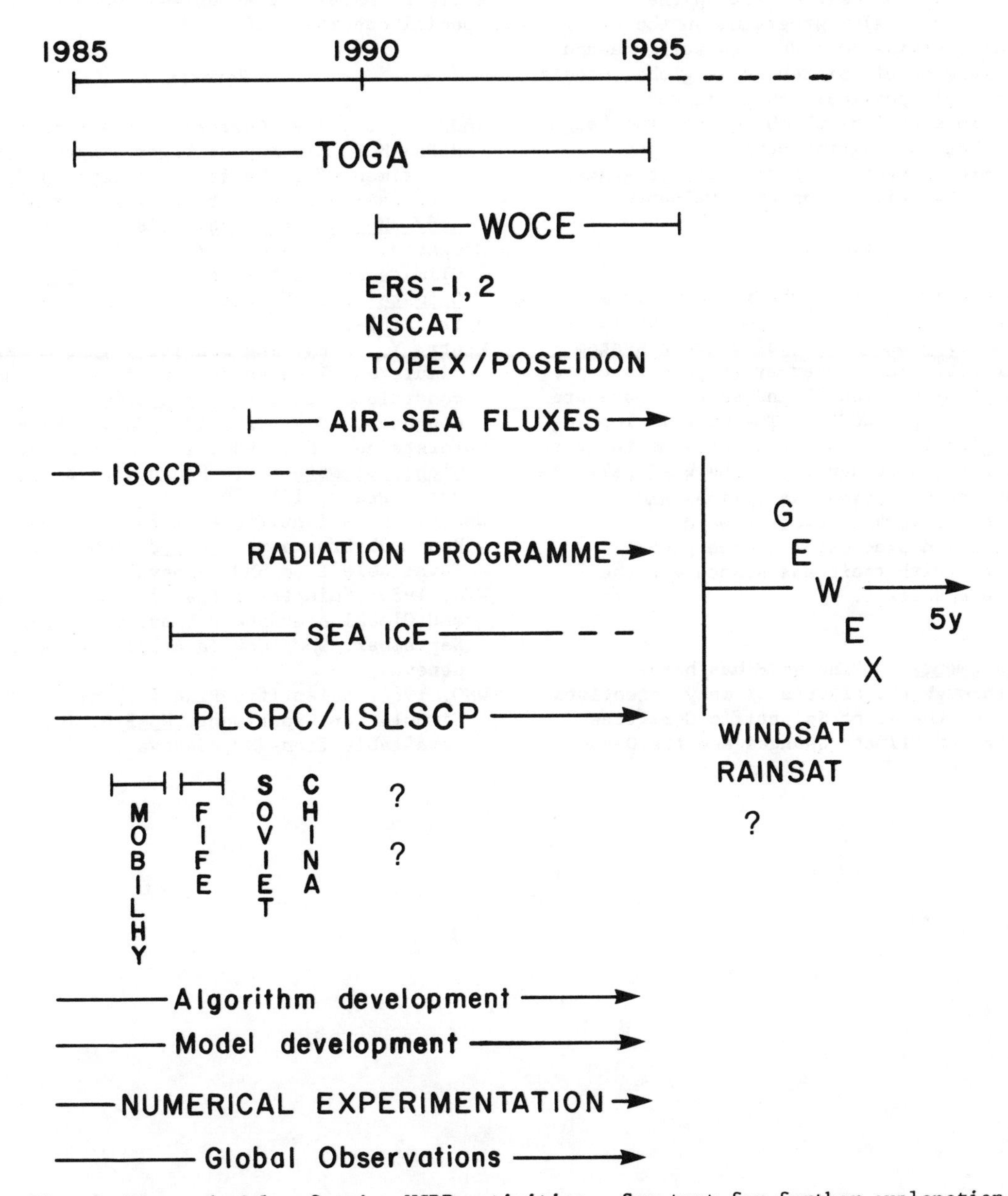


Fig. 2. Time schedule of major WCRP activities. See text for further explanation

The Joint Scientific Committee is now formulating a strategy for a Global Energy and Water Cycle Experiment, GEWEX, which would start in the period 1995+ and run about five years. The proposed objectives for GEWEX are:

I To describe and understand the transport of water (vapour, liquid and solid) and energy in the global atmosphere and at the underlying surface; and

II To develop methods of predicting changes in the distribution of water (vapour, liquid and solid) and energy within the global atmosphere and on the underlying surface, which may occur naturally or through the influence of man's activity.

Greenhouse Gases Programme

It is evident that changing concentrations of greenhouse gases (e.g., water vapour, carbon dioxide, ozone, methane, chloro-fluoro-carbons) will have a major impact on the earth's climate.

The WCRP is already developing models of radiative fluxes and their effect on the atmosphere. A new major programme on the water cycle is being envisaged. What is still needed is the development of comprehensive global models of the dynamical, physical, chemical and biochemical interactions which control the large-scale distribution of greenhouse gases and related chemical species. A specific programme towards this objective is under development.

## Summary

The World Climate Research Programme is a multi-component, endeavour to address the large-scale variability of the global climate system. Several activities are underway (Figure 2)(e.g., TOGA, HAPEX-MOBILHY, ISCCP) and several more are soon to start (e.g., WOCE). The WCRP is looking ahead to a global water cycle experiment to start before the end of the century. The WCRP calls upon scientists from many disciplines and nationalities to work together towards understanding and prediction of the global climate. Only with their assistance can the programme be a success.

Acknowledgments. The WCRP has been developed through the efforts of many scientists. The members of the Joint Scientific Committee, the Committee on Climate Changes and the Oceans of IOC/SCOR and their working groups and Prof. P. Morel, Director/WCRP have all made special contributions.

## References

Andre, J.C., J.P. Goutorbe and A. Perrier, 1986: HAPEX-MOBILHY, A hydrologic atmospheric pilot experiment for the study of water budget and evaporation flux at the climatic scale. Bull. Amer. Meteor. Soc., 67, 138-144.

Houghton, J.T., and P. Morel, 1984: The World Climate Research Programme. In The Global Climate (J.T. Houghton, ed.), Cambridge Univ. Press, 1-11.

Mintz, Y,. 1984: The sensitivity of numerically simulated climates to land-surface boundary conditions. In The Global Climate (J.T. Houghton, ed.), Cambridge Univ. Press, 79-105.

Untersteiner, N., 1984: The cryosphere. In The Global Climate (J.T. Houghton, ed.), Cambridge Univ. Press, 121-140.

WMO, 1984: Scientific Plan for the World Climate Research Programme. WCRP-2, September. WMO/ICSU (available from WMO, Geneva).

WMO, 1985: Scientific Plan for the Tropical Ocean and Global Atmosphere Programme. WCRP-3, September. WMO/ICSU (available from WMO, Geneva).

WMO, 1986: Scientific Plan for the World Ocean Circulation Experiment. WCRP-6, July, WMO/ICSU (available from WMO, Geneva).

# Section II

# PALEOCLIMATES AND ICE

# LONG-TERM CLIMATIC AND ENVIRONMENTAL RECORDS FROM ANTARCTIC ICE

[1]C. Lorius, [1]J-M. Barnola, [1]M. Legrand, [1]J. R. Petit, [1]D. Raynaud, [1]C. Ritz,
[2]N. Barkov, [2]Y. S. Korotkevich, [2]V. N. Petrov, [3]C. Genthon, [3]J. Jouzel,
[4]V. M. Kotlyakov, [5]F. Yiou, and [5]G. Raisbeck.

Abstract. Various records obtained from the Vostok (East Antarctica) ice core allow reconstruction of temperature, accumulation (precipitation), aerosol loading and atmospheric $CO_2$ concentration histories over the last climatic cycle (160 000 years). The results agree with those previously obtained from two other deep Antarctic ice cores going back to the Last Glacial Maximum.

The Vostok isotope-based temperature and $CO_2$ records show a large 100 ky signal with changes of the order of 10°C and 70 ppmv respectively. They are closely associated and show periodicities characteristic of the earth orbital parameters. These features suggest a fundamental link between the climate system and the carbon cycle and point out the possible role of $CO_2$, in addition to insolation inputs, in accounting for the observed temperature history.

The accumulation (precipitation) record appears to be governed by temperature with values during the coldest stages reduced to about 50 % of the current rate. Ice deposited during these coldest stages is also characterized by high concentrations of marine and terrestrial aerosols ; these peaks likely reflect strengthened sources and meridional transport during full glacial conditions, linked to higher wind speeds, more extensive arid areas on surrounding continents and the greater exposure of continental shelves. On the other hand there is no indication of a long term relationship between volcanism and climate.

## Introduction

Data on environmental and climatic changes having affected our planet in the past can be obtained from various marine and terrestrial sediments and from ice cores. Although ice-core studies have some inherent limitations (small number of available records, relative shortness of the period covered, difficulty of accurate dating), they also offer certain advantages and unique possibilities. In particular, they can provide high resolution records and access to the most important climatic parameters (temperature, precipitation, relative humidity and wind strength) as well as the past atmospheric composition including trace gases and aerosols of various origins.

This paper focuses mainly on the isotope, aerosol and $CO_2$ data from the Vostok ice core over the last 160 ky and their climatic interpretation. For the Last Glacial Maximum, these results essentially confirm those already obtained from the two other Antarctic deep ice cores (Byrd and Dome C) as shortly discussed for each of these parameters. Over the full glacial-interglacial cycle, the Vostok isotope-based temperature history is discussed in terms of spectral characteristics and in relation to $CO_2$, orbital forcing, precipitation and aerosol loading.

## The Vostok Ice Core

The Soviet Antarctic station of Vostok is located in East Antarctica (78°28'S and 106°48'E) at an elevation of 3490 m, with a mean annual temperature of - 55° C and a current snow accumulation of about 2.3 g $cm^{-2}$ $yr^{-1}$. An initial drilling went down to 950 m in successive steps between 1970 and 1974. In 1980, the drilling of a second deep hole was started. Data are now available on an almost continuous basis for the isotopic (deuterium) composition of the ice and on a discontinuous basis for impurities in ice and $CO_2$ contained in air bubbles down to a depth of 2083 m.

[1]Laboratoire de Glaciologie et de Géophysique de l'Environnement, B. P. 96 38402 St Martin d'Hères cedex (France).
[2]Arctic and Antarctic Research Institute, Beringa Street 38, 199236 Leningrad (USSR).
[3]Laboratoire de Géochimie Isotopique-LODYC (UA CNRS 1206) CEA/IRDI/DESICP/DPC 91191 Gif sur Yvette cedex (France).
[4]Institute of Geography, Academy of Science of the USSR, 29 Staronometny, Moscow 109107 (USSR).
[5]Laboratoire René Bernas, C. S. N. S. M. 91406 Orsay (France).

Establishing a reliable time scale is the first important step of data interpretation. For Vostok, this was established through a glaciological approach using a two-dimensional ice flow model, well suited to the Vostok area, and accounting for the change in snow accumulation with time [Lorius et al., 1985] ; this accumulation aspect is fully discussed in the following section. A key feature of the dating strategy is independence with respect to other paleo-series. This dating puts the 2083 m level at ~ 160 ky (with an accuracy of about 10-15 ky).

## The Climatic Record

The existence of a relationship between the isotope content of snow (deuterium or oxygen 18) and the temperature of the site is well documented over polar areas. Such a relationship results from the fractionation processes which take place during the atmospheric water cycle and is the basic tool for reconstructing past climatic conditions from ice cores. Our confidence in such temperature reconstruction for central East Antarctic deep ice cores relies on : i) a particularly well obeyed linear relationship observed between the annual averages of the surface temperature and of the snow isotope content in this part of Antarctica [Lorius and Merlivat, 1977], ii) the good agreement between observed slopes and the value derived from a one-dimensional isotope model [Jouzel and Merlivat, 1984]. For the deuterium content, upon which the following Vostok climatic interpretation is based, its value is 6 °/‥/°C (expressed in δD per mill with respect to S.M.O.W, the Standard Mean Ocean Water).

The deuterium profile is given in figure 1a with respect to depth (upper scale) and time (lower scale) with individual values representing averages on 1 to 2 m ice increments. Stages from A to H designate successive warm and cold periods. Stage A corresponds to the present Holocene interglacial ; stages B to F cover the last glacial with C and E being slightly warmer interstadials. Stage G characterizes the previous interglacial while H is the last part of the previous ice age.

Using a gradient of 6 °/‥/°C the deuterium record was interpreted in terms of temperature after correcting for the isotopic changes of sea water (fig. 1b). No corrections were applied for the possible influence of ice origin and ice cap thickness changes. The isotope temperature curve is given in figure 1c as a difference with respect to the present day value, after smoothing to filter out the very high frequency oscillations. The record obtained shows that the shift associated with the last deglaciation was around 9°C. This compares quite well with the 11°C change derived solely from the crystal size variation with depth [Petit et al., 1987], providing independent support for our temperature interpretation.

There is a very good agreement between Vostok, Byrd [Johnsen et al., 1972] and Dome C [Lorius et al., 1979] records over the last deglaciation and the Holocene periods. The temperature shift associated with this climatic transition is around 9°C both at Dome C and Vostok and its shape is remarkably similar for the two cores. A quantitative interpretation of the Byrd profile would lead to a quite comparable value as the associated $\delta^{18}O$ shift is very similar in the three cores. This supports the large geographical significance of the Vostok isotope temperature profile.

The main features of this record are in fact even of global significance, at least qualitatively. as suggested by the comparison with the marine

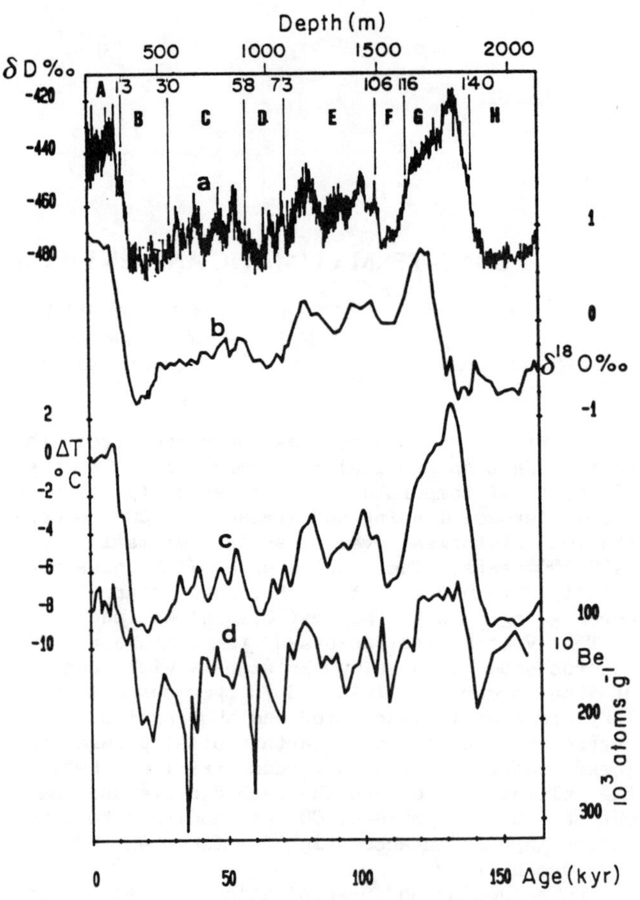


Fig. 1. a) Vostok isotope profile (deuterium content in °/‥ versus S.M.O.W with successive climatic stages A to $H_8$ as defined by Lorius et al. [1985] ; b) Marine $\delta^{18}O$ record of Martinson et al. [1986] ; c) Smoothed Vostok isotope temperature record expressed in °C as a difference with respect to current surface temperature value (from Jouzel et al.,1987] ; d) Vostok $^{10}Be$ concentration (from Raisbeck et al., 1987). Note inverted scale to facilitate com parison with climatic records. The upper scale gives the depth of the Vostok ice core ; the lower scale indicates the age (ky BP) of the various records.

$\delta^{18}O$ record (fig 1b) of Martinson et al. [1987] which is thought to essentially represent global continental ice volume changes. The two profiles correspond very closely down to 110 ky BP. However such consistency is no longer observed for the earlier part of the record (before 110 ky BP) as there is a major difference in the duration of the last interglacial estimated respectively from the ice and marine sediments. This likely results from relative uncertainties in both chronologies and does not affect the above conclusion on the representativeness of the Vostok record.

Aside from the large ~ 100 ky signal, visual inspection of the temperature curve (fig. 1c) also clearly shows a ~ 40 ky oscillation with four well marked temperature minima roughly in phase with the past total insolation at the Vostok latitude (governed by the obliquity cycle with a period of 41 ky). There are also similarities between this Vostok record and the 65°N July insolation changes which play a key role in the Milankovitch theory of the ice ages (this July 65° N insolation is largely influenced by precessional changes with periodicities of 19 and 23 ky). Spectral analysis [Jouzel et al., 1987] confirms these visual features showing that aside from the ~ 100 ky glacial-interglacial oscillation, the Vostok temperature record is dominated by a strong ~ 40 ky signal. It also suggests a lesser influence by a component slightly greater than 20 ky. The two frequency bands can be associated with the obliquity and precession cycles respectively, thus supporting the role of astronomical forcing in determining the late Pleistocene climate [Berger, in press], already convincingly demonstrated on the basis of deep sea core records [Hays et al., 1976].

Beyond this temperature record, Vostok data allow evaluation of past accumulation changes. As noted above, knowledge of this parameter is required for dating ; also, in a more global perspective growth and retreat of ice sheets largely depend on accumulation changes occurring over the glacial-interglacial time scale. One approach (the one used for dating) assumes [Robin, 1977] that the precipitation rate is governed by the amount of water vapor circulating above the inversion layer, itself controlled by the saturation vapor pressure, i.e. by the temperature [Lorius et al., 1985). From the Vostok temperature data we deduce that snow accumulation was quite similar during the Holocene and the previous Interglacial periods and reduced to about 50 % of the modern value during the coldest periods. Another approach [Raisbeck et al., 1981 ; Yiou et al., 1985] is based on the $^{10}Be$ profile (fig. 1d). At Vostok, concentrations of this cosmogenic isotope were quite similar during interglacials. The glacial-interglacial concentration changes have been interpreted as reflecting lower precipitation rates during cold periods. Indeed, apart from two beryllium peaks [Raisbeck et al., 1987], there is a good correlation between the precipitation rate estimated assuming a constant $^{10}Be$ deposition flux and that derived from temperature change. Although both may be affected by atmospheric circulation changes, the overall consistency between the two methods supports the estimate of past accumulation changes and also indirectly supports the deuterium temperature interpretation.

## $CO_2$ and Climate

Atmospheric $CO_2$ data obtained recently both from Greenland and Antarctic ice cores have shown in particular that concentrations during the Last Glacial Maximum were lower than Holocene values by 25-30 % and that the last deglaciation was characterized by a concentration increase from about 190-200 to 270-280 ppmv [Delmas et al., 1980 ; Neftel et al., 1982].

The Vostok ice core has made it possible to extend the record of past atmospheric $CO_2$ right through the glacial-interglacial cycle [Barnola et al., 1987]. Measurements performed at 66 different depth levels are separated by time intervals ranging from about 2 to 4.5 ky. Due to the gradual enclosure of atmospheric air in ice, air extracted from the core is younger than the age of the snow deposit and this difference (ranging from 2.5 to 4.3 ky depending upon climatic conditions) has been taken into account in establishing the $CO_2$ time scale. The best estimates of the $CO_2$ concentrations are plotted (fig. 2a) together with the associated uncertainty bands. The $CO_2$ concentration exhibits two very large changes between two levels centred around 190-200 and 260-280 ppmv with the low and high values associated with full glacial and interglacial conditions respectively. The low values previously recorded in other ice cores [Delmas et al., 1980 ; Neftel et al., 1982] are fully confirmed. The high level is comparable with the so-called "pre-industrial" $CO_2$ concentration which prevailed about 200 years ago. Indeed there is a remarkable correlation ($r^2$ = . 79) between the Vostok $CO_2$ and temperature records (figures 2a and 2b). Although some differences are observed between the two records, such as the absence of a low $CO_2$ value associated with the rather cold stage around 110 ky BP and the existence of a $CO_2$-temperature lag when going from warm to cold periods, these results nevertheless provide the first direct evidence of a close association between atmospheric $CO_2$ and climatic (temperature) changes on a glacial-interglacial time scale.

Although there are too few $CO_2$ data to study the phase relation with the isotope-temperature record, such a close association suggests that $CO_2$ may have, through its radiative effect and the associated feedbacks, participated in the glacial-interglacial change. Indeed a simple linear multivariate analysis suggests that $CO_2$ changes may have accounted for more than 50 % of the Vostok temperature variability, the remaining part being associated with orbital forcing [Genthon et al., 1987]. Qualitatively these results agree with the findings of Broccoli and Manabe [1987] who found that the low $CO_2$ concentration was the main cause

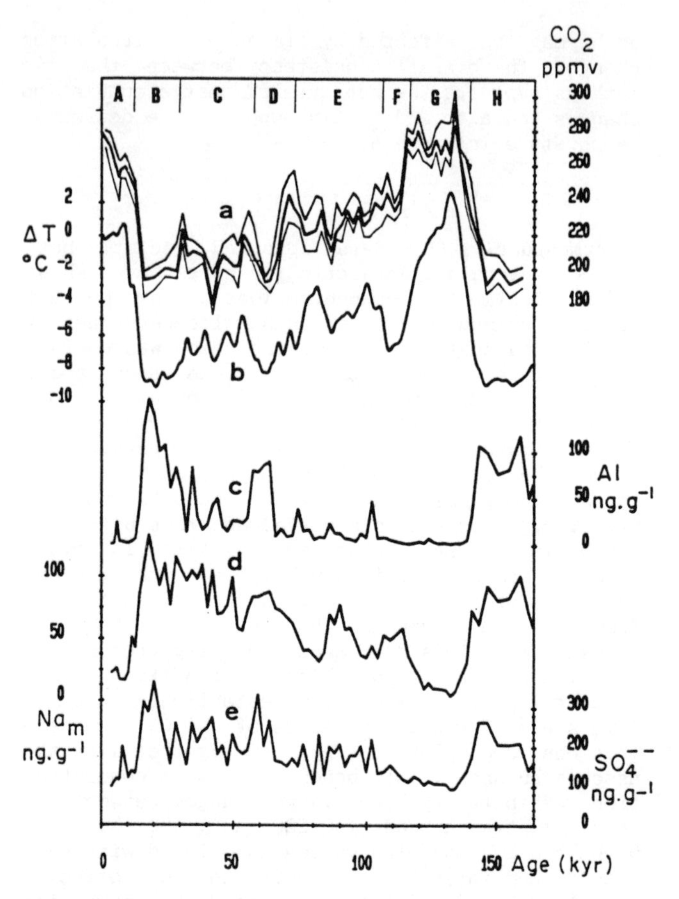


Fig. 2. Vostok ice core records; a) $CO_2$ concentrations (ppmv) with envelope of uncertainty (from Barnola et al., 1987) ; b) Smoothed Vostok isotope temperature record (as in Fig. 1c) ; c) Aluminium content (from Legrand et al., 1988) ; d) Marine Sodium content (from Legrand et al., 1988) ; e) Sulfate content (from Legrand et al., 1988).

of Last Glacial Maximum conditions in the southern hemisphere. A more quantitative agreement would likely be obtained by taking into account all the potential climatic feedbacks possibly combined with a polar amplification (see Genthon et al., 1987).

Our statistical approach [Genthon et al., 1987] is of course very simplistic. First non-linear interactions certainly exist between the different components of the climatic system [Imbrie and Imbrie, 1980 ; Le Treut and Ghil, 1983 ; Saltzman, 1987]. Second the role of the ocean, most likely important in linking Northern and Southern hemisphere climates, is not specifically taken into account, and neither are changes of sea ice extent which are also believed to have an important climatic role. It would also be useful to explore a wider range of possible forcings such as changes in the atmospheric optical depth resulting from variations in the aerosol loading [Rasool and Schneider, 1971] and in other trace gas concentrations [Wang et al., 1986] or, as recently suggested by Charlson et al. [1987], in the amount of cloud condensation nuclei which in the marine atmosphere are probably mainly formed from the oxidation products of the dimethylsulfide emitted by the ocean.

An important spectral characteristic of the $CO_2$ record is that, in addition to the 100 ky signal, it suggests a variance concentration around 20 ky [Barnola et al., 1987]. Beyond significant implications related to the understanding of modifications of the carbon cycle, this suggests that astronomical forcing is involved in both temperature and $CO_2$ changes, although the role of this process is complex and not yet understood.

Nevertheless the close correlation between $CO_2$ and temperature records and their spectral characteristics supports the idea that climatic changes could be triggered by an insolation input, with the relatively weak orbital forcing strongly amplified by possibly orbitally induced $CO_2$ changes. In particular these results suggest that the 100 ky glacial-interglacial oscillation may be related to the observed large $CO_2$ variations rather than to postulated non-linearities in ice sheet growth and decay forced by insolation changes. This interpretation can only be considered as tentative until backed by a clear understanding of the physical mechanisms involved. But in any case the results obtained from the Vostok ice core convincingly indicate that there is an interactive link between orbital forcing, $CO_2$ and climate, supporting Sundquist [1987] in his suggestion that further progress will require climate and the carbon cycle to be treated as part of the same global system rather than as separate entities.

## The Aerosol Record

The basic assumption for reconstructing past aerosol data from ice cores is that the concentration of impurities in the snow layers is directly related to the concentration in the atmosphere. A common characteristic of the Byrd [Cragin et al., 1977 ; Palais and Legrand, 1985], Dome C [Petit et al., 1981 ; Thompson and Mosley-Thompson, 1981] and Vostok [De Angelis et al., 1987 ; Legrand et al.,1988] cores is a much larger impurity concentration during the Last Glacial Maximum than during the Holocene. These changes, affecting both the continental and to a lesser extent the marine contributions, cannot be explained by accumulation variations.

For the input of dust mainly of continental origin the LGM/Holocene ratio is ~ 8 for Byrd (a core showing many ash layers which may have a local volcanic origin) ~ 27 for Dome C and up to 30 for Vostok. Marine aerosol concentrations were also higher for the LGM with LGM/Holocene ratios of about 2 to 3 for Byrd and ~ 5 both at Dome C and Vostok. Besides a possible effect connected with a lower accumulation these high glacial concentration values have been interpreted as resulting from strengthened sources and meridional transport linked to higher wind speeds (likely induced by higher temperature gradients with latitude), more extensive arid areas over the surrounding continents and the greater exposure of continental shelves due to a lower sea level [Petit et al., 1981]. More generally marine and terrestrial aerosol concentrations measured in ice are strongly dependent upon climatic conditions of global (source strength and atmospheric transport efficiency), regional (sea ice extent) and local (rate of snow accumulation) concern. Changes in continental and marine inputs and in all major ions is now documented over the full glacial-interglacial cycle [De Angelis et al., 1987 ; Legrand et al., 1988]. As an illustration, the Vostok aluminium record, an indicator of continental aerosol, is given in figure 2c as evaluated by Legrand et al., in press. This record is characterized by sharp increases during cold stages B, D and H and also by a slight increasing trend in the Al background over the last glacial period which may be explained by the emergence of large parts of the continental shelves due to sea level lowering [De Angelis et al., 1987] and also connected to the snow accumulation trend. As for the LGM, the sharp increase of continental input at the end of the previous ice age and during the cold stage D may be explained by expansion of arid areas, higher emissions due to higher wind speed over continents and more efficient atmospheric transport. There is not yet a clear explanation of the absence of large concentration values during stage F, but this suggests that global climatic conditions may have been different from full glacial during this stage at the beginning of the last glacial period.

The marine aerosol record (as illustrated by the Na marine concentration profile in fig. 2d) shows also higher values during cold climatic conditions superimposed on a well marked increasing trend for the concentrations during the last glacial period. In this case we observe a moderate increase of marine deposits during stage F, possibly related to climatic changes around Antarctica as depicted in the Vostok temperature record. In comparison, the $SO_4^{--}$ profile (fig. 2e) exhibits rather weak variations (up to a factor 3). Different sources contribute to the observed inputs ; continental dust and sea salt inputs are higher during cold climatic stages but the acidic-gas ($H_2SO_4$) derived contribution remains relatively stable over the whole climatic cycle, indicating the absence of a long-term relationship between volcanism and climate. More generally a comprehensive study of all major ions present in the ice points out significant changes in the aerosol loading associated with climate which may have some impact on the radiation balance of the earth atmosphere system. The mineral acid contribution represents a large part (over 50 %) of ice impurities deposited during interglacials. For glacial ice the inputs of marine and terrestrial salts become preponderant, representing up to 75 % of total soluble impurities [Legrand et al., 1988].

Acknowledgments. This work was supported in France by several institutions (Terres Australes et Antarctiques Françaises, CNRS-Institut National des Sciences de l'Univers and C.E.A) and in the USSR by the Arctic and Antarctic Research Institute. It is based on a drilling programme performed by Soviet Antarctic Expeditions. We are also grateful to the NSF-Division of Polar Programs who provided logistic support.

## References

De Angelis, M. , N.I. Barkov and V.N. Petrov, Aerosol concentrations over the Last Climatic Cycle (160 ky) from an Antarctic ice core. Nature, 325, 318-321, 1987.

Barnola J.M., D. Raynaud, Y. S. Korotkevich and C. Lorius, Vostok ice core : a 160,000 year record of atmospheric $CO_2$, Nature, 329, 408-414, 1987.

Berger A., Milankovitch theory and climate, Review of Geophysics, in press.

Broccoli, A.J. and S. Manabe, The influence of continental ice, atmospheric $CO_2$ and land albedo on the climate of the Last Glacial Maximum, Climate Dynamics, 1, 87-99, 1987.

Charlson, R.J., J. E.Lovelock, M. O. Andrae and S.G. Warren, Oceanic phytoplankton, atmospheric sulfur cloud albedo and climate, Nature, 326, 655-661, 1987.

Cragin, J. H., M. M. Herron, C. C. Jr. Langway, G. Klouda, Interhemispheric comparison of changes in the composition of atmospheric precipitation during the late cenozoic era. in Dunbar Maxwell J. (ed.), Polar oceans, Proceedings of the polar oceans conference, Montreal, 1974, Calgary, Arctic Institute of North America, 617-631, 1977.

Delmas, R.J., J. M. Ascencio and M. Legrand, Polar ice evidence that atmospheric $CO_2$, 20,000 yr BP was 50 % of the present, Nature, 284, 155-157, 1980.

Genthon C., J.M. Barnola, D. Raynaud, C. Lorius, J. Jouzel, N.I. Barkov, Y. S. Korotkevich and V. M. Kotlyakov, Vostok ice core : the climate response to $CO_2$ and orbital forcing changes over the last climatic cycle (160,000 years), Nature, 329, 414-418, 1987.

Hays, J.D., J. Imbrie and N.J. Shackleton, Variations in the Earth orbit : pacemaker of the ice ages, Science, 194, 1121-1132, 1976.

Imbrie, J. and J.Z. Imbrie, Modelling the climatic response to orbital variations, Science, 207, 243-253, 1980.

Johnsen, S. J., W. Dansgaard, H. B. Clausen and C. C. Langway, Oxygen isotope profiles through the Antarctic and Greenland ice sheets, Nature, 235, 429-434, 1972.

Jouzel,J. and L. Merlivat, Deuterium and oxygen 18 in precipitation. Modelling of the isotopic effect during snow formation, J. Geophys. Res., 89, 11749-11757, 1984.

Jouzel C., C. Lorius, J. R. Petit, C. Genthon, N.I. Barkov, V.M. Kotlyakov and V. N. Petrov, Vostok ice core : a continuous isotope temperature record over the last climatic cycle (160,000 years), Nature, 329, 403-408, 1987.

Legrand M., C. Lorius, N.I. Barkov, V. N. Petrov, Vostok (Antarctica) ice core : atmospheric chemistry changes over the last climatic cycle (160,000 yr), Atmospheric Environment, 22, 2, 317-331, 1988.

Le Treut, H. and M. Ghil, Orbital forcing, climatic interactions and glaciation cycles, J. Geophys. Res., 88, 5167-5190, 1983.

Lorius, C. and L. Merlivat, Distribution of mean surface stable isotope values in East Antarctica observed changes with depth in a coastal area. In : Isotopes and Impurities in Snow and Ice. Proc. Grenoble Symp. August-September 1975, IAHS N° 118, 127-137, 1977.

Lorius, C., L. Merlivat, J. Jouzel and M. Pourchet A 30,000 yr isotope climatic record from Antarctic ice, Nature, 280, 644-48, 1979.

Lorius, C., J. Jouzel, C. Ritz, L. Merlivat, N.I. Barkov, Y.S. Korotkevich and V.M. Kotlyakov A 150,000 year climatic record from Antarctic ice, Nature, 316, 591-596, 1985.

Martinson, D. G., N. G. Pisias, J. D. Hays, J. Imbrie, T. C. Moore and N. J. Shackleton, Age dating and the orbital theory of the ice ages : development of a high resolution 0 to 300 000 y. chronostratigraphy, Quaternary Research, 27, 1, 1-30, 1987.

Neftel, A., H. Oeschger, J. Schwander, B. Stauffer and R. Zumbrunn, Ice core sample measurements give atmospheric $CO_2$ content during the past 40,000 years, Nature, 295, 220-223, 1982.

Palais, J.M. and M. Legrand, Soluble impurities in the Byrd station ice core, Antarctica : their origin and sources, J. Geophys. Res., 90, 1143-1154, 1985.

Petit, J.R., M. Briat and A. Royer, Ice age aerosol content from East Antarctic ice core samples and past wind strength, Nature, 293, 391-394, 1981.

Petit, J. R., P. Duval and C. Lorius, Long-term climatic changes indicated by crystal growth in polar ice, Nature, 326, 62-64, 1987.

Raisbeck, G. M., F. Yiou, M. Fruneau, J. M. Loiseaux, M. Lieuvin, J.C. Ravel and C. Lorius Cosmogenic $^{10}$Be concentrations in Antarctic ice during the past 30 000 years, Nature, 292, 825-826, 1981.

Raisbeck G.M.,= F. Yiou, D. Bourles, C. Lorius, J. Jouzel, and N. I. Barkov, Evidence for two intervals of Enhanced $^{10}$Be Deposition in Antarctic Ice during the Last Glacial period, Nature, 326, 273-277 ,1987.

Rasool, S.I. and S. H. Schneider, Atmospheric carbon dioxide and aerosols : effects of large increases on global climate, Science, 173, 138-141, 1971.

Robin, G. de Q., Ice cores and climatic changes, Phil.Trans.R.Soc. Land., B 280, 143-168, 1977.

Saltzman, B., Carbon dioxide and the $\delta^{18}O$ record of late-quaternary climatic change : a global model, Climate Dynamics, 1, 77-85, 1987.

Sundquist, E., Ice core links $CO_2$ to climate, Nature, 329, 389-390, 1987.

Thompson, L. and E. Mosley-Thompson, Microparticle concentration variations linked with climatic change : evidence for polar ice cores, Science, 212, 4496, 812-815, 1981.

Wang, W.C., D. J. Wuebbles, V.M. Washington, R. G. Isaac and G. Molinar, Trace gases and other potential perturbations to global climate, Reviews of Geophysics, 24, 110-140, 1986.

Yiou, F., G. Raisbeck, D. Bourles, C. Lorius and Barkov N.I., $^{10}$Be at Vostok Antarctica during the last climatic cycle, Nature, 316, 616-617, 1985.

# THE ROLE OF LAND ICE AND SNOW IN CLIMATE

Michael H. Kuhn

Institut für Meteorologie und Geophysik, Universität Innsbruck
A-6020 Innsbruck, Austria

## Introduction

Some of the most distinct climatic contrasts like winter and summer or ice ages and interglacials are associated with changes in snow and ice at the earth's surface. The presence of ice on land has effects of a magnitude similar to or greater than those of vegetation or soil moisture as far as the boundary conditions for atmospheric circulation are concerned. In the hydrological cycle snow has a remarkable seasonal storing capacity. Ice sheets can increase surface elevation by kilometers, their development leads to sea level changes and on a 100 000 years time scale they despress the earth's crust and trigger lithospheric feedbacks.

Apart from their direct and indirect climatic influence glaciers and ice sheets are of interest to climatologists both in view of predicting future and unriddling past climatic situations: their slow motion at meters to kilometers per year, long turnover times and large thermal inertia make long term forecasts in the cryosphere more meaningful than in the atmosphere. On the other hand, moraines and other geomorphic traces of past ice extent have significantly promoted paleoclimatic reconstruction. Finally, the polar ice sheets contain ice that originated in the atmosphere hundreds of thousands of years ago. With the appropriate technology ice cores are now being used as archives containing information on temperatures, precipitation, dust and trace gas content of the atmosphere as far back as 150 000 years.



## An Inventory of Ice and Snow

Table 1 contains a summary of global ice masses and snow covered areas. The mass in long term storage is roughly $30 \cdot 10^{18}$ kg and if it were melted and evenly distributed over today's oceans would stand 83 m high. It would take roughly $10^{25}$ J to melt it, which, under the hypothetical assumption that this melting would be accomplished in 10 000 years, would require a heat flux density of 0.06 W $m^{-2}$ when averaged globally, or about 2 W $m^{-2}$ when averaged over the land area permanently ice covered today.

Of the global average annual precipitation of nearly 1 m (1000 kg $m^{-2}$) 6 per cent (60 kg $m^{-2}$) fall as snow. A correspondingly much higher amount of water is taken into seasonal storage in those areas actually affected by snow fall. A spring snow pack of 1 m water equivalent, which is not uncommon in midlatitude mountains, requires 64 W $m^{-2}$ in order to melt in a period of two months, while on a global, annual average solid precipitation turns over 0.6 W $m^{-2}$.

Both sea ice and seasonal snow cover extent fluctuate from year to year. Figure 1 shows a 1973-85 mean northern hemisphere snow cover of $32 \cdot 10^6$ $km^2$ with maxima ranging from 42 to $52 \cdot 10^6$ $km^2$. Note that November values are not well correlated with the seasonal maxima.

## Boundary Conditions

As a boundary of atmospheric circulation and of energy and mass exchange between air and soil, snow and ice covered surfaces have the following characteristics that distinguish them from bare soil:
- High albedo
- Potential evaporation
- Stable surface boundary layers

TABLE 1. Ice on Land

| | Area $10^6 km^2$ | Volume $10^6 km^3$ | Mass $10^{18} kg$ | Mean thickness km |
|---|---|---|---|---|
| Northern hemisphere | | | | |
| Greenland | 1.7 | 2.7 | 2.4 | 1.6 |
| Arctic islands | 0.35 | 0.2 | 0.2 | 0.6 |
| Mountain glaciers | 0.2 | 0.03 | 0.03 | 0.2 |
| Southern hemisphere | | | | |
| Mountain glaciers | 0.003 | 0.01 | 0.01 | 0.3 |
| Antarctica including ice shelves | 13.6 | 30.1 | 27.1 | 2.2 |
| Earth | 15.9 | 33.0 | 29.7 | 2.1 |
| Permanently snow or ice covered land | 16 | 11 % of continents | | |
| Maximum simultaneous cover on land (Northern hemisphere winter) | 53 | 35 % of continents | | |

- Strong thermal insulation, suppression of soil heat flux.

All of these properties make snow and ice colder than bare soil under comparable external conditions. Their combined climatic effect is:
- local temperature inversions
- baroclinic boundary layers and associated small scale circulation
- large scale differential heating and associated synoptic effects.

## The Surface Energy Balance of Snow and Ice

We shall treat the surface energy balance in the conventional way, i.e. express the distribution of energy gained from net radiation R to sensible heat flux H, latent heat flux of evaporation LE and heat fluxes associated with temperature changes S or melting M in the snow and underlying ground.

$$R = H + LE + S + M \qquad (1)$$

where the individual terms are most conveniently taken as energy flux densities ($W\ m^{-2}$).

Table 2 presents examples for melting snow and for Antarctic mid-winter conditions.

In these examples, the turbulent flux of sensible heat is directed towards the surface, supplying energy for snow melt (a and b) together with positive net radiation. In examples c and d, sensible heat replaces the energy loss of net radiation. In all four examples energy fluxes into the snow and evaporation from the snow surface play a subordinate role.

The formulation chosen here applies to the snow surface only. Since the snow pack is translucent to shortwave radiation and permeable to both air and liquid

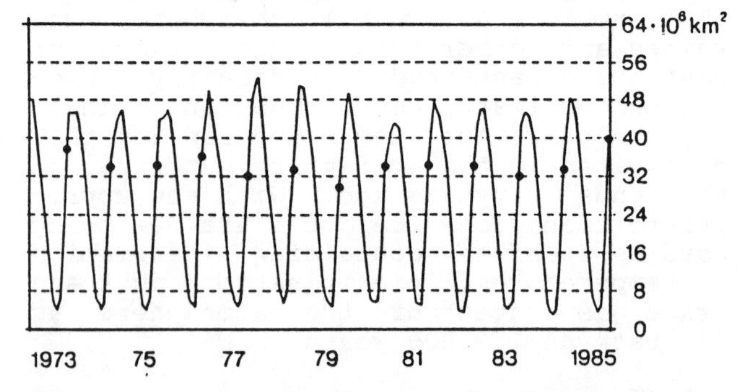


Figure 1. Time series of Northern Hemisphere snow cover area, in $10^6$ $km^2$. The dots mark the November snow cover for each year. After Ropelewski (1986).

TABLE 2. Examples of Snow Surface Energy Balances. R Net Radiation, H Sensible Heat, LE Latent Heat of Evaporation, and M, of Melting, S Heat Flux into Snow, all Expressed as Flux Densities in W $m^{-2}$.

| | R | = | H | + LE | + S + | M |
|---|---|---|---|---|---|---|
| a) Equilibrium line at Hintereisferner, 2960 m, 15 July - 18 August 1971 | 66 | = | -32 + | 3 | + 0 + | 95 |
| b) Bad Lake, Saskatchewan, 27-30 March 1976 | 14 | = | - 8 + | 1 | + 2 + | 19 |
| c) Plateau Station (80°S, 40°E, 3625 m), June + July 1967 | -15 | = | -13 + | 0 | -2 | |
| d) Maudheim (75°S, 11°E, sea level), June + July 1951 | -23 | = | -15 | -4 | -4 | |

Sources: b) adapted from Male and Gray, 1981
d) Liljequist, 1957

water, energy and mass is being exchanged internally to the effect that the size of snow grains increases with age.

## Albedo

Increasing grain size, liquid water content and impurities lead to a lowering of snow albedo. As the snow metamorphism progresses faster at higher temperatures, an absorption-temperature feedback will accelerate the ripening of the snow pack. Reflection from snow does not take place in a specular fashion at the surface but mainly by multiple scattering within the snow. The emerging flux shows selective absorption by ice and liquid $H_2O$ bands and displays anisotropic spatial distribution.

Albedo changes accompanying the decay of an alpine snow pack were monitored with a spectral radiometer as shown in Figure 2. The curves show the change from high visible to low near infrared albedo associated with absorption bands in the vicinity of 1, 1.3 and 1.6 μm. As the grain size increases from less than 1 mm (27.3.) to nearly 5 mm (14.6.) diameter, near infrared albedo drops to values comparable to that of rock.

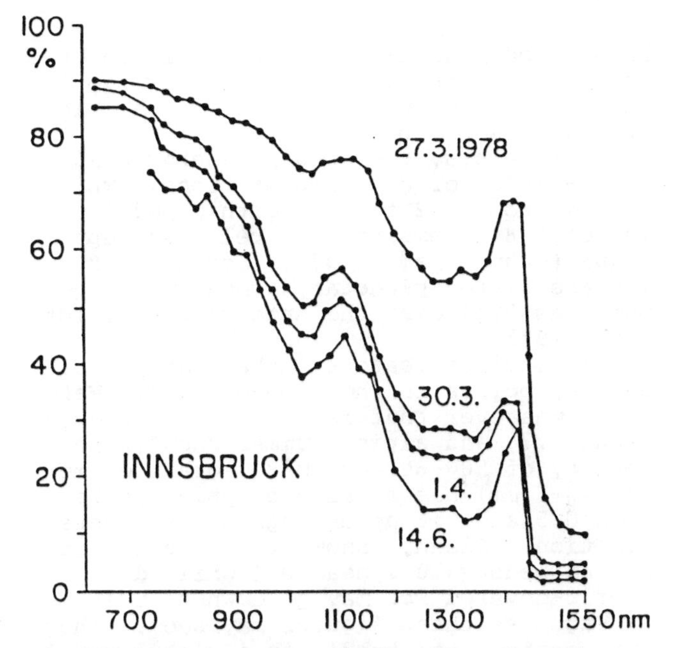


Figure 2. Spectral albedo of a decaying alpine snow field near Innsbruck at 1900 m. Grain diameters increased from less than 1 to nearly 5 mm from March to June.

In the dry snow facies of polar ice caps the scattering geometry at low solar incidence is far more important than metamorphism and grain size. Figure 3 shows the increase of albedo with sinking sun at four antarctic stations, Plateau and Charcot being situated in the dry snow facies while the other two are coastal stations with seasonal melting episodes.

The curves in fig. 3 are to be understood as hemispherical integrals over an anisotropic reflectance field. Measurements of bidirectionl reflectance (Kuhn 1985) of antarctic and alpine snow show reflectance enhancement by a factor of two, approximately, in the solar meridian.

Low sun and high albedo make measurements and modelling of polar radiation fluxes highly susceptible to errors: at 10 degrees solar elevation and an albedo of 0.8, the value of the direct component

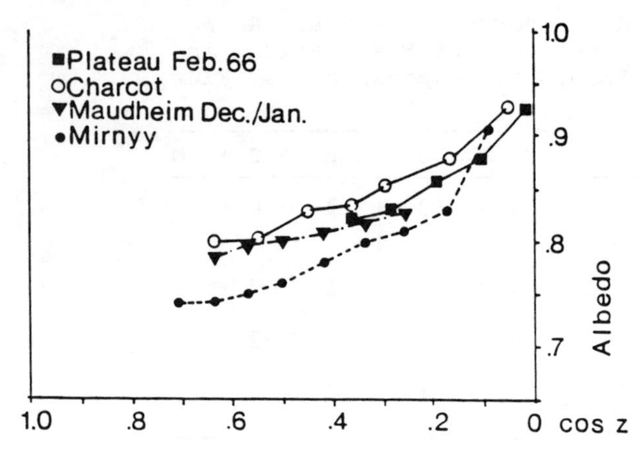


Figure 3. Increase of hemispherical albedo with increasing solar zenith distance Z, for four antarctic stations (Kuhn et al., 1977).

of absorbed shortwave radiation is in error by 10 per cent if either albedo is wrong by ± 0.02 or the level of the instrument is off by ± 1 deg.

Nonetheless, the annual, energy weighted albedo of dry snow stations, which ranges from 0.82 to 0.85 with a modal value of 0.84 seems to be a reliable upper value (Kuhn et al. 1977, Koerner 1980). The associated planetary albedo was computed as 0.74 over the Antarctic Interior (Kuhn 1975).

For obvious reasons, there are no other fix points in the albedo scale. "Wet" snow, whatever that is, will range widely about 0.7, and alpine summer snow (with a density of 500-600 kg $m^{-3}$, grain diameter 3-5 mm) has an albedo typically between 0.5 and 0.6 depending on dust contamination. Clean, snow free glacier ice has albedo values near 0.4 while dust or debris-covered ice may go down to 0.15.

Vague as these figures may seem, they are easier to handle in a global model than the spatially and temporally highly variable albedo of midlatitude model grid points, representing an area, say, 100 by 100 km. The work of Robinson and Kukla (1986) from which Figure 4 is taken, concentrated on late winter development of albedo of Eastern US surfaces. Figure 4 shows farmland albedo changing from dry snow values in February to a snow free 0.2, while forest albedo declines from 0.2 to 0.1 and a variety of surfaces range inbetween.

## Thermal Insulation by the Snow Cover

The equations of molecular heat conduction and ensuing temperature changes are sufficiently known. For given density, heat capacity and conductivity, the thermal diffusivity and the downward progression of a harmonic temperature change at the surface can be computed. Table 3 summarizes relevant values for snow, ice, sand and rock. The depth $D_a$ at which the amplitude of the annual temperature wave is damped to 1/e of its surface value illustrates the insulating properties of the material, low density snow being the best insulator of the examples given in Table 3. Higher frequency fluctuations penetrate less efficiently, the daily wave, for instance, to a depth $D_d = 365^{-1/2} D_a = 0.05 D_a$. Dry winter snow, which generally is of density less than 300 kg $m^{-3}$ thus is a very efficient insulator between soil and atmosphere. Compared to wet soil, rock or ice, low density snow effectively reduces subsurface heat storage. This situation can be grossly summarized by saying that the dry

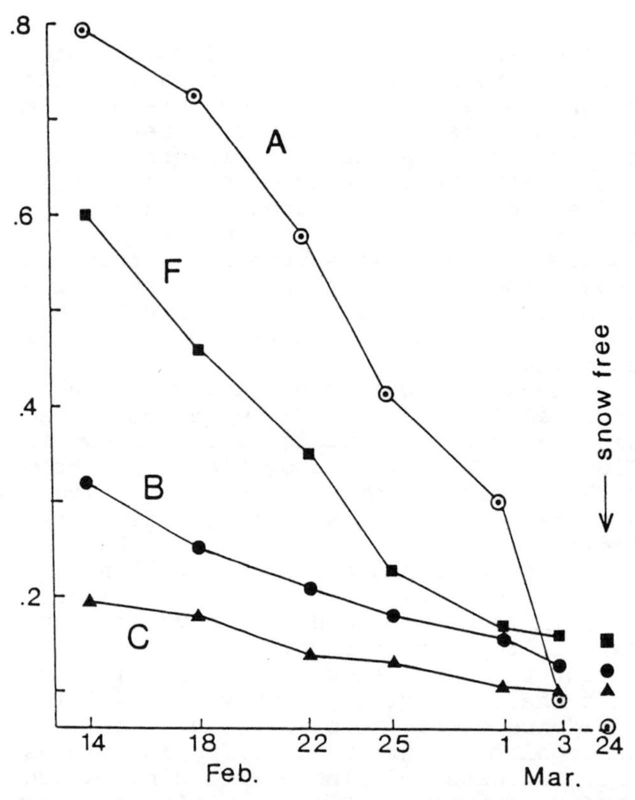


Figure 4. Albedo of major surface elements in southeastern New York and northern New Jersey under a variety of snow cover conditions: (A) dark soiled farmland, (B) deciduous forest, (C) mixed coniferous forest and (F) shrubby grassland. ( From Robinson and Kukla, 1984).

TABLE 3. Penetration of Annual Temperature Wave Damping of Amplitude to 0.37 of its Surface Value at Depth $D_a$

| Density $kg\ m^{-3}$ | Conductivity $W\ m^{-1}\ K^{-1}$ | Diffusivity $m^2\ s^{-1}$ | $D_a$ m |
|---|---|---|---|
| Snow | | | |
| 50 | 0.007 | $0.07 \cdot 10^{-6}$ | 0.8 |
| 100 | 0.03 | 0.15 | 1.2 |
| 300 | 0.27 | 0.45 | 2.1 |
| Ice | | | |
| 917 | 2.5 | 1.28 | 3.6 |
| Sand | | | |
| dry | 0.29 | 0.22 | 1.5 |
| 40 % water | 2.2 | 0.76 | 2.7 |
| Rock | | | |
| granite | 3.7 | 1.9 | 4.4 |

snow energy balance is an unbuffered, quasi-instant conversion of sensible heat gain into radiative heat loss (see Table 2, c and d) and vice versa.

To give an arbitrary example, I found the February heat content of an alpine snow pack at 3000 m elevation to be only 15 MJ $m^{-2}$ less than that of a zero-degree snow pack of same mass, a difference that can be restored by the energy gain of only two summer days.

## Stability of Cryospheric Boundary Layers

The thermal decoupling of cold surfaces from the warmer atmosphere is enhanced by a temperature-stability feedback. Consider the midwinter situation of antarctic energy budgets where, approximately R = H (see Table 2), net radiation loss being a function of surface temperature $T_o$, and H being determined by the free atmosphere - surface temperature difference $T_a - T_o$ as well as by a stability dependent a transfer coefficient $\alpha_H$.

$$R = L\downarrow - \sigma T_o^4 = \alpha_H (T_o - T_a) \qquad (2)$$

A disturbance in the radiative forcing (where $L\downarrow$ is the longwave downward flux) will lead to a change in $T_o$ that includes a gain of several per cent from the stability dependence of $\alpha_H$.

## Synoptic and Large Scale Effects

While local modification of the surface energy budget can be expressed in simple, analytical terms, the synoptic and planetary wave scale effects of snow cover have to be simulated in general circulation models. In the following, two examples of large scale simulations will be presented.

In Figure 5 Dewey (1986) demonstrates that in winters with extensive snow cover the number of cyclones increases over the southern part of the US. Increasing cyclonic activity accompanies the southward extension of the snow cover. In Eastern Canada in turn, cyclones become less frequent when the margin of the snow cover moves south.

A far reaching experiment was recently carried out by Barnett et al. (1988) who investigated the effect of Eurasian snow cover on global climate, especially on the Indian monsoon. When doubling the Eurasian snow cover in their simulation they found that it retarded the warming of the Asian land masses, thereby increasing surface pressure by up to 8 hPa, cooling the surface to 200 hPa layer by 1.7 to 3.6 K. Since ocean temperatures changed little, this caused a weaker meridional temperature gradient and a weaker subtropical easterly jet. The results of these simulations, especially the reduction of Indian precipitation by about 300 mm, agree well with what one observes in poor monsoon years. Reduction of Eurasian snow cover to half yielded similar results with opposite sign.

The authors found teleconnections via Rossby waves to the pressure field in the West Pacific and in Western Tropical

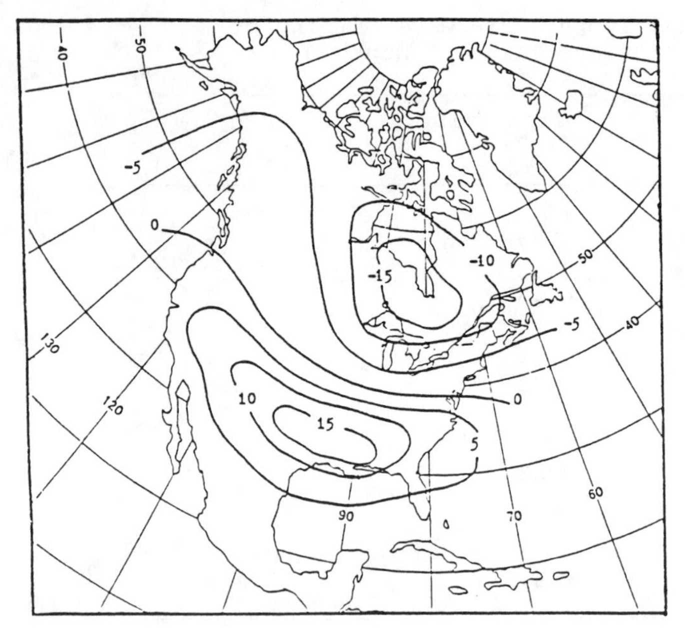


Figure 5. The difference in the total number of cyclones ( as calculated for 5 ° squares) between winters with extensive snow cover and winter with reduced snow cover. Positive values indicate that there is greater cyclonic activity during winters with extensive snow cover. (From Dewey, 1987).

Africa. The similarity with winter seasons of ENSO events lead them to remark that "snow fall perturbations may act as trigger for some ENSO events. They cannot directly force an ENSO since their characteristic time scale is about one season ...".

While the two reports quoted above treat seasonal snow cover, the growth of ice sheets to heights above 3000 m adds a third dimension to the forcing of atmospheric motion by snow and ice. By inserting ice age (18 ky BP) topography into a simple two-layer model Lindemann and Oerlemans (1987) found that orographic forcing was more important than thermal forcing. Their results are reproduced in Fig. 6 showing how the Laurentide and Fennoscandian ice sheets produce a wave number three anomaly in the 500 hPa height. Note that the Greenland and Ural ice sheets do not have any significant influence on this pattern. The negative anomaly over the eastern part of the large ice sheets promotes snow accumulation so that large ice sheets develope further by a planetary wave feedback.

The fact that in the wave number three pattern of Fig. 6 a third, large ice sheet is missing while Greenland, the Ural Mountains and central Asian highlands bear ice covers underlines the importance of land surface elevation for ice sheet development.

## Glacier Mass Balance

The mass balance of an existing glacier or ice sheet is dominated by atmospheric processes such as solid precipitation P, redistribution by snow drift D, deposition or erosion of snow by avalanches A, Evaporation E, melt M and calving C. P, D, C and A are strongly influenced by topographic features, E and M only to the extent that surface aspect governs the energy budget. The mass budget of an entire ice body is

$$B = P + D + A + E + M + C \qquad (3)$$

where P, D, A are usually positive (accumulation) and C, E, M negative (ablation).

Specific mass balance terms are referred to unit horizontal surface area and customarily denoted by lower case letters. The line connecting all points where the annual value b = 0 is called the equilibrium line. Further details are given in an earlier paper (Kuhn 1981).

A stationary or steady state glacier keeps its shape by annually transporting net specific mass gain from the accumulation area across the equilibrium line to the ablation area. By dividing annual net mass gain in the accumulation area by the cross-sectional area underneath the equilibrium line, a characteristic velocity of the ice flow is derived which has typical values of 10 to $10^2$ m $yr^{-1}$ in mountain glaciers and $10^3$ to $10^4$ m $yr^{-1}$ in polar outlet glaciers. Considering the sizes of glaciers and ice sheets one finds mean residence times of ice of less than 1000 years in mountain glaciers and of 10 000 years in polar ice caps, with extremes probably reaching 200 000 years.

From the areas and volumes given in Table 1, the typical thickness of glaciers has the order of 100 m, that of ice sheets is in the km-range. Comparing the life times and dimensions of these ice bodies to those of seasonal snow one finds a remarkable gap in the size and age spectrum. This impression is confirmed by glacier inventories and by individual observations.

While the seasonal snow cover may be gradually dissolved in smaller patches and vanish in spring as is evident in Figure 4, the majority of glaciers and ice sheets have a sharp boundary. In the time domaine this is paralleled by the rapid transitions from glacials to interglacials or the switching back and forth at

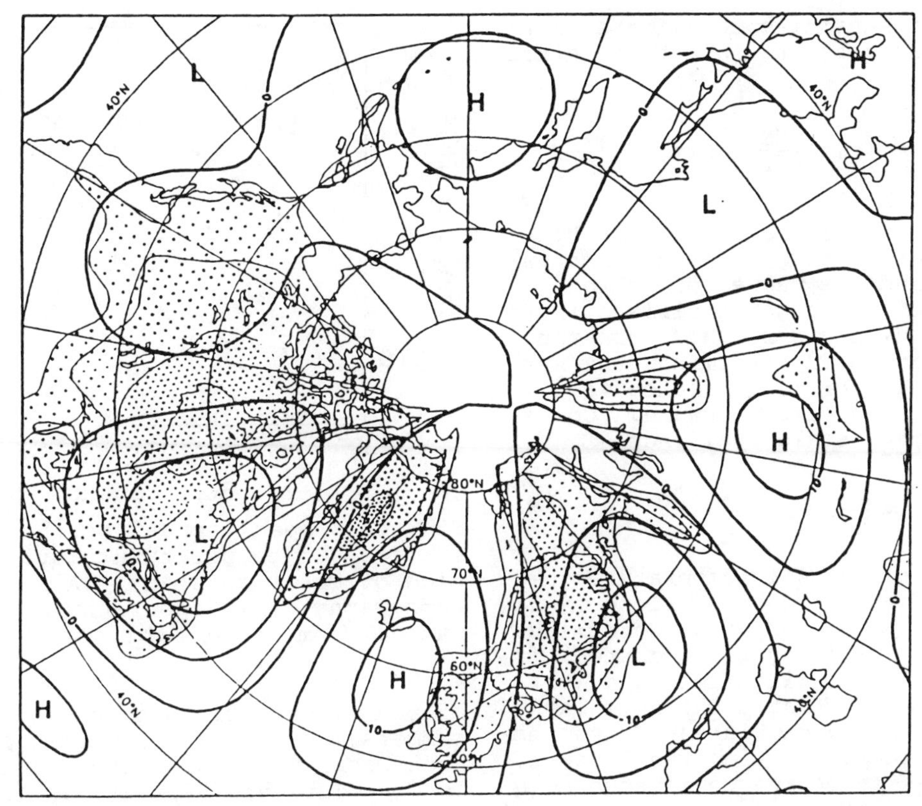


Figure 6. Perturbation of 500 hPa height in gpdam (heavy lines) due to topographic forcing by the 18 000 years BP ice sheets (stippled areas). From Lindeman and Oerlemans (1987).

higher frequency between less long-lived stable states of the climate system.

An obvious requirement for such sharp or rapid transitions is the existence of feedbacks involving snow and ice, and it is helped by the potential of ice to suffer catastrophic, dynamic changes.

## Feedbacks and Instabilities

### The Albedo-Temperature Feedback

We have mentioned so far two feedbacks, one involving surface layer instability and one involving orographic forcing of planetary waves. The one most often quoted in connection with snow and ice is the albedo-temperature feedback. With rising global temperature, planetary albedo is diminished by the melting of ice thus adding to the original forcing. This feedback was the central theme of the classical papers of Sellers (1969) and Budyko (1969).

Hansen et al. (1984) have modelled the effects of doubling $CO_2$ or increasing the solar constant by 2 % and found that an original temperature change of about 1.2 K (similar in both cases) was amplified by various feedbacks to a total of more than 4 K, about 0.4 K of this being due to surface albedo changes.

As was mentioned, planetary albedo is higher than surface albedo over most of the globe but lower (about 0.75) than surface albedo (0.84) over the dry snow facies of the Antarctic Plateau. This masking of surface contrasts by atmospheric backscattering and absorption decreases the efficiency of the albedo feedback. Shine et al. (1984), among others, have pointed out that cloudiness is higher over snow free than over snow covered surfaces, a fact that further contributes to reduce the surface albedo feedback.

### Topographic Feedbacks Promoting Ice Sheet Development

It would be more precise to speak of an energy budget-albedo feedback rather than of a temperature effect. However, both temperature and energy balance display a strong negative correlation with surface elevation, both $dT/dz$ and $db/dz$ are negative as is change of T and b with latitude $\phi$ . It was long known to alpine glaciologists that a glacier with a flat part at the altitude of its equilibrium line is most sensitive to climate changes, for instance to a temperature change, since

$$\frac{dB}{dT} = \frac{d(bS)}{dz}\frac{dz}{dT} = \left(b(z)\frac{dS}{dz} + S(z)\frac{db}{dz}\right)\frac{dz}{dT} \quad (4)$$

S being the glacier surface area up to altitude z. A flat part means a large value of dS/dz and the value of db/dz is characteristic of a given climate (Kuhn 1984), while dT/dz is roughly constant. This shows that dS/dz determines the sensitivity of glacier mass balance to climatic changes on a local scale. Considering the global scale one finds

$$\frac{dB}{dT} = \frac{dB}{dz}\frac{dz}{d\phi}\frac{d\phi}{dT} \quad (5)$$

For a given meridional temperature gradient $dT/d\phi$ the continental slope $dz/d\phi$ is thus the parameter that determines mass balance sensitivity to a change in global temperature. Birchfield and Wertman (1983) noted that "The Himalaya-Alpine Belt, the Tibet Plateau, and the Colorado Plateau appear in the zonally averaged elevation of the continents as a single plateau at an altitude of more than 1600 m centered between 30° and 40°N". Any climatic forcing that moves the snow line (equilibrium line) southward will thus be enhanced when a further southward shift means an increase of surface above sea level. This effect is illustrated in Figure 7 which shows Northern Hemisphere temperatures following a change in solar constant for sea level and for a realistic topography.

Koerner (1980) has given an interesting contribution to the problem of the initiation of the Labrador-Ungava Ice Sheet. In defeat of the earlier hypothesis that this ice sheet had been initiated by the buildup and subsequent spread of mountain glaciers, Koerner points at the situation of the Baffin Island Plateau just below today's equilibrium line. In the terms used above this plateau means high climatic sensitivity due to high values of dS/dz.

Barry et al. (1975) found Little Ice Age conditions in Labrador approaching ice sheet initiation and stated that (today) "the climatic changes required to initiate the necessary snow line lowering may involve only a minor summer cooling". Koerner (1980) stresses the point that a reduction of annual mass balance variance is just as important for ice sheet initiation as a lowering of the mean equilibrium line, since occasional, extreme summers may waste the ice accumulated under nearly average conditions.

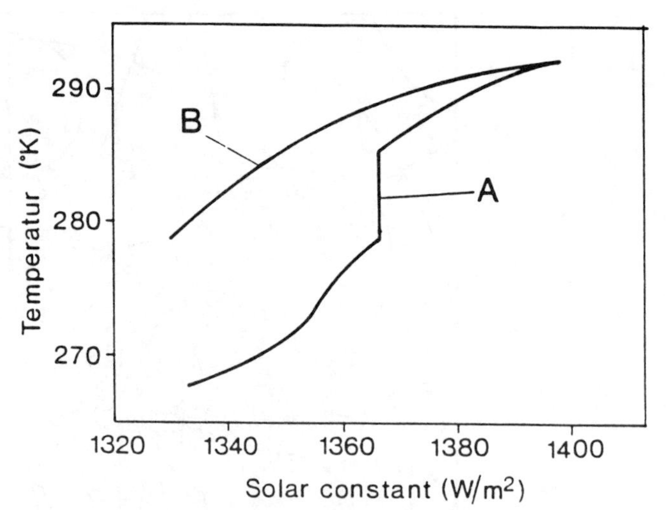


Figure 7. Mean annual Northern Hemisphere surface air temperature plotted as a function of solar constant. (A) with realistic topography, (B) topography has been eliminated. From Birchfield and Wertman, 1983.

The Tibetan Plateau is in a similar situation with respect to present equilibrium lines and may well have experienced a rapid glaciation in the quaternary. Kuhle (1986) believes to have found morphological evidence for a substantial pleistocene ice cover in spite of the low latitude position of the plateau, his associated climatic speculations, however, have not been confirmed.

## Reversible and Irreversible Changes

Annual snow cover extent, equilibrium line lowering and other processes involving snow masses of the order of $10^3$ kg $m^{-2}$ have an almost instantaneous adjustment and are reversible on a time scale of years. A mass balance disturbance is transmitted downglacier at a speed several times that of mean mass flow so that most ice bodies are continuously in a transient state, lagging behind smaller climatic changes.

A developing ice sheet soon reaches a size large enough to modify atmospheric temperatures or circulation. For example, Braithwaite (1977) found surface air temperatures over Ellesmere Island glaciers 1 to 2 degrees lower than over adjacent bare ground. This cooling effect is magnified when an ice sheet builds up to several hundred meters thickness attaining a surface climate significantly different from that governing its initial growth. When sea level temperature or sea level energy balance have returned to their original, no-snow values, the top of the new ice sheet is still in a healthy glacial condition.

This, by definition, is an unstable reaction to climatic forcing. It is not entirely irreversible, however, and on a large enough time scale the ice sheet surface temperature will follow a hysteresis loop around its original point when forced by a negative and subsequent positive sea level temperature excursion.

## Surges, a Periodic Instability

Glacier surges are a dynamical instability that is due to internal causes, not triggered by climatic forcing. Only a minor fraction of glaciers are capable of surging and few have been observed, one outstanding example being the 1982/83 surge of Variegated Glacier (Kamb et al. 1985). When basal water pressure comes near ice overburden pressure, basal sliding greatly accelerates producing ice velocities of up to 50 m per day. Under usual conditions, higher water pressure enlarges the subglacial tunnel system which controls water outflow. Only at high enough speeds frictional melting enables a subglacial cavity network without major tunnels to be established that in turn promotes further pressure to build up. The conditions suitable for this second mode of subglacial water flow seem to be fulfilled cyclically with periods of several decades. West Greenland outlet glaciers are believed to be permanently in the surging mode.

Since surges are not climate-related their moraines should not be used indiscriminantly in paleoclimatic reconstructions. The role of surges in the climate system lies in their potential of draining large ice basins in extraordinarily short time.

## The Grounding Line Instability and the Possible Drainage of West Antarctica

A survey of climate-related dynamical behavior of terrestrial and marine ice sheets was given by Bentley (1984). He shows that changes in the surface mass balance (precipitation or melting) are not nearly as effective in accelerating ice flow as are shifts in the grounding line position. At the grounding line the hydrostatic pressure of the ice load equals that of the sea water, an equilibrium that may be disturbed either by sea level changes or by thinning or thickening of the ice shelf. When ice at the old grounding line becomes afloat due to either sea level rise or ice sheet thinning the grounding line retreats. Many marine ice sheets are terminating on sills, bedrock sloping downward inland due to a combination of glacial erosion and isostatic depression (Fastook 1984). Under such conditions the grounding line retreats into deeper water, greatly amplifying the response to the climatic signal and possibly becoming unstable as was recently the case with Columbia Glacier in Alaska (Meier et al. 1985).

With long distance grounding line retreat the ice sheet surface becomes steeper and ice outflow is accelerated, the ice sheet becomes thinner and requires further grounding line retreat. Numerous speculations have been made about a possible catastrophic drainage or collapse of the West Anatarctic Ice Sheet. After detailed consideration of all dynamical processes involved Bentley (1984) comes to the conclusion that, if it drains, it will take at least 500 years for doing so.

In a classical paper Wilson (1964) discusses global cooling due to the albedo increase by ice bergs spreading over the Southern Ocean following an Antarctic surge.

# Snow and Ice in the Hydrological Cycle

Apart from the various ways in which ice and snow cover modifies the climate via the surface energy budget, some climatically relevant processes involve mass transport or mass storage in the cryosphere. These range from the daily freeze and thaw cycle to glacial sea level changes.

## The Seasonal Cycle

Figure 1 shows that an appreciable amount of water is seasonally stored on the land surface. Compared to no-snow conditions on the same soil this means delayed runoff and prolonged evaporation at the potential rate not only from the snow but also from the soil moisture that is charged by melt water. The hydrologic effect is not restricted to the local scale: many semi-arid areas receive an important fraction of their water balance from ice and snow melt in distant mountain. This is particularly important in summer-dry regions with sufficient winter precipitation that would otherwise drain unused.

While the peak of annual runoff occurs during the spring in seasonally snow covered lowland it is farther shifted to July and August in glacierized mountain regions. Figure 8 illustrated this effect with data from an alpine basin (Ötztal, 47°N, 11°E).

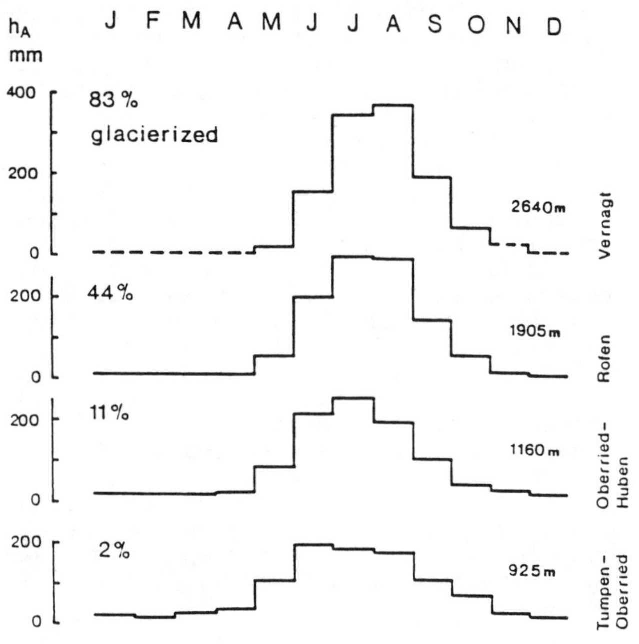


Fig. 8. Monthly runoff expressed in mm (kg $m^{-2}$ $mo^{-1}$) for four alpine gauging stations that are successively lower and drain less glacierized catchment areas in the basin of Ötztaler Ache.

## A Possible Runoff Change in a Warming Climate

To a useful approximation, the limit of snowfall is determined by the 0°C-isotherm. Following a climatic rise of mean temperatures, then, considerable fractions of a region may lose their seasonal snow cover and consequently change their annual river runoff characteristics.

Figure 9 gives an idea of the possible change in the Alps. Mean monthly temperatures in Austria presently favor a seasonal snow cover over the entire country, frequent valley inversions cause nearly isothermal conditions from 1000 to 1500 m elevation. A five-degree warming is equivalent to replacing the present -5°C isotherm (heavy line in Figure 9) by the new 0°C isotherm so that the lower limit of the winter snow cover would then be at 1600 m. Similar as in equation (4) the

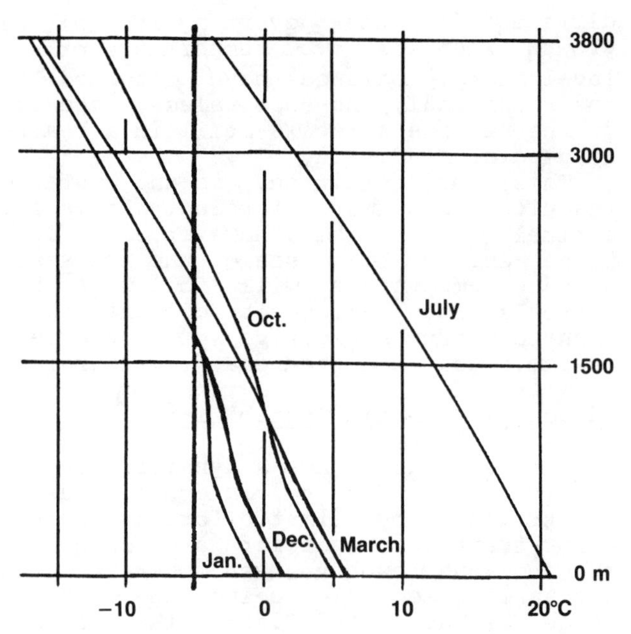


Fig. 9. Monthly mean tempertures of Austrian stations (1851-1950) as function of evaluation. Isothermal winter conditions between 1000 and 1500 m are due to inversions in valleys and basins.

sensitivity of this effect to climatic change is highest where the change of area with altitude is most rapid and where temperature changes most slowly with altitude.

## Sea Level Changes

Difficult as the determination of recent sea level changes may be on a global scale, there is evidence that since 1900 sea level has risen about 80 mm (Gornitz, in press). Of this rise more than half is caused by thermal expansion of the oceans, the rest being attributed to ice melt.

From a survey of the restricted number of mass balance measurements of sufficient length Meier (1984) found that about 30 mm could be attributed to this century's melting of extrapolar glaciers alone. The possible future contribution of these glaciers to a four-degree global warming is expected to be limited to another 120 mm (Kuhn, in press).

The additional contributions to sea level rise from Greenland and Antarctica are still largely unknown. Whether South Greenland surface melt, wasting of antarctic ice shelves from below and storage of possibly increased snow fall in polar regions are all insignificant compared to glacier melt or whether they simply have cancelled in this century is still a matter of stimulating discussions.

References


Barnett, T. P., L. Dümenil, U. Schlese, and E. Roeckner, The effect of Eurasian snow cover on global climate, Science, 239, 504-507, 1988.

Barry, R. G., J. T. Andrews, and M. A. Mahaffy, Continental ice sheets: conditions for growth, Science, 190, 979-981, 1975.

Bentley, C. R., Some aspects of the cryosphere and its role in climatic change, AGU Geophysical Monograph, 29, Maurice Ewing Volume 5, 207-220, 1984.

Birchfield, G., and J. Wertman, Topography, albedo-temperature feedback, and climate sensitivity, Science, 219, 284-285, 1983.

Braithwaite, R. J., Air temperature and glacier ablation, a parametric approach, Ph. D. Thesis, McGill University, Montreal, 146 pp., 1977.

Budyko, M. I., The effect of solar radiation variations on the climate of the Earth, Tellus, XXI (5), 611-619, 1969.

Dewey, K. F., The relationship between snow cover and atmospheric thermal and circulation anomalies, Snow Watch 1985, Glaciological Data, Report GD-18, World Data Center A for Glaciology, Boulder, 37-53, 1986.

Fastook, J. L., West Antarctica, the sea-level controlled marine instability: past and future, AGU Geophysical Monograph, 29, Maurice Ewing Volume 5, 275-287, 1984.

Gornitz, V., Mean sea level changes in the recent past. Paper given at the International Workshop on the Effects of Climatic Change on Sea Level, Severe Tropical Storms and their Associated Impacts. Norwich, UK, 1-4 Sep. 1987 (in press).

Hansen, J., A. Lacis, D. Rind, G. Russel, P. Stone, I. Fung, R.Ruedy, and J. Lerner, Climate sensitivity: Analysis of feedback mechanisms, AGU Geophysical Monograph, 29, Maurice Ewing Volume 5, 130-163, 1984.

Kamb, B., C. F. Raymond, W. D. Harrison, H. Engelhardt, K. A. Echelmeyer, N. Humphrey, M. M. Brugman, and T. Pfeffer, Glacier surge mechanism: 1982-1983 surge of Variegated Glacier, Alaska, Science, 227, 469-479, 1985.

Koerner, R. M., Instantaneous glacierization, the rate of albedo change, and feedback effects at the beginning of an ice age, Quaternary Research, 13, 153-159, 1980.

Kuhle, M., Die Vergletscherung Tibets und die Entstehung von Eiszeiten, Spektrum der Wissenschaften, 1986 (9), 42-54, 1986.

Kuhn, M., Climate and glaciers. Proceedings of the Canberra Symposium on sea level, ice and climatic change, International Association of Hydrological Sciences Publication, 131, 3-20, 1981.

Kuhn, M., Mass budget imbalances as criterion for a climatic classification of glaciers, Geografiska Annaler, 66 A (3), 229-238, 1984.

Kuhn, M., The bidirectional reflectance of polar and alpine snow surfaces, Annals of Glaciology, 6, 164-167, 1985.

Kuhn, M., A. J. Riordan, and I. A. Wagner, The climate of Plateau Station, In The Climate of the Arctic, (G. Weller, and S. A. Bowling, eds.). AMS, Fairbanks, 255-267, 1975.

Kuhn, M., L. S. Kundla, and L. A. Stroschein, The radiation budget at Plateau Station 1966/67, AGU, Antarctic Research Series, 25, 41-73, 1977.

Kuhn, M., Possible future contributions to sea level change from small glaciers. Paper given at the International Workshop on the Effects of Climatic Change on Sea Level, Severe Tropical Storms and their Associated Impacts. Norwich, UK, 1-4 Sep. 1987 (in press).

Lindeman, M., and J. Oerlemans, Northern Hemisphere ice sheets and planetary waves: a strong feedback mechanism, Journal of Climatology, 7, 109-117, 1987.

Liljequist, G. H., Energy Exchange of an Antarctic snow-field, Norwegian-British-Swedish Antarctic Expedition, 1949-52, Scientific Results, Vol. II, Part 1 D, Norsk Polarinstitutt, Oslo, 1957.

Male, D. H., and D. M. Gray, Snowcover ablation and runoff, In Handbook of Snow (D. M. Gray, and D. H. Male, eds.), Pergamon Press, 360-436, 1981.

Meier, M. F., Contribution of small glaciers to global sea level rise. Science, 226, 1418-1421, 1984.

Meier, M. F., L. A. Rasmussen, and D. S. Miller, Columbia Glacier in 1984: Disintegration underway. U.S. Geological Survey Open-File Report, 85-81, 17pp, 1985.

Moser, H., H. Escher-Vetter, H. Oerter, O. Reinwarth, D. Zunke, Abfluß in und von Gletschern, GSF-Bericht 41/86 (ISSN 0721-1694), Gesellschaft für Strahlen- und Umweltforschung München, 1986.

Robinson, D. A., and G. Kukla, Albedo of a dissipating snow cover, Journal of Climate and Applied Meteorology, 23, 1626-1634, 1984.

Ropelewski, C. F., Snow cover in real time climate monitoring, Snow Watch 1985, Glaciological Data, Report GD-18, World Data Center A for Glaciology, Boulder, 105-108, 1986.

Sellers, W. D., A global climatic model based on the energy balance of the earth-atmosphere system, Journal of Applied Meteorology, 8, 392-400, 1969.

Shine, K. P., A. Henderson-Sellers, and R. G. Barry, Albedo-climate feedback: the importance of cloud and cryospheric variability, In: A. Berger, and C. Nicolis (eds.), New Perspectives in Climate Modelling, 135-155, 1984.

Wilson, A. T., Origin of ice ages: an ice shelf theory for Pleistocene glaciation, Nature, 201, 147-149, 1964.

# Section III

# VOLCANOES AND CLIMATE

# PETROLOGIC EVIDENCE OF VOLATILE EMISSIONS FROM MAJOR HISTORIC AND PRE-HISTORIC VOLCANIC ERUPTIONS

Julie M. Palais* and Haraldur Sigurdsson

Graduate School of Oceanography, University of Rhode Island, Kingston, R.I. 02881

*Abstract.* Estimates of volcanic volatile composition and mass release of sulfur, chlorine and fluorine to the atmosphere from twelve large Recent and Quaternary volcanic eruptions has been made on basis of pre-eruption volatile composition, as determined by electron microprobe in glass inclusions trapped in phenocrysts in tephra. These estimates extend our knowledge of atmospheric loading by volcanic gases to include events more than two orders of magnitude larger than recent eruptions observed with modern methods. Results for several events agree well with other independent estimates, based on ice cores and atmospheric studies. The results show, that yield of sulfur, chlorine and fluorine to the atmosphere is not only dependent on total erupted mass, but largely determined by the composition of the erupting magmas. Thus volcanic volatile yield from high-silica or rhyolitic explosive eruptions is one or two orders of magnitude lower than during eruption of equal mass of basaltic or trachytic magmas. Sulfur yield during individual events is up to $3 \times 10^{13}$ g, such as in the case of the basaltic fissure eruptions of Laki and Eldgja in Iceland. However, in certain trachytic eruptions the yield of halogens may exceed sulfur output, such as during the great 1815 Tambora eruption in Indonesia, when chlorine and fluorine yield to the atmosphere is estimated as $2 \times 10^{14}$ and $1.7 \times 10^{14}$ g, respectively. Petrologic estimates of sulfur yield correlate closely with northern hemisphere annual temperature anomalies observed following the eruptions, in agreement with the findings of Devine et al [1984] on a smaller data base.

## Introduction

Volcanic activity is often cited as one of the major causes of climate cooling, on time scales ranging from a few years to millions of years [Lamb, 1970; Kennett and Thunell, 1975, 1977; Porter, 1981; Rampino et al., 1985]. While early work in this field attributed the climatic effects of volcanic eruptions to dust veils of volcanic ash [Lamb, 1970; 1971], more recent studies have shown that acidic volcanic aerosols are more important in producing a climatic impact, because of their longer atmospheric residence time [Pollack et al., 1976; Rampino and Self, 1982, 1984; Devine et al., 1984; Rampino et al., 1985]. In fact, small sulfur-rich eruptions (both explosive and non-explosive) may produce similar or even greater atmospheric-climatic effects, as measured by the stratospheric optical thickness and deviations in mean hemispheric temperature, as large, explosive silicic sulfur-poor ash-producing eruptions [Self et al., 1981; Sigurdsson, 1982; Rampino and Self, 1982, 1984; Rose et al., 1983; Devine et al., 1984; Sigurdsson et al., 1985; Rampino et al., 1985]. It is now apparent that the potential climatic impact of a volcanic eruption is not only governed by the intensity or magnitude of the eruption, but probably more importantly by the chemical composition of the magma, i.e. the concentration and type of volatile components, which are degassed during eruption and may generate a volcanic aerosol.

The determination of pre-eruption volatile concentration of the magma and the yield of volcanic gases to the atmosphere during eruption is therefore of major interest in the study of the possible relationship between volcanic aerosols and climate. The potential of trapped glass inclusions in phenocrysts as recorders of pre-eruption volatile content of magmas was first recognized by Anderson [1974], who applied this method in estimating the volcanic volatile contribution to the sulfur and chlorine budget of the oceans. The method was also applied to the 1976 Mt. St. Augustine eruption by Johnston [1980], who demonstrated the potentially great contribution of volcanic eruptions to the chlorine budget of the Earth's stratosphere. These studies paved the way for the petrologic estimates of volcanic degassing during earlier historic and pre-historic eruptions.

Petrologic estimates of volatile emissions from a number of recent historic and several Quaternary eruptions were made by Sigurdsson [1982], Devine et al.[1984], and Sigurdsson et al. [1985] (Table 1). Devine et al. [1984] showed that, in general, basaltic magmas release about an order of magnitude more sulfur than silicic eruptions of similar magnitude. These authors also found that surface temperature decreases after four historic volcanic eruptions (Laki 1783, Tambora 1815, Krakatau 1883, Agung 1963) were positively correlated with the estimated mass of sulfur

*Present address: Glacier Research Group, University of New Hampshire, Durham, N.H. 03824

TABLE 1. Estimated Mass of Volcanic Volatiles Released in Major Historic and Pre-Historic Eruptions (Devine et al., 1984; Sigurdsson et al., 1985)

| Eruption | Age | Total Erupted Mass Kg | $H_2SO_4$ metric tons | HCl metric tons | HF metric tons | Total Acids metric tons |
|---|---|---|---|---|---|---|
| Roseau Tuff | 28,000 BP. | 7.5 x $10^{13}$ | n.d. | 1.5 x $10^{7}$ | n.a. | 1.15 x $10^{7}$ |
| Minoan | 3,500 BP. | 6.3 x $10^{13}$ | 3.86 x $10^{6}$ | n.d. | n.a. | 3.86 x $10^{6}$ |
| Hekla -3 | 2,800 BP. | 5.5.x $10^{12}$ | 4.84 x $10^{5}$ | 2.55 x $10^{5}$ | n.a. | 7.39 x $10^{5}$ |
| Hekla-1 | 1104 AD. | 1.3 x $10^{12}$ | 2.24 x $10^{5}$ | n.d. | n.a. | 2.24 x $10^{5}$ |
| Katla | 1357 AD. | 4.2 x $10^{11}$ | 7.22 x $10^{5}$ | n.d. | n.a. | 7.22 x $10^{5}$ |
| Laki | 1783 AD. | 3.4 x $10^{13}$ | 9.03 x $10^{7}$ | 1.60 x $10^{6}$ | n.d. | 9.19 x $10^{7}$ |
| Tambora | 1815 AD. | 2.4 x $10^{14}$ | 5.24 x $10^{7}$ | 2.16 x $10^{8}$ | 1.26 x $10^{8}$ | 3.94 x $10^{8}$ |
| Krakatau | 1883 AD. | 2.7 x $10^{13}$ | 2.94 x $10^{6}$ | 3.75 x $10^{6}$ | n.d. | 6.69 x $10^{6}$ |
| Agung | 1963 AD. | 2.4 x $10^{12}$ | 2.84 x $10^{6}$ | 1.53 x $10^{6}$ | 8.0 x $10^{5}$ | 5.17 x $10^{6}$ |
| Surtsey | 1963 AD. | 2.8 x $10^{12}$ | 3.52 x $10^{6}$ | 1.44 x $10^{5}$ | n.a. | 3.66 x $10^{6}$ |
| Heimaey | 1973 AD. | 3.6 x $10^{11}$ | 4.68 x $10^{5}$ | n.d. | n.a. | 4.68 x $10^{5}$ |
| Soufriere | 1979 AD. | 4.0 x $10^{10}$ | 1.53 x $10^{4}$ | 3.50 x $10^{4}$ | n.d. | 5.03 x $10^{4}$ |
| Mt. St. Helens | 1980 AD. | 6.5 x $10^{11}$ | 7.90 x $10^{4}$ | 3.52 x $10^{4}$ | 5.3 x $10^{4}$ | 1.67 x $10^{5}$ |
| Krafla | 1981 AD. | 5.6 x $10^{10}$ | 1.01 x $10^{5}$ | 3.77 x $10^{3}$ | n.a. | 1.05 x $10^{5}$ |

n.a.= not analyzed; n.d.= not detected

released by the eruptions, confirming the suggestion that sulfate aerosols from volcanic eruptions have a greater climatic impact than silicate dust. Temperature decreases were found to be related to the estimates of sulfur released by a power function, where the power to which the sulfur masses are raised is 0.345 (r=0.971) [Devine et al., 1984].

In this study we have applied the same techniques as those used by Devine et al. [1984] to examine twelve other important volcanic events. These include the 35,000 yr. B.P. Campanian eruption (Phlegrean Fields, Italy), the 130 A.D. Taupo eruption (North Island, New Zealand), the 1400 yr. B.P. Rabaul eruption (Papua, New Guinea), the 934 A.D. Eldgja eruption (Iceland), the 1362 Oræfajokull eruption (Iceland), the ~1500 A.D. and ~1800 A.D. Mount St. Helens eruptions (Wash., U.S.A.), the 1835 Coseguina eruption (Nicaragua), the 1886 Tarawera eruption (North Island, New Zealand), the 1902 Santa Maria eruption (Guatemala), the 1912 Katmai eruption (Alaska, U.S.A.), and the 1956 Bezymianny eruption (Kamchatka, U.S.S.R.). Each eruption is discussed in detail below, following a brief description of the analytical and statistical methods we have used to obtain our results.

## Analytical Methods and Significance Tests

Chemical analyses were obtained using a JEOL JXA-50A electron microprobe. The procedures that we followed for major element analyses have been discussed in detail by Devine and Sigurdsson [1983] and Devine et al. [1984]. Sodium and potassium are analyzed separately using the decay curve method of Nielsen and Sigurdsson [1981]. For sulfur and chlorine analyses we use a trace element program which measures counts on the background, at a fixed offset, for 100 seconds on each side of the peak (which is determined during standardization), and twice on the peak for 100 seconds each. Operating conditions for the trace element analyses (S and Cl) include a 1 to 5 μm beam diameter, 15 kv accelerating voltage, 1.65 kv detector voltage, and a $2.5x10^{-8}$ amp beam current. We also use pulse height analysis to filter out background radiation.

Fluorine analyses were performed with a similar procedure as that employed by the trace element routine used for the S and Cl analyses. Step scans were performed on the standard (KE-12, see Table 2) as well as for each sample to determine the shape of the fluorine peak and the best offset on each side of the peak to measure background counts. We measured the count rate twice on the low side of the peak (300 steps off the peak, 100 s each), twice on the peak (100 s each) and once on the high side of the peak (100 s), at between 50 to 90 steps off the peak. This procedure enabled us to better define the background levels so that they could be subtracted from the average measured peak rate.

Table 2 gives the results of test analyses that we made on soda-lime and natural glass standards (NBS-610/620 and

TABLE 2. Accuracy and Precision in Microprobe Analysis of Sulfur, Chlorine and Fluorine.

| Standard | NBS-620(1) | NBS-610(2) | KE-12(3) |
|---|---|---|---|
| Sulfur ppm | 1154 | 565 | |
| 1 Std. Dev. | 67 (5 %) | 55 (10 %) | |
| Accepted Value | 1155 | 500 | |
| No. of Analyses | 15 | 231 | |
| Chlorine ppm | | | 3228 |
| 1 Std. Dev. | | | 164 (5 %) |
| Accepted Value | | | 3300 |
| No. of Analyses | | | 20 |
| F ppm | | | 4338 |
| 1 Std. Dev. | | | 548 (13%) |
| Accepted Value | | | 4400 |
| No. of Analyses | | | 9 |

(1) Average of test analyses of NBS-620, National Bureau of Standards (NBS), Standard Reference Material 620 soda-lime glass; accepted value is average of two bulk analyses done by the Leco (TM) induction titration method; analyzed by W. Stockwell [Leco Corp, St. Joseph MI, quoted in Devine et al., 1984].
(2) Average of test analyses of NBS-610, soda-lime glass; accepted value from NBS (1970).
(3) Average of test analyses of KE-12, pantelleritic obsidian from Eburru, Kenya; collected by D.K. Bailey and R. Macdonald; accepted Cl analysis by S.A.Malik and D.A. Bungard [quoted in Devine et al., 1984]. Standard deviation is one standard deviation, number in parentheses is coefficient of variation.

KE-12) to check our trace element standardizations. We estimate a detection limit of 50 ppm and a precision of 5 to 10 % for S and Cl . The detection limit for fluorine is estimated 300 ppm and a precision of 5 to 10 %.

Calculation of the mass of sulfur, chlorine and fluorine emitted to the atmosphere is made by measuring the content of these elements in glassy melt inclusions which have been trapped in magma phenocrysts (pre-eruption volatile content), and that in matrix glasses (degassed magma). The minimum masses of S, Cl and F that have been degassed in volcanic eruptions are determined by subtracting the average matrix glass concentration from the inclusion mean and the difference is then scaled to the mass of magma erupted.

A number of assumptions made in this approach put certain limitations on the accuracy of the estimates [Devine et al., 1984]. First, we assume that only minor crystallization takes place between the time that the melt inclusions are trapped and the time of eruption. Second, we assume that little or no degassing of the magma occurs before trapping of the melt inclusions and that all melt inclusions are trapped more or less simultaneously. A third assumption requires us to infer that the crystal content of the erupted material was relatively low and thus that the entire volume of erupted material was effectively liquid, so that we can scale the melt inclusion- matrix glass difference to the total mass of erupted material. Fourth, we assume that all gases released in the eruption were derived from the erupted material and that intrusive magma contributed little to the volatile emissions. Finally, we assume that there were no sulfur- or chlorine-bearing minerals in the magma that could have decomposed and released volatiles in the process. As discussed by Devine et al. [1984], when all possible errors due to invalid assumptions are considered together, the tendency is to underestimate the volatile yield by this method, and thus the petrologic estimates are minimum estimates.

A number of measurements of S, Cl and F concentration (ppm) are made on each sample (inclusions and matrix glass), and the average and standard deviation are calculated. In order to take into account the standard deviation of the analyses, we have assessed the statistical significance of the inclusion matrix glass difference by student's T-test.

## Results

Data on the petrologic estimates of volatile emissions for the twelve eruptions studied are presented below. The results of our glass inclusion analyses (major and trace elements) and the calculations of the mass of volatiles degassed for each eruption are summarized in Table 3 (see Appendix 1 for sample information). Pertinent information on each eruption is presented, including observations on the effects of the eruptions from optical, climatological and other atmospheric studies. Proxy data from Greenland and Antarctic ice core studies is also discussed in order to assess the reliability of the different methods (e.g. petrologic, atmospheric, ice core) for estimating the amount of volatiles degassed from volcanic eruptions.

### *Bezymianny Eruption, 1956 A.D.*

The March 30, 1956 eruption of Bezymianny volcano (Kamchatka Peninsula, U.S.S.R.) is best known for the gigantic cataclysmic explosion that accompanied the event. Before the renewed activity, which began with a period of volcanic earthquakes on September 29, 1955, Bezymianny volcano was generally regarded as extinct [Gorshkov, 1959]. Volcanic earthquakes continued until October 21, 1955 when the first gas outburst and a series of vulcanian-type ash eruptions took place. Explosions continued until March 30, 1956, when the "gigantic paroxysmal explosion" occurred [Gorshkov, 1959].

The eruption produced a directed blast, inclined at 30-40$^o$ to the horizon, which removed the upper 200 m of the volcano, producing a crater about 2 km wide and an eruption column up to 45 km height. Following the explosions two pyroclastic flows and several lahars were produced. Between April and late autumn of 1956, two endogenous domes were extruded in the new crater of the volcano, accompanied by weak to moderate explosions [Gorshkov, 1959].

Locally the ashfall was in a narrow zone (100-150 km wide) to the north and east, up to 400 km from Bezymianny. In addition, a strong mist of sulfurous gases was reported to the west and south of the volcano, although no ashfall was reported in that region [Gorshkov, 1959]. However, the atmospheric effects of the eruption were very widespread and a stratospheric dust layer was reported over Great Britain on April 3-4, 1956 [Bull and James, 1956].

We have analyzed the major and trace element composition of glass inclusions in plagioclase crystals and matrix glass in tephra collected 80 km north of the volcano (columns 41 to 44; table 3). Two compositionally distinct

TABLE 3. Electron microprobe analyses of glass inclusions and matrix glasses from volcanic eruptions, and estimates of sulfur, chlorine and fluorine yield to the atmosphere.

| Eruption | Eldgja 934 AD | | Oraefajokull 1362 AD | | Taupo-Hatepe 131 AD | | Taupo-Rotongaio | |
|---|---|---|---|---|---|---|---|---|
| Anal. No. | 1 | 2 | 3 | 4 | 5 | 6 | 7 | 8 |
| Glass | Inclusions | Matrix | Inclusions | Matrix | Inclusions | Matrix | Inclusions | Matrix |
| SiO2 | 47.12(144) | 46.06(62) | 71.74(94) | 73.07(94) | 71.81(57) | 74.92(96) | 72.17(143) | 74.85(90) |
| TiO2 | 3.91(43) | 4.77(26) | 0.16(2) | 0.27(2) | 0.22(4) | 0.23(8) | 0.29(8) | 0.30(3) |
| Al2O3 | 13.86(181) | 13.22(19) | 12.95(38) | 12.9(29) | 12.34(36) | 13.01(23) | 13.02(136) | 12.79(16) |
| FeO* | 14.57(183) | 15.4(32) | 3.40(16) | 3.11(25) | 2.04(27) | 1.77(28) | 2.22(61) | 1.80(19) |
| MnO | 0.26(6) | 0.22(5) | 0.10(7) | 0.02(2) | 0.02(4) | 0.01(2) | 0.07(6) | 0.01(2) |
| MgO | 5.14(54) | 5.41(4) | 0 | 0.01(1) | 0.10(2) | 0.12(1) | 0.16(28) | 0.56(119) |
| CaO | 9.8(43) | 10.17(15) | 0.93(6) | 1.01(7) | 1.33(5) | 1.49(7) | 1.32(10) | 1.16(52) |
| Na2O† | 2.42(51) | 2.91(12) | 5.74(59) | 5.71(63) | 4.75(0) | 4.76(0) | 4.02(0) | 4.02(0) |
| K2O | 0.59(30) | 0.72(3) | 3.60(15) | 3.61(16) | 2.59(0) | 2.69(0) | 2.93(0) | 2.93(0) |
| P2O5 | 0.5(27) | 0.43(5) | 0.13(8) | 0 | 0.06(7) | 0.12(3) | 0.17(10) | 0.11(6) |
| Total | 98.17(66) | 99.31(90) | 98.75(101) | 99.71(116) | 95.24(89) | 99.00(102) | 96.40(156) | 98.50(107) |
| No. Anal. | 6 | 8 | 8 | 4 | 16 | 6 | 9 | 6 |
| Mineral | Pyroxene | | Pyroxene | | Pyroxene | | Pyroxene | |
| S(ppm) | 1266(319) | 178(75) | 77(21) | 50(8) | 42(22) | 48(28) | 44(22) | 38(25) |
| No. Anal. | 11 | 17 | 9 | 5 | 4 | 8 | 8 | 5 |
| Cl(ppm) | 482(131) | 331(26) | 2084(63) | 1944(136) | 1729(60) | 1734(66) | 1773(192) | 1303(213) |
| No. Anal. | 12 | 6 | 7 | 5 | 10 | 6 | 7 | 7 |
| F(ppm) | 764(583) | 534(171) | 1589(521) | 1111(407) | 327(101) | 322(44) | 406(181) | 280(105) |
| No. Anal. | 5 | 5 | 5 | 4 | 8 | 5 | 8 | 4 |
| Erupted Volume ($km^3$DRE): | 9 | | 2 | | 1.4 | | 0.7 | |
| Yield (metric tons): | | | | | | | | |
| $H_2SO_4$ | $8.4 \times 10^7$ | | $3.8 \times 10^5$ | | 0 | | $3 \times 10^4$ | |
| HCl | $3.9 \times 10^6$ | | $6.6 \times 10^5$ | | 0 | | $7.8 \times 10^5$ | |
| HF | $6.2 \times 10^6$ | | $2.4 \times 10^6$ | | $1.7 \times 10^4$ | | $2.2 \times 10^5$ | |
| Total Acids (metric tons): | $9.4 \times 10^7$ | | $3.4 \times 10^6$ | | $1.7 \times 10^4$ | | $1 \times 10^6$ | |
| Yield (grams): | | | | | | | | |
| Sulfur | $2.7 \times 10^{13}$ | | $1.2 \times 10^{11}$ | | 0 | | $9.7 \times 10^9$ | |
| | 1088±98 ppm§ | | 27±8ppm | | | | 6±14 ppm | |
| | (99.9%) | | (97.7%) | | | | (85.5%) | |
| Chlorine | $3.8 \times 10^{12}$ | | $6.4 \times 10^{11}$ | | 0 | | $7.8 \times 10^{11}$ | |
| | 151±39 ppm§ | | 140±65 ppm | | | | 470±108ppm | |
| | (98.4%) | | (94.9%) | | | | (99.8%) | |
| Fluorine | $5.8 \times 10^{12}$ | | $2.2 \times 10^{12}$ | | $1.6 \times 10^{10}$ | | $2 \times 10^{11}$ | |
| | 230±272 ppm§ | | 478±309 ppm | | 5±41 ppm | | 126±83ppm | |
| | (51.8%) | | (78.5%) | | (0%) | | (74.2%) | |

See Appendix for sample identification. Values in parentheses represent the standard deviation in terms of least units cited for the value to their immediate left, thus 47.12(144) indicates a standard deviation of 1.44 wt. %. n.a. indicates not analysed. * All iron calculated as FeO. †All analyses corrected for sodium loss by decay curve method of Nielsen and Sigurdsson [1981]. § Volatile difference between melt inclusion mean and matrix glass mean and percent confidence in parentheses. n.d. = not detected.

groups of inclusions and corresponding matrix glasses were discovered. Dacitic inclusions and matrix glass contain from 64 to 66 % $SiO_2$ and in the rhyolitic group of glasses $SiO_2$ ranges from 75.5 to 76.5 %. In addition to the major element variation between these two groups, the sulfur content of the inclusions is also dramatically different. Inclusions in the low-silica group have 514±128 ppm sulfur and the co-existing dacitic matrix glass has 47± 41 ppm sulfur. Rhyolitic glass inclusions have much lower sulfur content than the dacitic inclusions, or an average of 49± 23 ppm S.

These results show that Bezymianny volcano erupted a mixed magma in 1956, composed mainly of dacitic magma, with minor amounts of rhyolite. Gorshkov [1959] reports whole-rock $SiO_2$ of about 58 to 60 % for ejecta from the 1956 eruption. Recently, however, the tephra being emitted from Bezymianny has become more acid (61 to 62 % $SiO_2$;

TABLE 3 cont. Electron microprobe analyses of glass inclusions and matrix glasses from volcanic eruptions, and estimates of sulfur, chlorine and fluorine yield to the atmosphere.

| Eruption | Taupo-Plinian | | Katmai 1912 Early | | Katmai-Mid | | Katmai-Late | |
|---|---|---|---|---|---|---|---|---|
| Anal. No. | 9 | 10 | 11 | 12 | 13 | 14 | 15 | 16 |
| Glass | Inclusions | Matrix | Inclusions | Matrix | Inclusions | Matrix | Inclusions | Matrix |
| SiO2 | 71.15(81) | 72.92(55) | 67.16(212) | 75.40(194) | 68.06(143) | 74.92(87) | 70.83(111) | 75.39(146) |
| TiO2 | 0.24(5) | 0.2(4) | 0.56(14) | 0.39(12) | 0.6(10) | 0.28(8) | 0.4(9) | 0.33(4) |
| Al2O3 | 12.09(32) | 12.75(28) | 14.12(109) | 13.14(53) | 14.27(68) | 12.47(44) | 12.04(130) | 12.06(45) |
| FeO* | 2.23(19) | 1.88(14) | 3.37(89) | 1.83(51) | 3.02(34) | 1.64(17) | 2.24(45) | 1.59(31) |
| MnO | 0.14(5) | 0.12(8) | 0.11(6) | 0.12(6) | 0.11(6) | 0.09(3) | 0.12(7) | 0.06(5) |
| MgO | 0.08(1) | 0.09(2) | 0.82(40) | 0.39(18) | 0.71(18) | 0.28(5) | 0.34(8) | 0.25(3) |
| CaO | 1.36(9) | 1.33(8) | 3.14(77) | 1.38(75) | 2.81(47) | 1.40(15) | 1.83(29) | 1.30(17) |
| Na2O† | 4.01(42) | 4.01(42) | 4.16(55) | 4.29(57) | 4.16(40) | 4.2(49) | 4.93(30) | 4.37(61) |
| K2O | 2.51(12) | 2.51(12) | 2.44(12) | 2.48(28) | 2.68(21) | 2.69(27) | 2.85(9) | 2.85(9) |
| P2O5 | 0.04(3) | 0.22(5) | 0.25(62) | 0 | 0.24(10) | 0 | 0.13(5) | 0.14(11) |
| Total | 93.85(56) | 96.03(67) | 96.13(228) | 99.42(77) | 96.66(92) | 97.97(117) | 95.71(79) | 98.34(119) |
| No. Anal. | 6 | 8 | 11 | 6 | 9 | 7 | 6 | 10 |
| Mineral | Pyroxene | | Pyroxene | | Plagioclase | | Pyroxene | |
| S(ppm) | 46(19) | 38(4) | 359(176) | 53(19) | 116(72) | 92(54) | 93(55) | 51(18) |
| No. Anal. | 6 | 3 | 6 | 7 | 9 | 4 | 8 | 9 |
| Cl(ppm) | 1750(102) | 1656(59) | 1739(142) | 1847(73) | 1712(166) | 1671(32) | 1959(469) | 1808(89) |
| No. Anal. | 9 | 7 | 10 | 10 | 6 | 5 | 9 | 12 |
| F(ppm) | 338(92) | 259(39) | 382(167) | 399(70) | 416(182) | 428(90) | 422(314) | 413(86) |
| No. Anal. | 7 | 3 | 5 | 8 | 4 | 5 | 8 | 8 |
| Erupted Volume ($km^3$DRE): | 5.1 | | 3 | | 3 | | 3 | |
| Yield (metric tons): | | | | | | | | |
| $H_2SO_4$ | $2.9 \times 10^5$ | | $6.5 \times 10^6$ | | $5.1 \times 10^5$ | | $8.9 \times 10^5$ | |
| HCl | $1.1 \times 10^6$ | | 0 | | $2.9 \times 10^5$ | | $1.1 \times 10^6$ | |
| HF | $9.9 \times 10^5$ | | 0 | | 0 | | $6.6 \times 10^4$ | |
| Total Acids (metric tons): | $2.4 \times 10^6$ | | $6.5 \times 10^6$ | | $8 \times 10^5$ | | $2 \times 10^6$ | |
| Yield (grams): | | | | | | | | |
| Sulfur | $9.4 \times 10^{10}$<br>8±8 ppm<br>(45.5%) | | $2.1 \times 10^{12}$<br>306±72 ppm<br>(99.8%) | | $1.7 \times 10^{11}$<br>24±36 ppm<br>(26.6%) | | $2.9 \times 10^{11}$<br>42±20ppm<br>(94.8%) | |
| Chlorine | $1.1 \times 10^{12}$<br>94±41 ppm<br>(94.6%) | | 0 | | $2.8 \times 10^{11}$<br>41±69 ppm<br>(25.2 %) | | $1 \times 10^{12}$<br>151±158 ppm<br>(67.8%) | |
| Fluorine | $9.3 \times 10^{11}$<br>79±41 ppm<br>(77.6%) | | 0 | | 0 | | $6.2 \times 10^{10}$<br>9±115ppm<br>(0%) | |

SEAN Bull. 1986, 11:4, p. 20) indicating the presence of a more evolved batch of magma under the volcano, which may have the same source as the rhyolitic component erupted in 1956. The low-silica dacite glass inclusions and matrix glasses are more representative of the composition of the main Bezymianny tephra, and we have based our estimates of volatile release in the eruption on sulfur, chlorine and fluorine data from the dacite glass inclusions (table 3). We estimate a total of $7 \times 10^6$ metric tons of acids were released in the Bezymianny eruption, of which 52 % was $H_2SO_4$ and 48 % was HCl.

*Katmai Eruption, 1912 A.D.*

The Katmai explosive eruption in Alaska on June 6 to 8, 1912, was the most voluminous eruption of this century, producing about 15 $km^3$ (D.R.E.) of magma, of which about 9 $km^3$ were emitted during the plinian phase [Curtis, 1968; Hildreth, 1983]. The eruption is best known for the pyroclastic and debris flows in the Valley of Ten Thousand Smokes. In this study we are mainly concerned with the tephra deposits formed in the plinian phases of activity. Historical records of three distinct periods of ashfall at

TABLE 3 cont. Electron microprobe analyses of glass inclusions and matrix glasses from volcanic eruptions, and estimates of sulfur, chlorine and fluorine yield to the atmosphere.

| Eruption | Katmai-Late | | Katmai-Total | Rabaul Plinian | | | Rabaul Ignimbrite | |
|---|---|---|---|---|---|---|---|---|
| Anal. No. | 17 | 18 | 19 | 20 | 21 | 22 | 23 | 24 |
| Glass | Inclusions | Matrix | | Inclusions | Inclusions | Matrix | Inclusions | Matrix |
| SiO2 | 72.92(154) | 75.39(146) | | 65.79(126) | 65.48(164) | 68.22(119) | 66.93(34) | 67.35(43) |
| TiO2 | 0.27(9) | 0.33(4) | | 0.81(8) | 0.89(6) | 0.78(11) | 0.77(3) | 0.80(12) |
| Al2O3 | 11.72(23) | 12.06(45) | | 14.77(47) | 15.39(42) | 15.42(50) | 15.02(33) | 14.95(43) |
| FeO* | 1.56(42) | 1.59(31) | | 3.73(49) | 3.87(61) | 3.51(62) | 2.87(69) | 3.69(43) |
| MnO | 0.09(5) | 0.06(5) | | 0.04(5) | 0.07(6) | 0.08(10) | 0.11(9) | 0.03(2) |
| MgO | 0.20(8) | 0.25(3) | | 1.01(16) | 1.00(12) | 1.09(13) | 0.87(9) | 1.05(7) |
| CaO | 1.28(14) | 1.30(17) | | 2.77(25) | 3.04(56) | 2.87(35) | 2.53(18) | 3.07(21) |
| Na2O† | 3.5(37) | 4.37(61) | | 4.89(17) | 4.62(3) | 4.76(14) | 4.6(21) | 4.81(40) |
| K2O | 2.91(15) | 2.85(9) | | 3.23(8) | 2.96(14) | 3.11(14) | 3.32(11) | 3.11(37) |
| P2O5 | 0 | 0.14(11) | | 0.19(7) | 0.27(12) | 0.28(14) | 0.31(6) | 0.17(6) |
| Total | 94.45(159) | 98.34(119) | | 97.23(109) | 97.59(93) | 100.14(141) | 97.33(98) | 99.03(91) |
| No. Anal. | 9 | 10 | | 11 | 4 | 7 | 2 | 3 |
| Mineral | Plagioclase | | | Plagioclase | Pyroxene | | Plagioclase | |
| S(ppm) | 48(33) | 51(18) | | 314(83) | 615(222) | 279(113) | 218(41) | 570(205) |
| No. Anal. | 9 | 9 | | 7 | 6 | 23 | 3 | 5 |
| Cl(ppm) | 2221(585) | 1808(89) | | 2932(368) | 2274(555) | 2678(292) | n.a. | n.a. |
| No. Anal. | 11 | 12 | | 10 | 8 | 11 | | |
| F(ppm) | 218(138) | 413(86) | | 243(129) | n.a. | 264(92) | n.a. | n.a. |
| No. Anal. | 5 | 8 | | 6 | | 11 | | |
| Erupted Volume ($km^3$DRE): | 3 | | 9 | 0.6 | | 0.6 | | 3.5 |
| Yield (metric tons): | | | | | | | | |
| $H_2SO_4$ | 0 | | $7.9 \times 10^6$ | $1.5 \times 10^5$ | | $1.5 \times 10^6$ | | 0 |
| HCl | $2.9 \times 10^6$ | | $3.2 \times 10^6$ | $3.8 \times 10^5$ | | 0 | | n.a. |
| HF | 0 | | $6.6 \times 10^4$ | 0 | | n.a. | | n.a. |
| Total Acids (metric tons): | $2.9 \times 10^6$ | | $1.1 \times 10^7$ | $5.3 \times 10^5$ | | $1.5 \times 10^6$ | | 0 |
| Yield (grams): | | | | | | | | |
| Sulfur | 0 | | | $5 \times 10^{10}$<br>35±39ppm<br>(0%) | | $4.8 \times 10^{11}$<br>336±94ppm<br>(99.9%) | | 0 |
| Chlorine | $2.9 \times 10^{12}$<br>413±178 ppm<br>(97.4%) | | | $3.7 \times 10^{11}$<br>254±146 ppm<br>(90.6%) | | 0 | | n.a. |
| Fluorine | 0 | | 0 | n.a. | | n.a. | | n.a. |

Kodiak, Alaska, described by Griggs [1922] and Curtis [1968], have been correlated with the plinian tephra layers A, C-D and F-G [Hildreth, 1983; Fierstein and Hildreth, 1986].

Major element analyses of glass and phenocrysts from pumice collected near Novarupta (the source vent of all the ejecta) show that the initial fall unit (A) was rhyolitic and contains 1-2 % phenocrysts. The later fall units (C-D and F-G), on the other hand, were found to be > 98 % dacitic and contain 30-45 % crystals. The sequence of events which produced the Katmai (Novarupta) deposits, described in detail by Hildreth [1983], suggest that the eruption was the result of magma-mixing and fractionation in a shallow, zoned magma chamber.

Recent studies on ejecta dispersal and dynamics of the 1912 plinian eruptions at Novarupta by Fierstein and Hildreth [1986] have provided estimates of the eruption rate and column height for different phases of the eruption. Column heights were calculated using both volumetric eruption rate estimates and maximum lithic vs. area isopleth maps determined from studies of the plinian fall deposits. These calculations suggest maximum column heights for the initial "A" rhyolitic phase of 28 to 30 km. For the dacitic eruptions, column heights of 18 to 25 km (phase C-D) and 15 to 20 km (phase F-G) were calculated. Isopach maps of the plinian fall tephra studied by Fierstein and Hildreth [1986] suggest that strong winds occurred during some phases of the eruption and may explain the observed area vs. thickness and axis of dispersal patterns. Tephra collected in abyssal piston cores southeast of Kodiak island and from a site on Kodiak island [Federman, 1984] confirm the

TABLE 3 cont. Electron microprobe analyses of glass inclusions and matrix glasses from volcanic eruptions, and estimates of sulfur, chlorine and fluorine yield to the atmosphere.

| | Campanian Plinian | | | | Tarawera Plinian | | | |
|---|---|---|---|---|---|---|---|---|
| | 25 | 26 | 27 | 28 | 29 | 30 | 31 | 32 |
| | Inclusions | Matrix | Inclusions | Matrix | Inclusions | Matrix | Inclusions | Matrix |
| SiO2 | 59.88(137) | 60.43(93) | 58.47(64) | | 52.95(59) | 51.79(138) | 53.74(70) | 52.32(128) |
| TiO2 | 0.28(6) | 0.36(7) | 0.35(5) | | 0.91(7) | 0.92(40) | 1.08(33) | 0.76(23) |
| Al2O3 | 19.4(63) | 19.31(30) | 18.63(49) | | 19.15(68) | 20.65(2.52) | 20.86(206) | 18.96(413) |
| FeO* | 2.88(60) | 3.21(11) | 2.79(35) | | 9.56(85) | 7.38(116) | 5.35(167) | 7.53(242) |
| MnO | 0.19(7) | 0.12(8) | 0 | | 0.21(3) | 0.18(4) | 0.10(7) | 0.19(8) |
| MgO | 0.49(13) | 0.49(10) | 0.58(10) | | 3.14(36) | 2.96(132) | 2.04(123) | 4.23(168) |
| CaO | 1.91(28) | 2.04(22) | 2.17(19) | | 7.41(22) | 11.98(1.89) | 12.91(181) | 11.52(101) |
| Na2O† | 4.04(51) | 4.13(82) | 3.39(24) | | 0.79(55) | 3.05(73) | 3.92(105) | 3.29(77) |
| K2O | 7.68(93) | 6.55(70) | 8.69(87) | | 0.46(13) | 0.61(25) | 0.51(13) | 0.59(20) |
| P2O5 | .07(4) | 0.15(8) | 0.18(7) | | 0.13(10) | 0.18(11) | 0.05(5) | 0.14(11) |
| Total | 96.82(148) | 96.79(127) | 95.25(300) | | 94.72(151) | 99.69(189) | 100.58(49) | 99.71(160) |
| No. Anal. | 15 | 6 | 8 | | 7 | 6 | 3 | 8 |
| Mineral | Pyx/Plag? | | Pyx/Plag? | | Pyroxene | | Pyroxene | |
| S(ppm) | 313(109) | 380(176) | 565(189) | 263(217) | 1888(594) | 74(27) | 801(201) | 47(12) |
| No. Anal. | 11 | 8 | 11 | 6 | 5 | 7 | 3 | 7 |
| Cl(ppm) | 3649(407) | 5500(1736) | 3588(101) | 6539(900) | 1286(301) | 418(195) | 1857(797) | 377(228) |
| No. Anal. | 9 | 6 | 10 | 9 | 2 | 7 | 3 | 5 |
| F(ppm) | 1408(230) | 3553(389) | 1634(102) | 2735(405) | 772(37) | 728(138) | 1848(700) | 691(241) |
| No. Anal. | 6 | 5 | 7 | 6 | 4 | 4 | 4 | 4 |
| Erupted Volume ($km^3$DRE): | 23 | | 23 | | 0.35 | | 0.35 | |
| Yield (metric tons): | | | | | | | | |
| $H_2SO_4$ | 0 | | $5.5 \times 10^7$ | | $5.4 \times 10^6$ | | $2.3 \times 10^6$ | |
| HCl | 0 | | 0 | | $8.7 \times 10^5$ | | $1.5 \times 10^6$ | |
| HF | 0 | | 0 | | $4.6 \times 10^4$ | | $1.2 \times 10^6$ | |
| Total Acids (metric tons): | 0 | | $5.5 \times 10^7$ | | $6.4 \times 10^6$ | | $5 \times 10^6$ | |
| Yield (grams): | | | | | | | | |
| Sulfur | 0 | | $1.8 \times 10^{13}$ | | $1.78 \times 10^{12}$ | | $7.39 \times 10^{11}$ | |
| | | | 302±105 ppm | | 1814±266 ppm | | 754±116pp | |
| | | | (98.7%) | | (99.9%) | | (99.9%) | |
| Chlorine | | | | | $8.5 \times 10^{11}$ | | $1.5 \times 10^{12}$ | |
| | | | | | 868±225 ppm | | 1480±471ppm | |
| | | | | | (99.6%) | | (98.2%) | |
| Fluorine | | | | | $4.3 \times 10^{10}$ | | $1.1 \times 10^{12}$ | |
| | | | | | 44±71 ppm | | 1157±370ppm | |
| | | | | | (42.7%) | | (95.9%) | |

conclusions of Hildreth [1983] and suggest the zonation in the magma chamber may have been the dominant factor responsible for the compositional variability in the Katmai deposits.

We have determined the sulfur, chlorine and fluorine content of glass inclusions in plagioclase and pyroxene phenocrysts and of matrix glasses in three samples from early, middle and late stages of tephra fall deposition (columns 11 to 19, table 3). We have divided the volume of the entire plinian fall deposit (9 $km^3$ D.R.E.) equally among the three phases of the eruption in order to estimate volatile mass. Therefore the reported total volatiles released from the Katmai eruption (column 19) is the sum of the samples from the early, middle and late phases of the eruption. Volatile emission from the early phase of the Katmai eruption was dominated by sulfur ($6.45 \times 10^6$ metric tons $H_2SO_4$). The eruption appears to have become progressively enriched in halogens (mainly Cl), especially in late stages, when $1 \times 10^6$ (pyroxene) to $2.92 \times 10^6$ (plagioclase) metric tons of HCl were emitted to the atmosphere. Lyons et al. [in prep.] have identified the Katmai eruption in an ice core from Dye 3, Greenland on the basis of peaks in excess chloride and sulfate. They also find a pattern of increased sulfate followed by chloride during the course of the Katmai eruption.

TABLE 3 cont. Electron microprobe analyses of glass inclusions and matrix glasses from volcanic eruptions, and estimates of sulfur, chlorine and fluorine yield to the atmosphere.

| | Mt.St. Helens ~1530 | | Mt.St. Helens ~1800 | | Coseguina 1835 | |
|---|---|---|---|---|---|---|
| | 33 | 34 | 35 | 36 | 37 | 38 |
| | Inclusions | Matrix | Inclusions | Matrix | Inclusions | Matrix |
| SiO2 | 71.17(162) | 73.19(76) | 69.57(158) | 71.85(86) | 59.39(211) | 61.46(504) |
| TiO2 | 0.03(6) | 0 | 0.05(5) | 0.07(9) | 0.97(8) | 0.64(49) |
| Al2O3 | 14.33(64) | 14.45(32) | 15.19(164) | 14.75(54) | 16.10(53) | 17.63(656) |
| FeO* | 1.35(31) | 1.37(11) | 1.96(54) | 1.98(37) | 7.99(110) | 6.16(411) |
| MnO | 0.02(3) | 0.01(3) | 0.03(3) | 0.04(4) | 0.12(9) | 0.23(20) |
| MgO | 0.24(7) | 0.30(8) | 0.56(17) | 0.47(14) | 2.55(73) | 1.36(96) |
| CaO | 1.53(11) | 1.65(7) | 1.99(34) | 2.34(42) | 4.99(97) | 5.68(294) |
| Na2O† | 4.07(38) | 4.07(38) | 4.85(80) | 4.85(80) | 4.13(34) | 4.57(0) |
| K2O | 2.23(24) | 2.23(24) | 2.05(16) | 2.05(16) | 2.09(19) | 1.71(0) |
| P2O5 | 0.10(11) | 0 | 0.12(13) | 0.01(2) | 0.27(10) | 0.20(17) |
| Total | 95.07(190) | 97.27(104) | 96.37(102) | 98.41(115) | 98.6(85) | 99.64(77) |
| No. Anal. | 9 | 6 | 7 | 7 | 21 | 4 |
| Mineral | Plag | | Plag | | Plag/Pyrox | |
| S(ppm) | 71 | 38(15) | 48(24) | 46(22) | 355(63) | 357(81) |
| No. Anal. | 1 | 5 | 7 | 9 | 15 | 4 |
| Cl(ppm) | 767 | 732(107) | 1665(540) | 1217(51) | 1870(215) | 1408(288) |
| No. Anal. | 1 | 10 | 9 | 8 | 15 | 10 |
| F(ppm) | n.a. | | n.a. | | 0 | 0 |
| No. Anal. | | | | | | |

| | Mt.St. Helens ~1530 | Mt.St. Helens ~1800 | Coseguina 1835 |
|---|---|---|---|
| Erupted Volume ($km^3$DRE): | 1 | 0.25 | 10 |
| Yield (metric tons): | | | |
| $H_2SO_4$ | $2.3 \times 10^5$ | $3.5 \times 10^3$ | 0 |
| HCl | $8.3 \times 10^4$ | $2.7 \times 10^5$ | $1.2 \times 10^7$ |
| HF | n.a. | n.a. | 0 |
| Total Acids (metric tons): | $3.2 \times 10^5$ | $2.7 \times 10^5$ | $1.2 \times 10^7$ |
| Yield(grams): | | | |
| Sulfur | $7.6 \times 10^{10}$<br>$33 \pm 7$ ppm<br>(99.9%) | $1.2 \times 10^9$<br>$2 \pm 12$ ppm<br>(0%) | 0 |
| Chlorine | $8.1 \times 10^{10}$<br>$35 \pm 34$ ppm<br>(0%) | $2.6 \times 10^{11}$<br>$448 \pm 181$ ppm<br>(96.2%) | $1.2 \times 10^{13}$<br>$462 \pm 107$ ppm<br>(99.9%) |
| Fluorine | n.a. | n.a. | 0 |

See the Appendix for sample identification. Values in parentheses represent the standard deviation in terms of least units cited for the value to their immediate left, thus 47.12(144) indicates a standard deviation of 1.44 wt. %. n.a. indicates not analysed. *All iron calculated as FeO. †All analyses corrected for sodium loss by decay curve method of Nielsen and Sigurdsson [1981]. § Volatile difference between melt inclusion mean and matrix glass mean and percent confidence in parentheses. n.d. = not detected.

Although fluorine was reported to have been a major component of the fumarolic emanations in the Valley of Ten Thousand Smokes [Zies, 1929], our results suggest that much of the fluorine in the magma was retained in the matrix glass, and thus probably did not contribute greatly to the volcanic volatile input to the atmosphere during the eruption. Lyons et al. [in prep.], however, find fluoride in the ice at the time of the Katmai eruption, one of only two volcanic events (the other being Hekla, 1947) to exhibit fluoride in ice core acidity layers in the period from 1869 to 1984.

### *Santa Maria Eruption, 1902 A.D.*

The plinian eruption of Santa Maria volcano, Guatemala in October 1902 was one of the largest of this century, lasting 18 to 20 hours and producing an eruption column estimated to have reached well into the stratosphere, 28 km high [Williams and Self, 1983]. The eruption is estimated to have produced about 20 $km^3$ of white dacitic pumice (8.5 $km^3$ D.R.E.) containing 20-30 % crystals; mainly

TABLE 3 cont. Electron microprobe analyses of glass inclusions and matrix glasses from volcanic eruptions, and estimates of sulfur, chlorine and fluorine yield to the atmosphere.

| Eruption | Santa Maria 1902 | | Bezymianny 1956 | | Bezymianny 1956 | |
|---|---|---|---|---|---|---|
| Anal. No. | 39 | 40 | 41 | 42 | 43 | 44 |
| Glass | Inclusions | Matrix | Inclusions | Matrix | Inclusions | Matrix |
| SiO2 | 69.72(228) | 73.55(93) | 63.99(44) | 66.06(41) | 76.39(1.07) | 75.43(231) |
| TiO2 | 0.24(5) | 0.27(2) | 0.84(8) | 0.71(7) | 0.19(5) | 0.11(6) |
| Al2O3 | 13.83(56) | 13.86(29) | 14.54(66) | 15.23(20) | 10.66(26) | 12.34(151) |
| FeO* | 1.61(50) | 1.35(11) | 5.11(52) | 4.76(26) | 1.89(24) | 0.19(10) |
| MnO | 0.10(11) | 0.07(5) | 0.21(6) | 0.24(6) | 0.11(3) | 0.08(4) |
| MgO | 0.55(38) | 0.35(7) | 1.43(25) | 1.20(11) | 0.40(10) | 0.06(14) |
| CaO | 1.66(25) | 1.65(10) | 3.26(41) | 3.48(15) | 0.77(12) | 0.71(60) |
| Na2O† | 5.16(84) | 5.60(3) | 3.78(56) | 4.34(0) | 3.78(54) | 4.97(0) |
| K2O | 2.67(26) | 2.71(1) | 3.25(25) | 3.42(0) | 3.67(75) | 4.86(2.26) |
| P2O5 | 0.24(19) | 0.01(3) | 0.20(8) | 0.07(5) | 0.01(1) | 0.02(1) |
| Total | 95.77(220) | 99.42(112) | 96.62((89) | 99.51(45) | 97.86(115) | 98.78(63) |
| No. Anal. | 22 | 19 | 9 | 5 | 11 | 5 |
| Mineral | Pyroxene | | Plagioclase | | Plagioclase | |
| S(ppm) | 198(89) | 108(48) | 514(128) | 47(41) | 49(23) | 27(14) |
| No. Anal. | 17 | 12 | 10 | 4 | 6 | 3 |
| Cl(ppm) | 1388(119) | 1212(64) | 3984(495) | 2721(342) | n.a. | n.a. |
| No. Anal. | 7 | 7 | 8 | 5 | | |
| F(ppm) | 0 | 0 | 0 | 0 | 0 | 0 |
| No. Anal. | | | | | | |
| Erupted Volume ($km^3$DRE): | | 8.5 | | 1.0 | | 1.0 |
| Yield (metric tons): | | | | | | |
| $H_2SO_4$ | | | | $5.6 \times 10^6$ | | $3.7 \times 10^6$ |
| | | $1.8 \times 10^5$ | | | | |
| HCl | | $3.7 \times 10^6$ | | $3.4 \times 10^6$ | | n.a. |
| HF | | 0 | | 0 | | 0 |
| Total Acids (metric tons): | | $1.2 \times 10^7$ | | $7.06 \times 10^6$ | | $1.8 \times 10^5$ |
| Yield (grams): | | | | | | |
| Sulfur | | $1.84 \times 10^{12}$ | | $1.2 \times 10^{12}$ | | $5.7 \times 10^{10}$ |
| | | 90±26 ppm | | 467±45 ppm | | 22±12 ppm |
| | | (99.9%) | | (99.9%) | | (79%) |
| Chlorine | | $3.59 \times 10^{12}$ | | $3.28 \times 10^{12}$ | | n.a. |
| | | 176±51 ppm | | 1263±232 ppm | | |
| | | (99.2%) | | (99.9%) | | |
| Fluorine | | 0 | | 0 | | 0 |

plagioclase, hornblende, orthopyroxene, clinopyroxene and titanomagnetite [Williams and Self, 1983; Rose, in press]. Traces of dark gray, scoriaceous high-Al basalt occur both as discrete blebs within the dacitic pumice and as isolated fragments scattered throughout the deposit. Medium gray mixed pumices are also a minor component of the deposit and are evidence that the eruption may have been triggered by the injection of a high temperature mafic magma into an already vapor-saturated dacitic magma body [Sparks et al., 1977; Williams, 1979; Rose, in press].

Our results of the major and trace element compositions of glass inclusions and matrix glass from the 1902 Santa Maria eruption are given in columns 39 and 40, respectively, of Table 3. We estimate that the Santa Maria eruption emitted a total of $9.31 \times 10^6$ metric tons of acids, 61% of which was $H_2SO_4$ ($5.63 \times 10^6$) and 39 % ($3.68 \times 10^6$) of which was HCl. Legrand and Delmas [1987] report a small peak of sulfuric acid in the ice at Dome C, Antarctica which they attribute to the eruption of Santa Maria and calculate global acid fallout of $2.2 \times 10^7$ metric tons. No evidence of the Santa Maria eruption has yet been reported in ice cores from Greenland, but a small peak in the Crete ice core at this time interval may correspond to the event [Hammer et al. 1980].

### *Tarawera Eruption, 1886 A.D.*

The largest and most destructive volcanic eruption in New Zealand in historical times was the basaltic plinian fissure eruption of Tarawera in 1886. The activity was mainly along a 4 km long fissure, which erupted a total of 0.7 $km^3$ of high-alumina basaltic magma, with a mass eruption rate of $1.8x10^8$ kg/s [Walker et al., 1984]. The plinian column height is estimated to have been 28 km to 34 km, based on calculations of Walker et al. [1984] and Carey and Sparks [1986], respectively. The powerful nature of the Tarawera eruption is thought to be due to the interaction of the basaltic magma with groundwater from a nearby geothermal field [Walker et al., 1984]. Violent phreatic explosions associated with the Tarawera eruption suggest that a significant fraction of the total volatiles may have been derived from ground water. Calculations of the gas exit velocity by Walker et al. [1984] lead to an estimate of 1.5 to 3.0 % for the total volatile content of the erupted gas-clast mixture, including any groundwater which may have interacted with the erupting magma.

Petrologic study of glass inclusions in pyroxenes from two tephra samples from the eruption (early, TW-1 and late, TW-2 stage ejecta) suggest a total volatile yield of about $1.15x10^7$ metric tons (as acids) (early ejecta- $6.5x10^6$ metric tons; late ejecta- $5x10^6$ metric tons). While both phases of the eruption appear to have emitted significant quantites of sulfur, chlorine and fluorine, the early ejecta were enriched in sulfur relative to late-stage products, which were richer in halogens. The inclusions in the early erupted tephra (TW-1) contain the highest concentration of sulfur ($1888 \pm 594$; n=5) of any natural glass that we have analyzed so far. Furthermore, according to major element analyses, the magma in the early stage of the eruption contained about 4.9 % total volatiles, most of which were lost on eruption. The magma erupted in late-stage activity was virtually devoid of volatiles.

Because of its location in mid-latitude region of the southern hemisphere, it is unlikely that the Tarawera eruption would be recorded in Greenland ice cores or that it would have an effect on northern hemisphere climate. Legrand and Delmas [1987], however, report a peak of sulfuric acid in an ice core from Dome C, Antarctica which they attribute to the Tarawera eruption. No estimate of global acid fallout is feasible on basis of the ice core as the Tarawera peak overlaps with that of the 1883 Krakatau eruption, but Legrand and Delmas [1987] estimate a $H_2SO_4$ deposition flux of 3.7 $kg/km^2$ for the Tarawera eruption.

*Coseguina Eruption, 1835 A.D.*

The 1835 eruption of Coseguina volcano has been regarded as one of the largest and most violent volcanic eruptions occurring in historic times in the Americas. Coseguina is located on the southern end of the Gulf of Fonseca in the northwest corner of Nicaragua. The eruption was locally of great significance [Williams, 1952], with fine ash fall near the volcano and as far as Jamaica (1300 km to the east), and darkness lasted locally for 2 to 3 days. Composition of the ejecta is andesitic, with predominantly (~ 66%) pale to dark brown glass, containing minute grains of augite, iron oxide, and abundant plagioclase microlites [Williams, 1952]. Discrete crystals of plagioclase, augite, hypersthene and magnetite are also common in the 1835 ejecta, along with lithic fragments from the original summit cone.

One of the major difficulties in trying to reconstruct an accurate scenario of the eruption and total erupted volume is the lack of outcrops of the 1835 deposit [Williams, 1952]. Original estimates for the volume of the 1835 eruption products range from 150 $km^3$ (D.R.E.) [Penck, quoted in Williams, 1952] to 50 $km^3$ (D.R.E.) [Reclus, 1891]. Williams [1952] concluded that the total volume did not exceed 10 $km^3$ (D.R.E.), although this is only an order of magnitude estimate.

Our results of the major and trace element compositions of glass inclusions and matrix glass from the 1835 Coseguina eruption are given in columns 37 and 38, respectively, of Table 3. Although the Coseguina eruption may have been responsible for a large emission of fine ash to the atmosphere, our results suggest that the aerosol cloud from this eruption may have contained small amounts of sulfur but was composed mostly of HCl ($1.23x10^7$ metric tons). There is no evidence from Greenland ice cores of sulfuric acid deposition from the Coseguina eruption. Legrand and Delmas [1987] report a small sulfuric acid peak in the ice at Dome C, Antarctica, at the time of the Coseguina eruption and estimate a global acid yield of $2.3x10^7$ metric tons, only a factor of 2 greater than our petrologic estimate. It should be noted, however, that there is a discrepancy in the composition of the acids; our petrologic estimate suggesting a volcanic volatile emission composed dominantly of HCl, whereas ice core studies suggest an aerosol composed entirely of $H_2SO_4$.

*Mount St. Helens Eruptions 1500 A.D. and 1800 A.D.*

Sulfur emission from the 1980 Mount St. Helens eruption was low [Sigurdsson, 1982], as is typical of explosive eruptions of most dacite magmas [Devine et al., 1984]. In order to characterize further the volatile degassing of this volcano, we have studied dacite tephra fall from two previous eruptions of Mount St. Helens. They are the "T" tephra fall from the Goat Rocks eruptive period (ca. 1800 A.D.) and the "W" tephra from the Kalama eruptive period [ca. 1530 A.D. Mullineaux and Crandell, 1981; Hoblitt et al., 1980; Mullineaux, 1986]. The volume of erupted tephra has not been accurately determined, but we estimate volumes of about 0.25 and 1 $km^3$(DRE) for the "T" and "W" tephras, respectively, on the basis of published field observations [Crandell et al., 1975; Crandell and Mullineaux, 1978; Mullineaux, 1986].

Our results indicate the the 1800 A.D. eruption of the "T" tephra had a negligible sulfur yield ($1.15x10^9$ g S; based on $2 \pm 12$ ppm S inclusion-matrix glass difference) but emitted substantial quantities of chlorine ($2.58x10^{11}$ g Cl). Our data indicate that the 1530 A.D. "W" eruption may have emitted as much as $7.6x10^{10}$ g S and negligible Cl (inclusion-matrix glass difference only $35 \pm 34$ ppm Cl). As with the "T" eruption, we did not measure fluorine in the samples of the "W" tephra, however, based on the analyses of the 1980 tephra, the 1530 A.D. eruptive event may have produced as much as $2x10^{11}$ g F.

These results are very similar to those of Sigurdsson et al. [1985] who found S, Cl and F yields of $2.5x10^{10}$ g, $3.4x10^{10}$ g and $5x10^{10}$ g, respectively, for the 1980 Mount St. Helens eruptive event. These results confirm our data on Taupo, reported below, and the findings of others [Devine et al., 1984; Sigurdsson et al., 1985] that dacitic magmas are generally sulfur-poor and often halogen-rich.

*Oræfajokull Eruption 1362 A.D.*

The Oræfajokull eruption in Iceland in 1362 A.D. was a major rhyolitic explosive eruption, which ejected about 2 $km^3$ D.R.E. of magma and laid waste to several districts in southern Iceland [Thorarinsson, 1958]. We have made estimates of the volatile degassing during this eruption, on the basis of the data in columns 3 and 4 of Table 3. Our results indicate a total of $3.4x10^6$ metric tons of acid, of which 11 % was $H_2SO_4$, 20 % was HCl and 69 % was HF. Of all the eruptions that we have studied so far, the Oræfajokull eruption emitted the largest proportion of halogens to the atmosphere, or slightly more than the Tambora eruption, in which 87 % of the total gas emissions were halogens [55 % HCl and 32 % HF; Sigurdsson et al., 1985; see table 1]. The Oraefajokull volcano emitted comendite magma [Carmichael, 1967], but comendite and other peralkaline magmas are generally enriched in halogens [Bailey and Macdonald, 1970]. However, because of the large difference in the mass of these two eruptions ($2.4x10^{14}$ kg vs. $4.6x10^{12}$ kg), the Tambora eruption emitted nearly 2 orders of magnitude more HCl and HF than the Oræfajokull eruption.

Ice core evidence of aerosol deposition from the Oræfajokull eruption is unclear. An acidity layer appears in both the Crete and Dye 3 Greenland ice cores at this level, but it is unsure whether these acidities are due to a volcanic event or an artifact of surface melting on the ice sheet [Hammer, 1984]. Until another core is examined in this interval, we have no way of comparing the petrologic estimate with any proxy data.

*Eldgja Eruption, 934 A.D.*

The basaltic fissure eruption of Eldgja in Iceland in 934 A.D. was the second largest historic lava eruption on Earth (second only to Laki 1783), with an estimated volume of 9 $km^3$ (DRE) of magma [Larsen, 1979]. The eruption involved vigorous and high fire-fountains which formed large tephra deposits along margins of the eruptive fissure. Our results, shown in columns 1 and 2 of Table 3, indicate a volatile yield of $2.7x10^{13}$ S, $3.8x10^{12}$ g Cl and $5.8x10^{12}$ g F, corresponding to a mass of $9.39x10^7$ metric tons of acids ($H_2SO_4$+HCl+HF). By comparison, Hammer et al. [1980] estimate $1.65x10^8$ metric tons global acid fallout from this eruption, on basis of D.C. electrical conductivity measurements of Greenland ice cores. We consider these estimates in very good agreement when assumptions and uncertainties in both measurements are considered. The factor of two difference between these estimates may be related to the chlorine content. Hammer [1980] estimates that at least 65 % of the acid fallout was HCl as opposed to only 4 % in our petrologic estimate, where sulfur dominates (89 %). This discrepancy may be due to reactions between volcanic and other aerosols during atmospheric transport, as $H_2SO_4$ reacts with NaCl from sea salt to produce HCl and $Na_2SO_4$ [Legrand and Delmas, 1984; Palais and Legrand, 1985, Chuan et al., 1986] causing an overestimate of volcanic Cl in the ice cores.

Herron [1982] studied the concentration of anions ($SO_4^=$, $NO_3^-$, $Cl^-$ and $F^-$) in ice from the Eldgja eruption interval in the Dye 3 core from Greenland. Herron's results suggest a ten-fold increase in sulfate and a hundred-fold increase in fluoride concentration over background values in the ice associated with the eruption. Chloride concentrations were only slightly elevated in the Eldgja layer, consistent with the petrologic results. Hammer [1983] noted that the increase in fluoride concentration, associated with the Eldgja eruption began almost a full year prior to the sulfate increase. This is consistent with the idea that fluorine (as HF) may have been adhering to fine grained dust associated with the eruption [Oskarsson, 1980]. Fallout of this fine grained tephra, perhaps due to particle aggregation, could then explain the deposition of the HF prior to the $H_2SO_4$, which would have a longer atmospheric residence time. Few studies have examined whether there is any fine grained tephra associated with volcanic acidity layers in Greenland ice, however, because of the difficulties of working with small amounts of fine-grained tephra [Hammer, 1984].

*Rabaul Eruption, 1400 years B.P.*

Rabaul volcano, on New Britian island in Papua New Guinea, contains an 8x14 km caldera, which has been highly active in the past. The volcano showed evidence of renewed activity in 1983, when a dramatic short-term increase in seismicity and ground deformation took place within the caldera. The most recent eruption occurred in 1937 to 1943, but the last major eruption is $^{14}C$ dated 1400 yr. B.P. This event produced a wide-spread plinian dacite fall deposit, with a volume of 1.66 $km^3$ (0.6 $km^3$ DRE). Following the plinian phase, a further 8 $km^3$ (3.5 $km^3$ DRE) were erupted during the ignimbrite phase [Heming, 1974; Walker et al., 1981].

We have studied the volatile content of glass inclusions and matrix glass in samples of the plinian and ignimbrite phases of the 1400 yr. B.P. Rabaul eruption. The results are given in columns 20 to 24 of Table 3. Glass inclusions in pyroxene phenocrysts from the plinian phase are higher in sulfur and slightly lower in chlorine content than inclusions in plagioclase. This may be the result of degassing after the pyroxene crystallized, just prior to the formation of the plagioclase. We have used the pyroxene data ($336 \pm 94$ ppm S) to obtain a maximum petrologic estimate for sulfur degassing from this eruption. Our results suggest that the Rabaul plinian eruption had a maximum sulfur yield of $4.8x10^{11}$ g (~$1.5x10^6$ metric tons $H_2SO_4$).

The difference between the chlorine concentration in the plagioclase inclusions ($2932 \pm 368$ ppm) and the matrix glass ($2678 \pm 292$ ppm) is $254 \pm 146$ ppm Cl. This gives a Cl yield of about $3.7x10^{11}$ g Cl ($3.75x10^5$ metric tons HCl). Our fluorine analyses on glass inclusions in plagioclase ($243 \pm 129$ ppm) and associated matrix glass ($264 \pm 92$ ppm ) indicate no fluorine degassing from this eruption. No fluorine analyses were made on inclusions in pyroxene. Therefore, if we combine results from the plagioclase and pyroxene inclusions, we estimate that the plinian phase of the Rabaul eruption produced a maximum volcanic volatile yield (as acids) of about $1.9x10^6$ metric tons ($1.5x10^6$ metric tons $H_2SO_4$, $3.75x10^5$ metric tons HCl).

We have also analyzed samples from the 1400 yr B.P. Rabaul ignimbrite phase. Our analyses indicate no significant degassing of sulfur in the ignimbrite phase of the eruption. However, a higher concentration of sulfur in the matrix glass

(570 ± 205 ppm) than in the inclusions (218 ± 41 ppm) suggests that sulfur was taken up by the matrix glass by metasomatism after emplacement of the ignimbrite, as we have found in the Campanian ignimbrite (see below). A maximum estimate of volatile degassing from the entire Rabaul eruption can be made if we assume that the entire volume of the 1400 yr B.P. Rabaul eruption (plinian plus ignimbrite phase- 4.1 $km^3$ D.R.E.) degassed with the same yield as the plinian phase. This would give a maximum volcanic volatile yield (according to our petrologic estimate) of about $1.26 \times 10^7$ metric tons ($1 \times 10^7$ metric tons $H_2SO_4$, $2.56 \times 10^6$ metric tons HCl).

Our analyses of the Rabaul 1400 yr B.P. eruptive event indicate a minimum total volatile yield ($H_2SO_4$ + HCl) of $1.9 \times 10^6$ metric tons and a maximum yield (if the entire volume of the eruption is assumed to have degassed in the same manner as the plinian phase alone) of about $1.3 \times 10^7$ metric tons. It has been proposed that this eruption produced a wide-spread "dry fog" and other atmospheric phenomena which affected Europe and the Middle East in 536 A.D., with an estimated aerosol mass of $3 \times 10^8$ metric tons [Stothers, 1984]. This estimate was made from direct ancient observations of the optical depth perturbation of the stratosphere and by scaling the relative signal strengths for the Eldgja and Tambora eruptions to the 536 A.D. event using ice core data of Herron et al. [1982] and Hammer et al. [1980]. As we discuss below, ice core estimates of global acid fallout from tropical and high northern latitude eruptions cannot be compared, since the latter may be significantly overestimated in Greenland ice cores.

Our results therefore call into question the estimate of Stothers [1984]. Although the increases in anion concentration observed by Herron [1982] in the Dye 3 core (535 A.D.) and the strong acidity signal observed by Hammer et al. [1980] at about 540 ± 10 A.D. (global acid fallout estimate of $7 \times 10^7$ metric tons) may be the result of the same event discussed by Stothers [1984], we believe his aerosol loading is overestimated by one or two orders of magnitude. Future ice core studies may reveal additional information concerning the proportions of acids in the aerosol layer produced by this eruption. Our results indicate that approximately 80 % of the Rabaul volcanic volatile release was $H_2SO_4$ and the remainder was largely HCl.

*Taupo Eruption, 131 A.D.*

The Taupo eruption in New Zealand was one of the largest known explosive eruptions within the last 7000 years and falls in a class of its own as an "ultraplinian" eruption, because of the high mass eruption rate, high degree of fragmentation and consequently the wide dispersal of the tephra [Walker, 1980]. The statistical mean of 22 $^{14}C$ age determinations, yields 1819± 17 yrs. B.P. or 131 A.D. for the date of the eruption [Healy, 1964]. Two plinian pumice falls (Hatepe plinian pumice (1.4 $km^3$ D.R.E.) and Taupo plinian pumice (5.1 $km^3$ D.R.E.)) were generated, as well as three phreatomagmatic ash falls (initial phreatomagmatic ash (0.005 $km^3$ D.R.E.), Hatepe phreatoplinian ash (1.0 $km^3$ D.R.E.) and Rotongaio phreatoplinian ash (0.7 $km^3$ D.R.E.)) and several ignimbrite flow units (early intraplinian ignimbrite flow units (0.5 $km^3$ D.R.E.) and main Taupo ignimbrite (10 $km^3$ D.R.E.), [Wilson and Walker, 1985]). A total of $\geq 35$ $km^3$ D.R.E. of rhyolite magma erupted during this event includes all tephra fall deposits (8.2 $km^3$ D.R.E.), the ignimbrite units (10.5 $km^3$ D.R.E.), co-ignimbrite ash (~ 7 $km^3$ D.R.E.) and the primary material now under Lake Taupo [8 to 20 $km^3$ D.R.E.; Wilson and Walker, 1985 ]. Tephra dispersal from the eruption indicates a column height for the plinian phase of no less than 50 km [Walker, 1980 ; Carey and Sparks, 1986].

We have studied samples from the Hatepe plinian pumice, the Rotongaio phreatoplinian ash and the Taupo plinian pumice phases of the eruption. The concentrations of S, Cl and F in glass inclusions and matrix glasses from all three samples are very similar (columns 5 to 10 of Table 3). We have used the volumes estimated by Wilson and Walker [1985] to calculate the total volatile emission for the three phases of the eruption. Examination of Table 3 reveals very similar but low (close to our detection limit of 50 ppm) concentrations of sulfur in inclusions and matrix glasses from the three phases of the eruption. The inclusion-matrix glass difference for sulfur is not significant and for all intents and purposes we estimate that there was negligible sulfur emission from the Taupo eruption. Since our analyses reveal very similar S, Cl and F concentrations in the inclusions and matrix glasses of all three phases, we can extrapolate our analyses to the total volume of the eruption. Our results demonstrate that the Taupo eruption involved very sulfur-poor magma and caused negligible emission of sulfur to the atmosphere, despite the large volume of magma that was erupted (≥ 35 $km^3$ D.R.E.).

On the other hand, emissions of chlorine and fluorine were significant. Scaling the average inclusion-matrix glass differences for chlorine (~200 ppm) and fluorine (~100 ppm) to the estimated total volume for the eruption (~35 $km^3$) gives a total chlorine yield of $1.5 \times 10^{13}$ g and $8 \times 10^{12}$ g fluorine. The values quoted in table 3 for the yield of Cl and F are scaled only to the volume of magma erupted in the phases of the eruption that we studied and are thus minimum estimates while the above estimate is a maximum value. Judging from the low oxide totals of the glass inclusions compared to the matrix glass, we estimate that the pre-eruption total volatile content of the Taupo magma was between 3 to 6 wt. % and the total volatile loss on eruption was about 2 to 4 wt. %; $H_2O$ was probably a major component of the total volatiles.

Our results are similar to those of Dunbar et al. [1985] who found chlorine contents of melt inclusions in rhyolitic obsidian clasts in Taupo tephra ranging from 0.18-0.24 wt. % and sulfur contents below their detection limit of 200 ppm. They estimated an original magmatic water content of 2.5 to 4 wt. %. These results demonstrate that the Taupo eruption involved relatively sulfur-poor magma, and caused only minor emission of sulfur to the atmosphere. A significant degassing of both chlorine and fluorine occurred, however, and if one uses the total volume of the eruption and the average Cl and F yield estimates, the total volatile emissions from the eruption is estimated as $2.35 \times 10^7$ metric tons acids.

It has been proposed that the Taupo eruption was responsible for the atmospheric effects (red coloration of the sun and moon) observed in China and Rome in 186 A.D. [Wilson et al., 1980]. This date is, however, close to 3 standard deviations outside the mean of the $^{14}C$ age for the eruption. Furthermore, our results suggest that the volcanic volatile mass (especially $H_2SO_4$) was too low to have such far ranging effects, and that the $^{14}C$ age of 131 A.D. is probably a more reliable age for the eruption. Nevertheless, we do have evidence that significant amounts of HCl and HF were injected into the stratosphere (eruption column height > 50 km). Little is known about the climatic effects of the halogen acids, which are not known to form atmospheric aerosols, but they may have had a major effect on the chemistry of the stratosphere (e.g. ozone depletion) at that time [Crutzen, 1974; Mankin and Coffey, 1984].

Evidence for distant deposition from the Taupo eruption has been found in Greenland ice cores. Hammer [1984] documented a "dust" layer in the Dye 3 core from South Greenland, which is dated at about 174 to 175 A.D. The layer is composed of silicate particles of "high Si content" (precise composition unknown) and is not associated with high acidity. In Antarctica, closer to the source of the Taupo eruption, a layer at 104.2 to 104.5 m depth in the Dome C ice core (estimated age ~ 2000 yr B.P.) was found to have elevated levels of D.C. electrical conductivity, sulfate, and microparticles [Maccagnan et al., 1981; Benoist et al., 1982]. However, major element analyses of glass shards found in this layer ($SiO_2$ = 66.54 ±1.51%) by de Angelis et al.[1985] do not support the conclusion that the material is from the Taupo eruption. Therefore, confirmation that the Taupo eruption produced either acid fallout or ash deposition in Antarctica must await future ice core studies. Our petrologic data suggest, however, that the event may be only barely detectable in polar ice cores on the basis of acidity and D.C. conductivity measurements alone because of the low volcanic volatile yield.

*Campanian Eruption, 35,000 years B.P.*

A large explosive eruption in the Phlegrean Fields in Italy about 35,000 yrs. B.P. produced a total of 80 $km^3$ of magma and led to caldera collapse [Barberi et al., 1978; Rosi et al., 1983]. The Campanian eruption is the largest eruption documented in the Mediterranean region during the late Pleistocene. The products of the eruption include the Campanian ignimbrite deposit on land and a major tephra fall layer which extends over the eastern Mediterranean region, beyond Crete. The tephra fall deposit alone has a volume of 73 $km^3$ [about 23 $km^3$ D.R.E., Cornell et al., 1983].

We have studied two samples of tephra from a 50 cm thick pumice fall from the Lavorate Quarry, 50 km east of the Phlegrean Fields caldera, where the tephra fall deposit is overlain by incipiently welded pyroclastic flow. Our results (columns 25 to 28; table 3) show that the earliest phase of the eruption may have emitted $2x10^{13}$ g of S. Sulfur degassing in the second half of the eruption was apparently negligible; however, this may be due to metasomatic alteration of the upper tephra fall unit, as discussed below.

We find higher concentrations of chlorine and fluorine in the matrix glasses than in the glass inclusions of tephra fall samples, indicating no apparent degassing of chlorine or fluorine during the eruption. We believe, however, that Cl and F content in the matrix glass of these samples may have been affected by post-depositional hydrothermal alteration or metasomatism associated with cooling of the overlying ignimbrite deposit. Other studies of the S,Cl and F content of glass inclusions and matrix glass in samples of the Campanian ignimbrite [Devine et al., 1984; M. Rosi, pers. comm. 1986] reveal similar results. Furthermore, as discussed above, we have found a similar enrichment of halogens in matrix glass relative to glass inclusions in samples of the 1400 yr. B.P. Rabaul ignimbrite. These results suggest that the enrichment of halogens in the matrix glass may be related to either a syn-eruptive or post-depositional processes in the ignimbrite.

## Discussion

The results of petrologic estimates of degassing from the twelve eruptions presented in this paper and the seventeen events studied by Devine et al. [1984] can be compared with volcanic aerosol mass estimates based on ice core and aerosol optical depth studies for these events. These petrologic studies now provide the data base required to re-evaluate the potential climatic impact of volcanic degassing from large eruptions and a further test of the claim of Devine et al. [1984] and Rampino and Self [1984] of a direct relationship between mean northern hemisphere surface temperature decline after volcanic eruptions and the mass of sulfur released by the eruption.

*Aerosol Estimates*

The mass of total acids released by the eruptions which we studied (those for which aerosol data also exist) are plotted versus the total aerosol mass in figure 1, calculated from global peak optical depth estimates [Deirmendjian, 1973; Stothers, 1984] and direct measurements using airborne lidar and particle counters [McCormick et al., 1981; Kent, 1982; McCormick, 1982]. For most of the eruptions the agreement

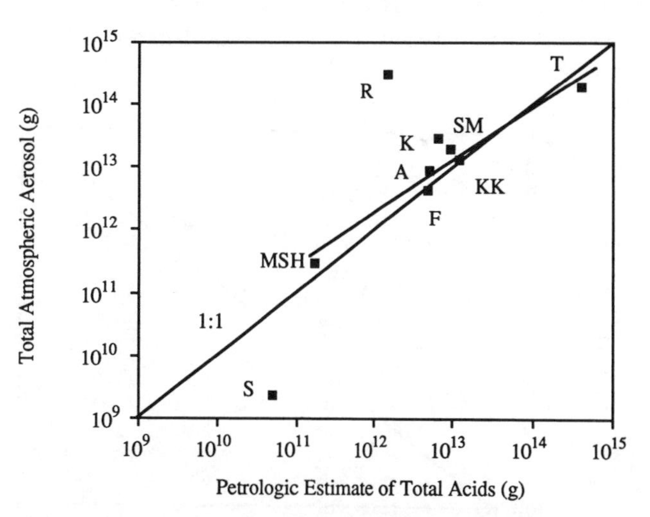


**Fig. 1. Plot of total acids (g) estimated by petrologic method versus atmospheric estimates of aerosol mass (g) calculated from optical depth perturbation and direct measurement for eruptions reported in this paper. Key: T - Tambora, SM- Santa Maria, K- Katmai, KK- Krakatau, R- Rabaul, A- Agung, F- Fuego, MSH- Mount St. Helens, S- Soufriere. Linear regression equation (y = 184.5065 * x^0.8372; R= 0.96) does not include R and S (see discussion in text). Line corresponding to 1:1 relationship also shown for comparison.**

between our petrologic estimates and the aerosol measurements is good, except Soufriere and Rabaul.

In case of the Rabaul eruption, the aerosol loading ($3x10^{14}$ g) estimated by Stothers [1984a] from observations in the literature at the time, is for an event which occurred around 536 A.D. and may not correspond to the Rabaul eruption. Because of the poor correlation between our petrologic estimate of total acids ($1.48x10^{12}$ g) and the total aerosol estimate of Stothers [1984a], we suspect that the "mystery cloud" event of A.D. 536 described by Stothers [1984a] is probably not due to the 1400 yr. B.P. Rabaul eruption, but derived from another event.

Sigurdsson et al. [1985] have discussed the discrepancy between the petrologic estimate and the volcanic aerosol estimate [McCormick et al., 1981] for the Soufriere 1979 eruption. They concluded that since the satellite-based aerosol extinction measurements probably included only about half of the erupted mass (only two of the eight eruption plumes), the two estimates are only off by an order of magnitude. In addition, a comparison of sulfur mass flux emission [measured by COSPEC; Hoff and Gallant, 1980] and petrologic estimates of sulfur yield per mass of erupted magma gives very similar results, supporting the validity of the petrologic method for the study of volatile release [Sigurdsson et al., 1985].

*Ice Core Estimates*

Petrologic estimates of volcanic degassing are compared in figure 2 with ice core estimates of total acids ($H_2SO_4$ + HX) from several eruptions [Hammer et al., 1980; Legrand and Delmas, 1987]. The estimates of Legrand and Delmas [1987] represent only $H_2SO_4$; however, they are comparable with the total acid estimates of Hammer et al.

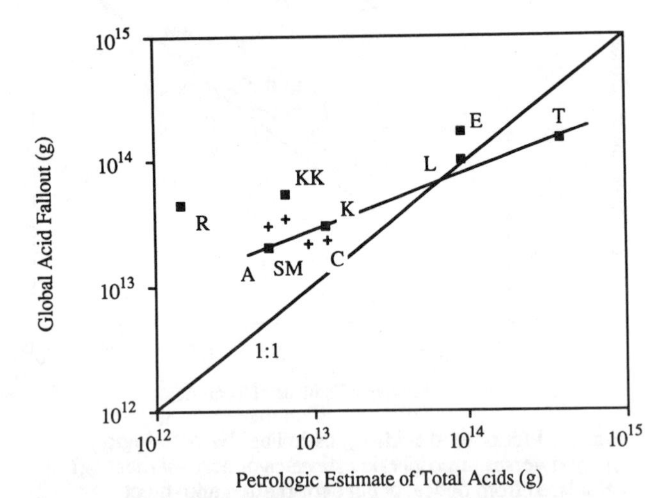


Fig. 2. Plot of total acids (g) estimated by petrologic method versus global acid fallout (g) estimated from Greenland [Hammer et al., 1980] and Antarctic [Legrand and Delmas, 1987] ice core studies. Key - same as in Fig. 1 including E- Eldgja, C- Coseguina. Linear regression equation (y = 1.366e+8*x^0.4111, R=0.92) calculated for all data points. Line corresponding to 1:1 relationship also shown for comparison.

[1980], as no hydrochloric acid or hydrofluoric acid (HX) were associated with any of the eruptions that they studied [Legrand, 1985].

The figure shows that the ice core estimate is greater than the petrologic estimate for many low-sulfur yield eruptions (Krakatau, Agung, Katmai, Coseguina and Santa Maria). In the case of eruptions with high sulfur yield (e.g. Laki, Eldgja and Tambora), the correspondance is good. The petrologic method tends to underestimate the amount of released volatile species in general, and therefore our results indicate minimum values [Devine et al., 1984].

The poor correspondance between the petrologic estimate for the Rabaul eruption and the ice core estimate of global acid fallout attributed by Stothers to this eruption, indicates that these estimates are for two different events. We therefore attribute the ice core acidity peak in 540 $\pm$ 10 A.D. [Hammer et al. 1980] and the 536 A.D. atmospheric event [Stothers, 1984] to another eruption of unknown origin, with volatile output over two orders of magnitude larger than the Rabaul event.

Our estimates of sulfuric acid yield are plotted in figures 3a and 3b versus the estimate of sulfuric acid deposition flux ($kg/km^2$; the time-integrated deposition of acid which takes into account the mean accumulation rate at the ice core site and the acid concentration in the peak) in Greenland [Hammer et al., 1980] and Antarctica [Legrand, 1985; Legrand and Delmas, 1987], respectively. The relationship between our estimates of sulfuric acid yield (g) and the ice core estimates of sulfuric acid deposition flux ($kg/km^2$) on the ice sheets (Greenland and Antarctica) (Figures 3a and 3b) is very good, with few exceptions. The order of magnitude difference in the sulfuric acid deposition flux ($kg/km^2$) in Greenland versus Antarctica is due to a combination of factors, including the order of magnitude difference in accumulation rates between the two sites, and differences in the amount of dry deposition and/or direct stratospheric injection of volcanic debris [Legrand and Delmas, 1987].

In Figure 3a, the Greenland ice core estimates for the Laki (64.08$^\circ$ N) and Eldgja (63.75$^\circ$ N) eruptions appear to be overestimated, compared the petrologic estimates, considering the good agreement for the other eruptions (Tambora, Agung, Krakatau and Katmai). The apparent overestimate from ice cores is probably due to the high northern latitude location of these volcanoes, their relative proximity to the Greenland ice sheet (~ 1500 km) and the nature of the atmospheric transport involved (tropospheric vs. stratospheric). The Antarctic ice core deposition flux for the Agung eruption may be an overestimate (figure 3b), whereas the deposition flux for the Santa Maria eruption may be underestimated, compared to the excellent correlation of petrologic and ice core estimates for the other three eruptions (Tarawera, Krakatau and Tambora). The discrepancy for the Agung eruption is readily accounted for by the observed two-thirds/one-third distribution of the 1963 Agung aerosol between the Southern and Northern Hemispheres, respectively [Castleman et al., 1974; Delmas et al., 1985; Legrand and Delmas, 1987]. The slight underestimate of the deposition flux in Antarcitca from the Santa Maria eruption is likely due to the northern latitude location (14$^\circ$ N) of the volcano.

*Halogens*

The results show very significant releases of HCl and HF gases to the atmosphere in many eruptions. Of the twenty-six

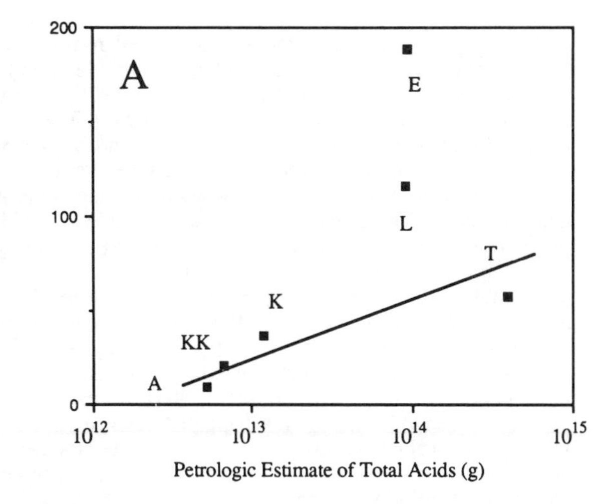


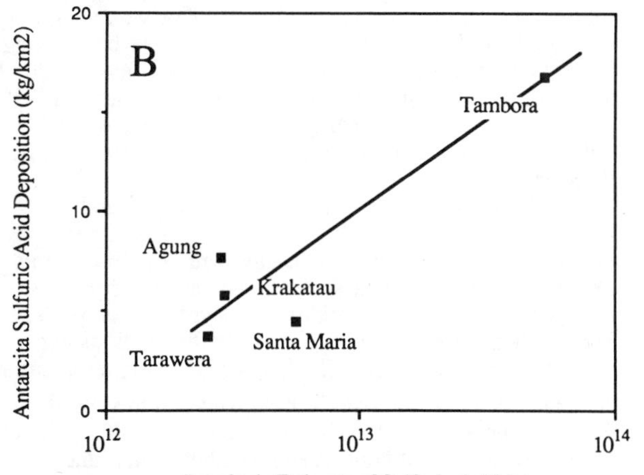


Fig. 3. (A): Plot of total acids (g) estimated by petrologic method versus deposition flux of total acids estimated by Hammer et al. [1980] from the Crete ice core, Greenland. Key same as in Figs. 1 and 2. Linear regression equation (y = 0.0015* x^0.3183, R = 0.80) calculated without the Laki (L) and Eldgja (E) points (see text).
(B): Plot of sulfuric acid (g) estimated by petrologic method vs. deposition flux of sulfuric aicd by Legrand and Delmas [1987] from the Dome C ice core, Antarctica. Key same as in Figs. 1 and 2. Linear regression equation (y = 4.0828 + 2.429e^-13x, R = 0.99) calculated without Santa Maria (SM) and Agung (A) points (see text).

events studied so far by the petrologic method, thirteen emitted sulfur-dominated volcanic gases, seven events emitted halogen-dominated gas clouds, while sulfur and halogens were emitted in roughly equal proportions in the remaining six events. The petrologic studies indicate that the halogen output may be truly exceptionally large in some events, such as during the great 1815 Tambora eruption in Indonesia, when chlorine and fluorine yield to the atmosphere is estimated as $2x10^{14}$ and $1.7x10^{14}$ g, respectively. However, it should not be assumed that these halogen gases form aerosols in the stratosphere, as physical and chemical models suggest that HCl and HF will not form liquid aerosols under normal stratospheric conditions [Miller, 1983; Solomon and Garcia, 1984].

Studies of the atmosphere following recent eruptions shows highly significant increase of chlorine. Thus spectroscopic observations of the total column amount of HCl, made shortly after the 1982 El Chichon eruption, show an increase of the stratospheric HCl burden of 40 per cent over a large part of the globe, with an estimated total increase of about $4x10^{10}$ g HCl, or about 9 per cent of the total global stratospheric HCl burden [Mankin and Coffey, 1984]. The hydrogen chloride may in part be derived from release of gaseous HCl during breakdown of halite particles observed in the aerosol cloud from the volcano [Wood et al, 1985], as well as from direct degassing of the magma. The atmospheric estimate of Cadle et al. [1977] of the mass of HCl released by the Agung 1963 eruption ($1.2x10^{12}$ g) is virtually identical to our petrologic estimate ($1.42x10^{12}$ g), whereas the estimate of Stolarski and Butler [1978] for the Krakatau 1883 eruption is one order of magnitude lower than the petrologic value. Stolarski and Cicerone [1974] were the first to consider that chlorine from volcanoes might perturb the stratospheric ozone layer. Based on very conservative estimates of chlorine degassing, the stratospheric photochemistry model of Stolarski and Butler [1978] for the Krakatau 1883 eruption estimates about 7 per cent ozone column depletion. Output of chlorine from the 1815 Tambora eruption is about three orders of magnitude higher on basis of petrologic data, and modeling of the impact on the ozone layer of such large volcanic chlorine emissions is urgently needed.

### *Temperature Data*

Devine et al. [1984] evaluated the possible effect of volcanic eruptions on climate and proposed a relationship between the mass yield of sulfur to the atmosphere from an eruption (as estimated by the petrologic method) and the observed decrease in mean northern hemisphere surface landmass temperature in the one to three years following the eruption, on basis of published temperature data [Rampino and Self, 1982, 1984]. Devine et al.[1984] found that the mean surface temperature decrease was related to the estimate of sulfur yield by a power function (r= 0.971), with the power to which the sulfur mass is raised being equal to 0.345.

We have combined new data on sulfur yield presented in this paper and data from the literature for the Fuego eruption with data base of Devine et al. [1984] in figure 4a. The resulting equation describing the relationship between estimated sulfur mass and observed surface landmass temperature decrease is similar to the previous one [Devine et al. 1984], with the power to which the sulfur mass is raised being equal to 0.3076 and r=0.92. The total yield of volcanic volatile components as determined by the petrologic method (sulfur, chlorine and fluorine) for these eruptions, calculated as acids, is plotted versus the observed northern hemisphere temperature decrease in figure 4b. The relationship is also a power function, with the power to which the total mass of acids are raised being equal to 0.274 and r= 0.87.

Although these results appear to confirm the relationship proposed by Devine et al. [1984], it should be emphasised that the sample is limited and the temperature deviations listed

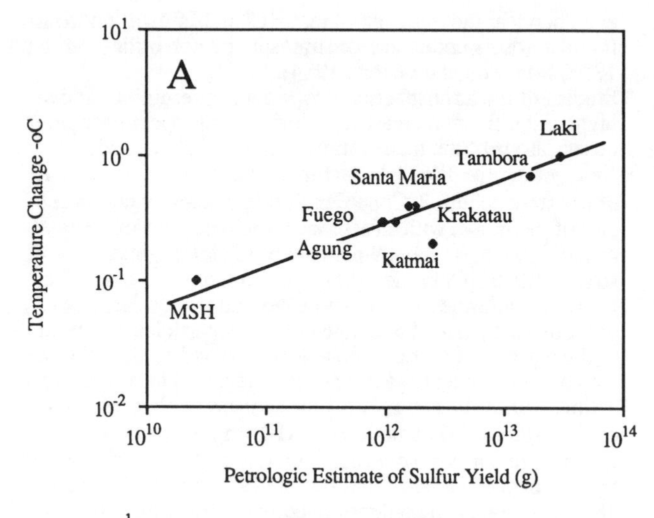


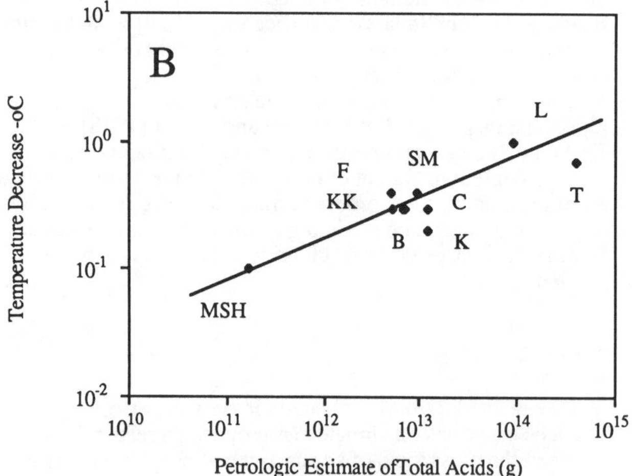


Fig. 4. (A): Plot of sulfur yield (g) estimated by petrologic method vs. estimated northern hemisphere temperature decrease. Linear regression equation (y = 5.891e-5*x^0.3076, R=0.92) calculated with all points. See text for further discussion.
(B): Plot of total acids (g) estimated by petrologic method vs. estimated northern hemisphere temperature decrease. Linear regression equation (y = 9.220e-5*x^0.2739, R = 0.87). calculated with all points. See text for further discussion.

in table 4 are all associated with large errors. This is especially true for the two earliest eruptions (Laki-1783 and Tambora-1815), as temperature records are less complete prior to about 1880 [Groveman and Landsberg, 1979; Jones et al., 1982; Angell and Korshover, 1985]. Furthermore, the northern hemisphere temperature deviations for the Tambora and Laki eruptions were calculated from a different data set and by a different method than for the other events [Rampino and Self, 1984].

The surface temperature changes following six major volcanic events between 1780 and 1980 (Laki-Asama, Tambora, Coseguina, Krakatau, Santa Maria-Pelee-Soufriere and Agung) were recently examined by Angell and Korshover [1985]. They calculated temperature deviations using data sets of Groveman and Landsberg [1979; prior to 1881] of Jones et al. [1982; 1881 through 1981]. The temperature deviations calculated by Angell and Korshover [1985] are based on the difference between the average temperature deviations for the five years immediately before and after the volcanic eruptions. Rampino and Self [1984], on the other hand, subtracted the lowest temperature anomaly in the three years after the eruption from the mean temperature of the year before the eruption. As stated by Angell and Korshover [1985] "...the relatively-long interval of five years before and after the eruption ...may be too long, based on the usual stratospheric lifetime of volcanic dust [e.g., Oliver, 1976]. Kelly and Sear [1984], for example, provide evidence for abrupt lowering of Northern Hemisphere surface air temperatures two to three months after eruptions in the past one hundred year data set, and a return to pre-eruption temperature level during the third year following eruption. We suspect that the five-year averaging of Angell and Korshover [1985] may therefore be too long to discern any effect from a short-lived volcanic aerosol cloud and have therefore used the data base of Rampino and Self [1984], based on three-year averages, to calculate the volcano-related Northern Hemisphere temperature anomalies listed in table 4.

As discussed by Handler [1984] and Angell and Korshover [1985], warming of the tropical (and perhaps the extratropical) troposphere associated with El Nino events might mask the cooling associated with an eruption if the timing of the El Nino coincides with a volcanic eruption. Such a phenomenon would complicate the relationships between volcanic aerosols ($H_2SO_4$ or HCl) and temperature decreases discussed above. Related to this is the possibility that volcanic aerosol clouds may even trigger (although not actually cause) an El Nino event. Handler [1984] and Schatten et al. [1984] have proposed models of such a triggering mechanism, involving feedbacks affecting the dynamics of the atmospheric circulation (shift in the ITCZ and/or decreases in the pole to equator temperature gradient which would affect zonal wind velocities and atmospheric motions). It is noteworthy that the last two anomalous El Nino events (those that have not followed the normal pattern) occurred in 1963 and 1982-83, or coinciding with the Agung and El Chichon eruptions. The strongest El Nino event in recent times occurred in 1940-41 and was apparently also an anomalous event, similar to the 1982-83 event. Although no known significant volcanic eruptions occurred at this time, Lyons et al. [in prep.] have documented a 1941 volcanic event on the basis of increased excess sulfate and chloride in an ice core from Dye 3, Greenland.

The occurrence of frost rings in trees exhibits a good relationship with the timing of known volcanic events and El Ninos, including a frost event in 1941, for which there is no known volcanic event [LaMarche and Hirschboeck, 1984]. A major volcanic eruption may have easily gone unreported during the second world war, as the number of volcanoes reported active in the western Pacific and Indonesia in the years 1941 to 1945 dropped by one-third from the preceeding five years [Simikin et al. 1981]. It is thus likely that a significant volcanic eruption was responsible for the volcanic aerosol in 1941 and may have been associated with the frost rings and El Nino event.

TABLE 4. Estimates of total volcanic aerosols from petrologic, atmospheric and ice core techniques, and estimated temperature decreases after the eruptions.

| Eruption | Petrologic[1] grams | Atmospheric [2] grams | Ice Core[3] grams | $\Delta^oC$[4] |
|---|---|---|---|---|
| Agung | $5.17 \times 10^{12}$ | 3 to $9 \times 10^{12}$ | $2 \times 10^{13}$ ($3 \times 10^{13}$) | 0.3 |
| Krakatau | $6.69 \times 10^{12}$ | 3 to $9 \times 10^{13}$ | $5.5 \times 10^{13}$ ( $3.8 \times 10^{13}$) | 0.3 |
| Tambora | $3.94 \times 10^{14}$ | $2 \times 10^{14}$ | $1.5 \times 10^{14}$ ($1.5 \times 10^{14}$) | 0.7 |
| Laki | $9.19 \times 10^{13}$ | | $1 \times 10^{14}$ | 1.0 |
| Eldgja | $9.39 \times 10^{13}$ | | $1.7 \times 10^{14}$ | |
| Katmai | $1.11 \times 10^{13}$ | $1.34 \times 10^{13}$ | $3 \times 10^{13}$ | 0.2 |
| Rabaul | $1.48 \times 10^{12}$ | $3 \times 10^{14}$ | $4.5 \times 10^{13}$(5) | |
| St Helens | $1.67 \times 10^{11}$ | $3 \times 10^{11}$ | | $<0.1$ |
| Coseguina | $1.23 \times 10^{13}$ | | ($2.3 \times 10^{13}$) | 0.4 |
| Tarawera | $1.13 \times 10^{13}$ | | | |
| Santa Maria | $9.31 \times 10^{12}$ | $\leq 2 \times 10^{13}$ | ($2.2 \times 10^{13}$) | 0.4 |
| Bezymianny | $7.06 \times 10^{12}$ | | | 0.3 |
| Fuego | $4.95 \times 10^{12}$ | 3 to $6 \times 10^{12}$ | | 0.4 |
| Oraefajokull | $3.39 \times 10^{12}$ | | | |
| Taupo | $2.41 \times 10^{12}$ | | | |
| Soufriere | $5.03 \times 10^{10}$ | $2.3 \times 10^{9}$ | | |

[1] Data from Devine et al. [1984], Sigurdsson et al. [1985], Rose et al [1982] and this work.
[2] Data from Deirmendjian [1973], Volz [1975], Stothers [1984a; 1984b], Kent [1982], McCormick et al [1981; 1982], Rampino and Self [1984], Cadle et al [1977].
[3] Data from Hammer et al [1980] ($H_2SO_4$); number in parentheses from Legrand and Delmas [1987] ($H_2SO_4$ only).
[4] Reported Northern Hemisphere surface temperature change assocaited with the eruption. Data from Sigurdsson [1982], Rampino and Self [1982], Robock [1981] and Jones et al [1981].
[5] Unknown eruption recorded as an acidity layer in Greenland Dye-3 core 540±10 AD. Possible Rabaul eruption (1400 yrs BP.; Hammer [1980].

## Conclusions

The petrologic estimate of volcanic volatile emissions from major volcanic eruptions is a viable method of determining the mass yield to the atmosphere of components such as sulfur, chlorine and fluorine. As this approach only requires sampling and study of the geologial deposit from the eruption, it can be applied to the estimation of volatile yield from any volcanic event in the geologic record. This is of particular value, as the recurrence interval of truly large volcanic events is several hundred years, and such events have not yet taken place since the rigorous monitoring of the atmosphere began. The method provides only a minimum estimate, but the results have helped us refine our understanding of a number of major eruptions for which little other information exits.

In general, the yield estimates of sulfur, chlorine and fluorine emissions on basis of the petrologic method compare well with other techniques (atmospheric and ice core methods) for estimating volcanic volatile emissions. The correspondance between the petrologic and atmospheric aerosol estimates is good while ice core estimates seem to compare best for the largest eruptions ($>10^{13}$ g). The data presented here support the observation of Devine et al. [1984] of a relationship between the sulfur yield of a number of major historic eruptions and a northern hemisphere temperature decrease. A similar correlation is seen for total acids versus temperature decrease, although the relationship is not as good.

The chemical composition of volcanic volatiles, i.e. the proportions of sulfur, chlorine and fluorine, differs systematically from volcano to volcano, and in general reflects the chemistry of the erupting magmas and the solubility of sulfur and halogens in these melts. Thus eruption of high-silica magmas, such as rhyolites and dacites, generally results in relatively low yield of sulfur and halogens to the atmosphere, in spite of the high mass eruption rate and large total erupted mass characteristic of many such events. Thus the very violent 131 A.D. Taupo eruption in New Zealand emitted 35 $km^3$ of silicic magma up to 51 km height, but produced a negligible yield of S, Cl and F. We observe, for example, that the atmospheric yield of $H_2SO_4$ from eruptions of silicic magmas is typically 100 to 500 ppm per erupted mass, whereas the $H_2SO_4$ yield of basaltic eruptions to the atmosphere is in the range of 1000 to 5000 ppm. Their explosive nature, along with the low content of associated acidic volatiles, suggests that silicic explosive eruptions may be more easily recognized in ice cores by the presence of fine silicate ash, rather than by the presence of acidity peaks.

At the other extreme are fissure eruptions of basaltic magmas, such as from Eldgja and Tarawera and the 1783 Laki eruption, which inject large amounts of sulfur, chlorine and fluorine into the atmosphere. While most fissure eruptions probably produce tropospheric aerosol clouds, their eruption clouds may reach the stratosphere under certain circumstances [Devine et al., 1984; Wolff et al., 1984; Walker et al., 1984; Carey and Sparks, 1986; Stothers et al. 1986].

A third compositional type of volcanic volatiles are exhibited during eruption of trachyte and trachyandesite magmas, such as the great Tambora eruption of 1815. Such magmas typically have very high yield of halogens to the atmosphere, in addition to sulfur. The relationship between magma composition and the petrologic estimates of volcanic yield of $H_2SO_4$, HCl and HF per $km^3$ of magma erupted are shown in figures 5a, 5b and 5c and listed in table 5. In general, there is a fairly systematic decrease in sulfur, chlorine and fluorine yield per $km^3$ with increasing $SiO_2$ of the magma. Most of the basaltic eruptions (Eldgja, Surtsey, Krafla, Laki), however, have a quite low yield of HCl per $km^3$ of magma (~ $10^5$ tons HCl/$km^3$). The only exception is the Tarawera basaltic eruption, with a yield of $3.4 \times 10^6$ tons HCl/$km^3$ magma. These results confirm the observation of Devine et al [1984] that eruptions of basaltic magma release an order of magnitude more sulfur, chlorine and fluorine than eruptions of silicic magma.

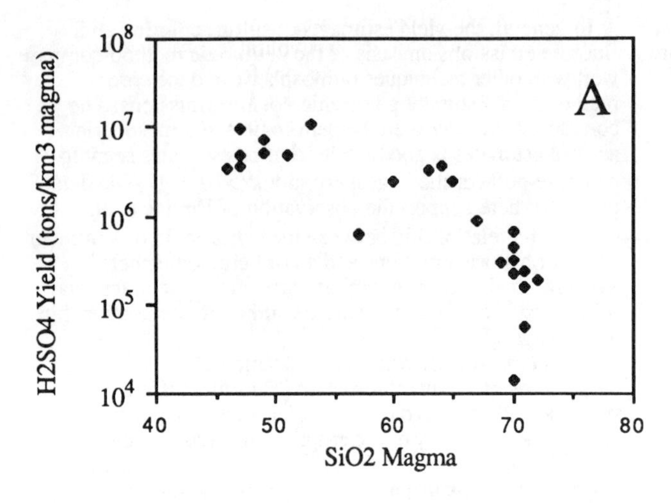


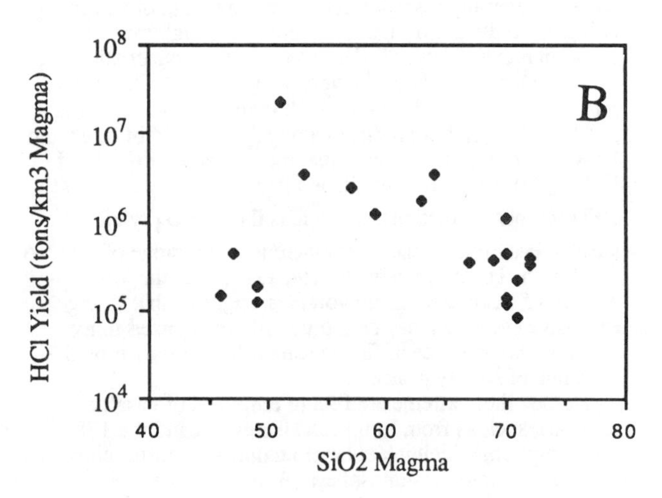


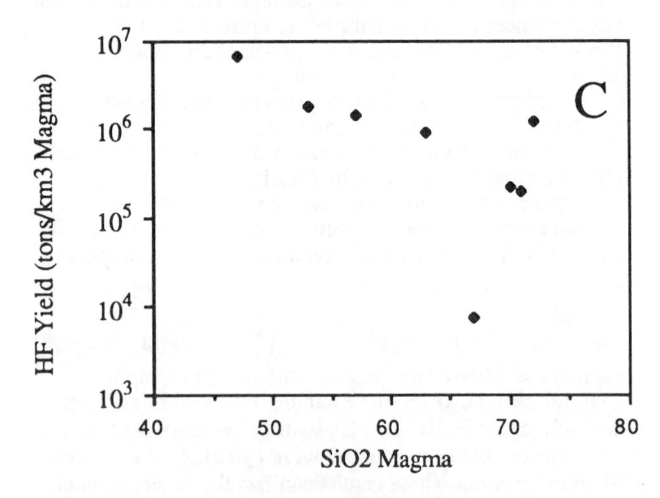


Fig. 5. Plot of sulfuric (A), hydrochloric (B) and hydrofluoric (C) acid yield (tons/km$^3$ magma) versus silica content of magma. See Table 5 for additional details.

TABLE 5. Eruptions studied by the petrologic method and their yield of acids ($H_2SO_4$, HCl, HF) per km$^3$ of magma erupted [Devine et al., 1984 and this work]. Eruptions are listed in order of increasing $H_2SO_4$ yield.

| Eruption | $H_2SO_4$/km$^3$ metric tons | HCl/km$^3$ metric tons | HF/km$^3$ metric tons |
|---|---|---|---|
| Tarawera | $1.10 \times 10^7$ | $3.38 \times 10^6$ | $1.79 \times 10^6$ |
| Eldgja | $9.31 \times 10^6$ | $4.34 \times 10^5$ | $6.89 \times 10^6$ |
| Laki | $7.17 \times 10^6$ | $1.27 \times 10^5$ | 0 |
| Krafla | $5.05 \times 10^6$ | $1.89 \times 10^5$ | n.a. |
| Soufriere | $4.82 \times 10^6$ | $2.32 \times 10^7$ | 0 |
| Katla | $4.81 \times 10^6$ | 0 | n.a. |
| Bezymianny | $3.70 \times 10^6$ | $3.36 \times 10^6$ | 0 |
| Heimaey | $3.61 \times 10^6$ | 0 | 0 |
| Surtsey | $3.52 \times 10^6$ | $1.44 \times 10^5$ | n.a. |
| Agung | $3.16 \times 10^6$ | $1.70 \times 10^6$ | $8.89 \times 10^5$ |
| Rabaul | $2.47 \times 10^6$ | 0 | n.a. |
| Campanian | $2.41 \times 10^6$ | 0 | 0 |
| Katmai | $8.72 \times 10^5$ | $3.57 \times 10^5$ | $7.38 \times 10^3$ |
| Santa Maria | $6.62 \times 10^5$ | $4.33 \times 10^5$ | 0 |
| Tambora | $5.99 \times 10^5$ | $2.47 \times 10^6$ | $1.44 \times 10^6$ |
| Hekla H-1 | $4.48 \times 10^5$ | 0 | n.a. |
| St Helens 1980 | $3.16 \times 10^5$ | $1.41 \times 10^5$ | $2.12 \times 10^5$ |
| Krakatau | $2.94 \times 10^5$ | $3.75 \times 10^5$ | 0 |
| St Helens 1530 | $2.32 \times 10^5$ | $8.26 \times 10^4$ | n.a. |
| Hekla H-3 | $2.20 \times 10^5$ | $1.16 \times 10^5$ | n.a. |
| Oraefajokull | $1.90 \times 10^5$ | $3.30 \times 10^5$ | $1.18 \times 10^6$ |
| Minoan | $1.54 \times 10^5$ | 0 | n.a. |
| Taupo Plinian | $5.63 \times 10^4$ | $2.22 \times 10^5$ | $1.94 \times 10^5$ |
| St Helens 1800 | $1.41 \times 10^4$ | $1.06 \times 10^6$ | n.a. |
| Roseau Tuff | 0 | $3.83 \times 10^5$ | n.a. |
| Coseguina | 0 | $1.23 \times 10^6$ | 0 |

n.a.= not analyzed

Analyses of both the Campanian and Rabaul ignimbrites have shown that metasomatism of ignimbrites and tephra-fall buried under ignimbrites may lead to anomalously high levels of S, Cl and F in the matrix glass of these deposits. Therefore, in the absence of unaltered ignimbrite samples, the amount of S, Cl and F degassed in ignimbrite-forming eruptions cannot be directly estimated. The petrologic estimate of the volatile yield from the plinian phase of the 1400 yr B.P. Rabaul eruption, on the other hand, is shown to be at least two orders of magnitude lower than the estimate of atmospheric aerosol mass present in the atmosphere in 536 A.D., as determined by Stothers [1984]. We therefore conclude that the event in 536 A.D. which perturbed the atmospheric optical depth was probably unrelated to the Rabaul eruption.

## Appendix

### Sample Identification

Sample numbers refer to numbers of columns in Table 3.

1. ELD-1, average composition of melt inclusion trapped by pyroxene phenocrysts in basaltic tephra from Eldgja

eruption in 934 A.D. Eldgja tephra fall collected by Thorvaldur Thordarson, 1.5 km SSW of Kambavatn. Sample no. 160883, section V.

2. ELD-1, average composition of matrix glass in basaltic tephra from Eldgja eruption in 934 A.D. Eldgja tephra fall collected by Thorvaldur Thordarson, 1.5 km SSW of Kambavatn. Sample no. 160883, section V.
3. O-1362, average composition of melt inclusions trapped by pyroxene phenocrysts in rhyolitic tephra fall from Oraefajokull, Iceland eruption in 1362 A.D. Collected by H. Sigurdsson, 29/7/83 at Fagurholsmyri, Iceland.
4. O-1362, average composition of matrix glass in rhyolitic tephra fall from eruption in Oraefajokull, Iceland 1362 A.D. Collected by H. Sigurdsson, 29/7/83 at Fagurholsmyri, Iceland.
5. TP-4, average composition of melt inclusions trapped by pyroxene phenocrysts in rhyolitic tephra, 130 A.D. Hatepe plinian phase of Taupo eruption, North Island, New Zealand. Collected by C.J.N. Wilson, locality at N103/580133.
6. TP-4, average composition of matrix glass in rhyolitic tephra, 130 A.D. Hatepe plinian phase of Taupo eruption, North Island, New Zealand. Collected by C.J.N. Wilson, locality at N103/580133.
7. TP-5, average composition of melt inclusions trapped by pyroxene phenocrysts in rhyolitic tephra, 130 A.D. Rotongaio phreatoplinian phase of Taupo eruption, North Island, New Zealand. Collected by C.J.N. Wilson, locality at N103/580133.
8. TP-5, average composition of matrix glass in rhyolitic tephra, 130 A.D. Rotongaio phreatoplinian phase of Taupo eruption, North Island, New Zealand. Collected by C.J.N. Wilson, locality at N103/580133.
9. TP-6, average composition of melt inclusions trapped by pyroxene phenocrysts in rhyolitic tephra, 130 A.D. Taupo plinian eruption, North Island, New Zealand. Collected by C.J.N. Wilson, locality at N103/549233.
10. TP-6, average composition of matrix glass in rhyolitic tephra, 130 A.D. Taupo plinian eruption, North Island, New Zealand. Collected by C.J.N. Wilson, locality at N103/549233.
11. K-192A, average composition of melt inclusions trapped by pyroxene phenocrysts in rhyolitic tephra (early tephra, basal 6 cm of a 19 cm thick section) from early phase of 1912 A.D. Katmai plinian ash fall, Alaska, U.S.A. Sample collected by J. Riehle, 9/85, 17 km NE of Novarupta Dome.
12. K-192A, average composition of matrix glass in rhyolitic tephra (early tephra, basal 6 cm of a 19 cm thick section) from early phase of 1912 A.D. Katmai plinian ash fall, Alaska, U.S.A. Sample collected by J. Riehle, 9/85, 17 km NE of Novarupta Dome.
13. K-192B, average composition of melt inclusions trapped by plagioclase phenocrysts in rhyolitic tephra (middle tephra, 6 cm of a 19 cm thick section) from middle phase of 1912 A.D. Katmai plinian ash fall, Alaska, U.S.A. Sample collected by J. Riehle, 9/85, 17 km NE of Novarupta Dome.
14. K-192B, average composition of matrix glass in rhyolitic tephra (middle tephra, 6 cm of a 19 cm thick section) from middle phase of 1912 A.D. Katmai plinian fall, Alaska, U.S.A. Sample collected by J. Riehle, 9/85, 17 km NE of Novarupta Dome.
15. K-192C, average composition of melt inclusions trapped by pyroxene phenocrysts in rhyolitic tephra (late tephra, top 7 cm of a 19 cm thick section) from late phase of 1912 A.D. Katmai plinian fall, Alaska, U.S.A. Sample collected by J. Riehle, 9/85, 17 km NE of Novarupta Dome.
16. K-192C, average composition of matrix glass in rhyolitic tephra (late tephra, top 7 cm of a 19 cm thick section) from late phase of 1912 A.D. Katmai plinian fall, Alaska, U.S.A. Sample collected by J. Riehle, 9/85, 17 km NE of Novarupta Dome.
17. K-192C, average composition of melt inclusions trapped by plagioclase phenocrysts in rhyolitic tephra (late tephra, top 7 cm of a 19 cm thick section) from late phase of 1912 A.D. Katmai plinian fall, Alaska, U.S.A. Sample collected by J. Riehle, 9/85, 17 km NE of Novarupta Dome.
18. K-192C, average compostion of matrix glass in rhyolitic tephra (late tephra, top 7 cm of a 19 cm thick section) from late phase of 1912 A.D. Katmai plinian fall, Alaska, U.S.A. Sample collected by J. Riehle, 9/85, 17 km NE of Novarupta Dome.
19. Total Katmai- Sum of the volatiles released in early, middle and late phase tephra
20. Rab-1, average composition of melt inclusions trapped by plagioclase phenocrysts in dacitic tephra from 1400 yr. B.P. Rabaul plinian fall deposit, Papua, New Guinea. Samples courtesy of Dr. P. Lowenstein.
21. Rab-1, average composition of melt inclusions trapped by pyroxene phenocrysts in dacitic tephra from 1400 yr. B.P. Rabaul plinian fall deposit, Papua, New Guinea. Samples courtesy of Dr. P. Lowenstein.
22. Rab-1, average composition of matrix glass in dacitic tephra from 1400 yr. B.P. Rabaul plinian fall deposit, Papua, New Guinea. Samples courtesy of Dr. P. Lowenstein.
23. Rab-3, average composition of melt inclusions trapped by plagioclase phenocrysts in dacitic tephra from 1400 yr. B.P. Rabaul pyroclastic flow deposit, Papua, New Guinea. Samples courtesy of Dr. P. Lowenstein.
24. Rab-3, average composition of matrix glass in dacitic tephra from 1400 yr. B.P. Rabaul pyroclastic flow deposit, Papua, New Guinea. Samples courtesy of Dr. P. Lowenstein.
25. IT-2053, average composition of melt inclusions trapped by pyroxene phenocrysts in trachytic tephra from upper half of plinian fall, Campanian eruption (~35,000 yrs. B.P.), Phlegrean Fields, Italy. Samples collected by H. Sigurdsson (5/84) in quarry near Lavorate, Italy.
26. IT-2053, average composition of matrix glass in trachytic tephra from upper half of plinian fall, Campanian eruption (~ 35,000 yrs. B.P.), Phlegrean Fields, Italy. Samples collected by H. Sigurdsson (5/84) in quarry near Lavorate, Italy.
27. IT-2054, average composition of melt inclusions trapped by pyroxene phenocrysts in trachytic tephra from lower half of plinian fall, Campanian eruption (~35,000 yrs. B.P.), Phlegrean Fields, Italy. Samples collected by H. Sigurdsson (5/84) in quarry near Lavorate, Italy.
28. IT-2054, average composition of matrix glass in trachytic tephra from lower half of plinian fall, Campanian eruption (~35,000 yrs. B.P.), Phlegrean Fields, Italy. Samples collected by H. Sigurdsson (5/84) in quarry near Lavorate, Italy.

29. TW-1, average composition of melt inclusions trapped by pyroxene phenocrysts in basaltic tephra from plinian fall, 1886 A.D. Tarawera eruption, North Island, New Zealand. Collected by C.J.N. Wilson, locality at N77/984913
30. TW-1, average composition of matrix glass in basaltic tephra from plinian fall, 1886 A.D. Tarawera eruption, North Island, New Zealand. Collected by C.J.N. Wilson, locality at N77/984913
31. TW-2, average composition of melt inclusions trapped by pyroxene phenocrysts in basaltic tephra from plinian fall, 1886 A.D. Tarawera eruption, North Island, New Zealand. Collected by C.J.N. Wilson, locality at N77/973936, exposure 1.5-2.0 m below top of 1886 scoria fall.
32. TW-2, average composition of matrix glass in basaltic tephra from plinian fall, 1886 A.D. Tarawera eruption, North Island, New Zealand. Collected by C.J.N. Wilson, locality at N77/973936, exposure 1.5-2.0 m below top of 1886 scoria fall.
33. MSH-201, average composition of melt inclusions trapped by plagioclase phenocrysts in dacitic tephra, Mount St. Helens Kalama eruptive period ~1530 A.D. ("W" tephra), Washington, U.S.A. Collected in Bear Meadow by H. Sigurdsson, 8/14/84.
34. MSH-201, average composition of matrix glass in dacitic tephra, Mount St. Helens Kalama eruptive period ~1530 A.D. ("W" tephra), Washington, U.S.A. Collected in Bear Meadow by H. Sigurdsson, 8/14/84.
35. MSH-202, average composition of melt inclusions trapped by plagioclase phenocrysts in dacitic tephra, Mount St. Helens Goat Rocks eruptive period ~1800 A.D. ("T" tephra), Washington, U.S.A. Collected in Bear Meadow by H. Sigurdsson, 8/14/84.
36. MSH-202, average composition of matrix glass in dacitic tephra, Mount St. Helens Goat Rocks eruptive period ~1800 A.D. ("T" tephra), Washington, U.S.A. Collected in Bear Meadow by H. Sigurdsson, 8/14/84.
37. COS-5, average composition of melt inclusions trapped by phenocrysts in dark grey-black scoria fall from the 1835 A.D. Coseguina, Nicaragua eruption. Sample collected by Steve Self on the E-SE flanks of the volcano.
38. COS-5, average composition of matrix glass in dark grey-black scoria fall from the 1835 A.D. Coseguina, Nicaragua eruption. Sample collected by Steve Self on the E-SE flanks of the volcano.
39. SM-1902, average composition of melt inclusions trapped in phenocrysts of pumiceous dacitic tephra from the 1902 A.D. eruption. Sample collected by Bill Rose on the rim of Siete Orejas volcano about 6 km NNW of the Santa Maria vent, Guatemala.
40. SM-1902, average composition of matrix glass of pumiceous dacitic tephra from the 1902 A.D. eruption. Sample collected by Bill Rose on the rim of Siete Orejas volcano about 6 km NNW of the Santa Maria vent, Guatemala.
41. BZ-1956, average composition of melt inclusions trapped in phenocrysts in andesitic tephra from the March 30, 1956 eruption of Bezymianny volcano, U.S.S.R. Sample collected by Dr. G.E. Bogoyavlenskaya at a distance of about 80 km north from the volcano (10 km north from the settlement of Klyuchi). At this site the tephra layer was about 2 cm thick.
42. BZ-1956, average composition of matrix glass from andesitic tephra from the March 30, 1956 eruption of Bezymianny volcano, U.S.S.R. Sample collected by Dr. G.E. Bogoyavlenskaya at a distance of about 80 km north from the volcano (10 km north from the settlement of Klyuchi). At this site the tephra layer was about 2 cm thick.
43. BZ-1956, average composition of melt inclusions trapped in phenocrysts in andesitic tephra from the March 30, 1956 eruption of Bezymianny volcano, U.S.S.R. Sample collected by Dr. G.E. Bogoyavlenskaya at a distance of about 80 km north from the volcano (10 km north from the settlement of Klyuchi). At this site the tephra layer was about 2 cm thick. Same sample as 41/42 but average of high-silica inclusions.
44. BZ-1956, average composition of matrix glass from andesitic tephra from the March 30, 1956 eruption of Bezymianny volcano, U.S.S.R. Sample collected by Dr. G.E. Bogoyavlenskaya at a distance of about 80 km north from the volcano (10 km north from the settlement of Klyuchi). At this site the tephra layer was about 2 cm thick. Same sample as 41/42 but average of high-silica matrix glass.

*Acknowledgements.* We wish to thank D. Browning for his help in sample preparation and microprobe analysis. S.N. Carey wrote the computer program for determining the statistical significance of our results. We gratefully acknowledge the following people who provided samples for this study: Thorvaldur Thordarson (Eldgja), C. J. N. Wilson (Taupo and Tarawera), J. Riehle (Katmai), P. Lowenstein (Rabaul), S. Self (Coseguina), W.I. Rose Jr. (Santa Maria), and G.E. Bogoyavlenskaya (Bezymianny). We also wish to thank S.N. Carey and W. Cornell for assistance with the microprobe and for helpful discussions. This work was funded by the National Science Foundation grant EAR-8503104.

## References

Anderson, A.T., Chlorine, sulfur, and water in magmas and oceans, Geol. Soc. Am. Bull., 85, 1485-1492, 1974.

Angell, J.K., The close relation between Antarctic total-ozone depletion and cooling of the Antarctic low stratosphere, Geophys. Res. Lett., 13, 1240-1243, 1986.

Angell, J.K., and J. Korshover, Surface temperature changes following the six major volcanic episodes between 1780 and 1980, Jour. Climate Appl. Meteorol., 24, 937-951, 1985.

Bailey, D.K., and R. Macdonald, Fluorine and chlorine in peralkaline liquids and the need for magma generation in an open system, Mineral. Mag., 40, 405-414, 1975.

Barberi, F., F. Innocenti, L. Lirer, R. Munno, and T. Pescatore, The Campanian ignimbrite: A major prehistoric eruption in the Neapolitan area (Italy), Bull. Volcanol., 41, 1-22, 1978.

Benoist, J.P., J. Jouzel, C. Lorius, L. Merlivat, and M. Pourchet, Isotope climatic record over the last 2.5 Ka from Dome C, Antarctica, ice cores, Annals Glaciol., 3, 17-22, 1982.

Bull G.A. and D.G.James, Dust in the stratosphere over

western Britain on April 3 and 4, 1956, Meteorol. Mag. G.B., 85, 1956.
Carey, S.and R.S.J. Sparks, Quantitative models of the fallout and dispersal of tephra from volcanic eruption columns, Bull. Volcanol., 48, 109-125, 1986.
Carmichael, I.S.E., The iron-titanium oxides of salic volcanic rocks and their associated ferromagnesian silicates, Contr. Min. Petrol., 14, 36-64, 1967.
Castleman, A.W., Jr., H.R. Munkelwitz, and B. Manowitz, Isotopic studies of the sulfur component of the stratospheric aerosol layer, Tellus, 26, 222-234, 1974.
Chuan, R. L., J. Palais, W.I. Rose, and P.R. Kyle, Fluxes, sizes, morphology and compositions of particles in the Mt. Erebus volcanic plume, December, 1983, J. Atmos. Chem., 4, 467-477, 1986.
Cornell, W.C., S.N. Carey, and H. Sigurdsson, Computer simulation of transport and deposition of the Campanian Y-5 ash, J. Volcanol. Geotherm. Res., 17, 89-109, 1983.
Crandell, D.R., D.R. Mullineaux, and M. Rubin, Mount St. Helens volcano: Recent and future behavior, Science, 187, 438-441, 1975.
Crandell, D.R. and D.R. Mullineaux, Potential hazards from future eruptions of Mt. St. Helens volcano, Washington, U.S. Geol. Surv. Bull., 1383-C, 1-26, 1978.
Crutzen, P., Photochemical reactions initiated by and influencing ozone in unpolluted tropospheric air, Tellus, 26, 47-57, 1974.
Curtis, G.H., The stratigraphy of the ejecta from the 1912 eruption of Mount Katmai and Novarupta, Alaska, in Studies in Volcanology, edited by R.R. Coats, R.L. Hay, and C.A. Anderson , Geol. Soc. Am. Mem., 116, 153-210, 1968.
De Angelis, M., L., Fehrenbach, C. Jéhanno, and M. Maurette, Micrometre-sized volcanic glasses in polar ices and snows, Nature, 317, 52-54, 1985.
Deirmendjian, D., On volcanic and other particulate turbidity anomalies, Advances in Geophysics, 16, 267-297, 1973.
Devine, J.D., and H. Sigurdsson, The liquid composition and crystallization history of the 1979 Soufriere magma, St. Vincent, W.I., J. Volcanol. Geotherm. Res., 16, 1-31, 1983.
Devine, J.D., H. Sigurdsson, A.N. Davis, and S. Self, Estimates of sulfur and chlorine yield to the atmosphere from volcanic eruptions and potential climatic effects, J. Geophys. Res., 89, 6309-6325, 1984.
Dunbar, N., and P. R. Kyle, $H_2O$ and Cl contents, and temperature of Taupo Volcanic Zone rhyolitic magmas (abstract), Inter. Volcanol. Congr., New Zealand, Feb., 1986.
Federman, A.N., Correlation and Petrological Interpretation of Abyssal and Terrestrial Tephra Layers, PhD Dissertation, Oregon State University, 241 pp. 1984.
Fierstein, J. and W. Hildreth, Ejecta dispersal and dynamics of the 1912 eruptions at Novarupta, Katmai National Park, Alaska, Eos Trans. AGU, 67, 1246, 1986.
Gorshkov, G.S., Gigantic eruption of the volcano Bezymianny, Bull.Volcanol., 20, 77-109, 1959.
Griggs, R.F., The Valley of Ten Thousand Smokes, National Geographic Society, 340 pp., 1922.
Groveman, B.S. and H.S. Landsberg, Reconstruction of Northern Hemisphere Temperature: 1579-1880, Meteorology Program, Univ. of Maryland, Publ. 79-181, 1979.
Hammer, C.U., Acidity of polar ice cores in relation to absolute dating, past volcanism, and radioechos, J. Glaciol., 25, 359-372, 1980.
Hammer, C.U., Initial direct current in the buildup of space charges and the acidity of ice cores, J. Phys. Chem., 87, 4099-4103, 1983.
Hammer, C.U., Traces of Icelandic eruptions in the Greenland ice sheet, Jokull, 34, 51-65, 1984.
Hammer, C.U., H. B. Clausen, and W. Dansgaard, Greenland ice sheet evidence of post-glacial volcanism and its climatic impact, Nature, 288, 230-235, 1980.
Handler, P., Possible association of stratospheric aerosols and El Nino type events, Geophys. Res. Lett., 11, 1121-1124, 1984.
Healy, J., Stratigraphy and chronology of late Quaternary volcanic ash in Taupo, Rotorua and Gisborne districts, Part 1, N.Z. Geol. Surv. Bull., 73, 88pp., 1964.
Heming, R.F., Geology and petrology of Rabaul caldera, Papua, New Guinea, Geol. Soc. Am. Bull., 85, 1253-1264, 1974.
Herron, M.M., Impurity sources of $F^-$, $Cl^-$, $NO_3^-$ and $SO_4^{2-}$ in Greenland and Antarctic precipitation, J. Geophys. Res., 87,3052-3060, 1982.
Hildreth, W., The compositionally zoned eruption of 1912 in the Valley of Ten Thousand Smokes, Katmai National Park, Alaska, J. Volcanol. Geotherm. Res., 18, 1-56, 1983.
Hoblitt, R.P., D.R. Crandell, D.R. Mullineaux, Mount St. Helens eruptive behavior during the past 1,500 yr., Geology, 8, 555-559, 1980.
Hoff, R.M. and A.J. Gallant, Sulfur dioxide emissions from La Soufriere volcano, St. Vincent, West Indies, Science, 209, 923-924, 1980.
Johnston, D.A., Volcanic contribution of chlorine to the stratosphere: More significant to ozone than previously estimated?, Science, 209, 491-493, 1980.
Jones, P.D., T.M.L. Wigley and P.M. Kelly, Variations in surface air temperatures: Part 1. Northern Hemisphere, 1881-1980, Mon.Wea. Rev., 110, 59-70, 1982.
Kelly, P.M., and C.B. Sear, Climatic impact of explosive volcanic eruptions, Nature, 311, 740-743, 1984.
Kennett, J.P. and R.C.,Thunell, Global increase in Quaternary explosive volcanism, Science, 187, 497-503, 1975.
Kennett, J.P., and R.C. Thunell, On explosive Cenozoic volcanism and climatic implications, Science, 196, 1231-1234, 1977.
Kent, G.S., SAGE measurements of Mount St. Helens volcanic aerosols, NASA Conf. Publ., 2240, 109-116, 1982.
LaMarche, V.C., Jr. and K.K. Hirshboeck, Frost rings in trees as records of major volcanic eruptions, Nature, 307, 121-126, 1984.
Lamb, H.H., Volcanic dust in the atmosphere, Phil. Trans. Roy. Soc. Lond., ser. A, 266, 425-533, 1970.
Lamb, H.H., Volcanic activity and climate, Paleogeograph., Paleoclimatol., Paleoecol., 10, 203-230, 1971.
Larsen, G. Um aldur Eldgjarhrauna, Natturufraedingurinn, 49, 1-25, 1979.
Legrand, M. and R.J. Delmas, A 220-year continuous record of volcanic $H_2SO_4$ in the Antarctic ice sheet, Nature, 327, 671-676, 1987.
Legrand, M., Chimie des neiges et glaces Antarctique: Un

reflet de l'environment, Universite Scientifique et Medicale de Grenoble, 1985.

Legrand, M.R. and R.J. Delmas, The ionic balance of Antarctic snow: A 10-year detailed record, Atmos. Environ., 18, 1867-1874, 1984.

Maccagnan, M. J.M. Barnola, R. Delmas, and P. Duval, Static electrical conductivity as an indicator of the sulfate content of polar ice cores, Geophys. Res. Lett., 8, 970-972, 1981.

Mankin, W.G. and M.T. Coffey, Increses stratospheric hydrogen chloride in the El Chichon cloud. Science, 226, 170-172, 1984.

McCormick, M.P., Ground-based and airborne measurements of Mount St. Helens stratospheric effluents, NASA Conf. Publ., 2240, 125-130, 1982.

McCormick, M.P., G.S. Kent, G.K. Yue, and D.M. Cunnold, SAGE measurements of the stratospheric aerosol dispersion and loading from the Sourfriere volcano, NASA Tech. Pap., 1922, 1-20, 1981.

Miller, E., Vapor-liquid equilibira of water-hydrogen chloride solutions below $0^{o}C$. J. Chem. Eng. Data, 28, 363-367, 1983.

Mullineaux, D.R., Summary of pre-1980 tephra-fall deposits erupted from Mount St. Helens, Washington state, USA, Bull. Volcanol., 48, 17-26, 1986.

Mullineaux, D.R. and D.R. Crandell, The eruptive history of Mt. St. Helens, In; P.W. Lipman, D.R. Mullineaux (eds.) The 1980 eruptions of Mt. St. Helens,Washington, U.S. Geol. Surv. Prof. Pap., 1250, 3-15, 1981.

Nielsen, C. H. , and H. Sigurdsson, Quantitative methods for electron microprobe analysis of sodium in natural and synthetic glasses, Am. Mineral., 66, 547-552, 1981.

Oliver, R.C., On the response of hemispheric mean temperature to stratospheric dust: An empirical approach, Jour. Appl. Meteor., 15, 933-950, 1976.

Oskarsson, N., The interaction between volcanic gases and tephra: Fluorine adhering to tephra from the 1979 Hekla eruption, J. Volcanol. Geotherm. Res., 8, 251-266, 1980.

Palais, J.M. and M. Legrand, Soluble impurities in the Byrd Station ice core, Antarctica: Their origin and sources, J. Geophys. Res., 90, 1143-1154, 1985.

Pollack, J.B., O.B. Toon, C. Sagan, A. Summers, B. Baldwin, and W. Van Camp, Volcanic explosions and climate change: A theoretical assessment, J. Geophys. Res., 81, 1071-1083, 1976.

Porter, S.C., Recent glacier variations and volcanic eruptions, Nature, 291, 139-142, 1981.

Rampino, M.R. and S. Self, Historic eruptions of Tambora (1815), Krakatau (1883), and Agung (1963), their stratospheric aerosols, and climatic impact, Quat. Res., 18, 127-143, 1982.

Rampino, M.R. and S. Self, Sulphur-rich volcanic eruptions and stratospheric aerosols, Nature, 310, 677-679, 1984a.

Rampino, M.R., and Self, S., The atmospheric effects of El Chichon, Sci. Am., 48, 1984b.

Rampino, M.R., R.B. Stothers, and S. Self, Climatic effects of volcanic eruptions, Nature, 313, 272, 1985.

Reclus, E., Nouvelle Géographie universelle, 17, 488-489, Paris, 1891.

Robock, A., The Mount St. Helens volcanic eruption of 18 May 1980: Minimal climatic effect, Science, 212, 1383-1384, 1981.

Rose, W.I., Jr., R.E. Stoiber, and L.L. Malinconico, Eruptive gas compositions and fluxes of explosive volcanoes: budget of S and Cl emitted from Fuego volcano Guatemala, in Andesites, edited by R.S. Thorpe, pp. 669-676, John Wiley, New York, 1982.

Rose, W.I., R.L. Wunderman, M.F. Hoffman, and L. Gale, A volcanologist's review of atmospheric hazards of volcanic activity, Fuego and Mt. St. Helens, J. Volcanol. Geotherm. Res., 17, 133-157, 1983.

Rose, W.I., Santa-Maria, Guatemala: Bimodal soda-rich calcalkalic stratovolcano, J. Volcanol. Geotherm. Res., in press.

Rosi, M., A. Sbrana and C. Principe, The Phlegraean fields: Structural evolution, volcanic history and eruptive mechanisms, J. Volcanol. Geotherm. Res., 17, 273-288, 1983.

Schatten, K.H., H.G. Mayr, I. Harris, and H.A. Taylor, Jr., A zonally symmetric model for volcanic influence upon atmospheric circulation, Geophys. Res. Lett., 11, 303-306, 1984.

Self, S., M.R. Rampino, and J.J. Barbera, The possible effects of large 19th and 20th century volcanic eruptions on zonal and hemispheric surface temperatures, J. Volcanol. Geotherm.Res., 11, 41-60, 1981.

Sigurdsson, H., Volcanic pollution and climate: The 1783 Laki eruption, Eos Trans. AGU, 63, 601-602, 1982.

Sigurdsson, H. J. D. Devine, and A. N. Davis, The petrologic estimation of volcanic degassing, Jokull, 35, 1-8, 1985.

Simkin, T., L. Seibert, L. McClelland, W.G. Melson, D. Bridge, C.G. Newhall, and J. Latter, Volcanoes of the World, Smithsonian Institution, Wash. D.C., 1981.

Solomon, S. and R.R. Garcia, On the distribution of long-lived tracers of chlorine species in the middle atmosphere. J. Geophys. Res. 89, 11633-11644, 1984.

Sparks, R.S.J., H. Sigurdsson, and L. Wilson, Magma mixing: a mechanism for triggering acid explosive eruptions, Nature, 267, 315-318, 1977.

Stolarski, R.S., and D.M. Butler, Possible effects of volcanic eruptions on stratospheric minor constituent chemistry. Pageoph, 117, 486-497, 1978.

Stolarski, R.S., and R.J. Cicerone, Stratospheric chlorine: a possible sink for ozone. Can. J. Chem. 52, 1610-1615, 1974.

Stothers, R.B., Mystery cloud of AD 536, Nature, 307, 344-345, 1984a.

Stothers, R.B., The great Tambora eruption in 1815 and its aftermath, Science, 224, 1191-1198, 1984b.

Stothers, R.B., J.A. Wolff, S. Self and M.R. Rampino, 1986: Basaltic fissure eruptions, plume heights and atmosperic aerosols. Geophys. Res. Lett. 13, 725-728.

Thompson, L.G., E. Mosley-Thompson, W. Dansgaard, and P.M. Grootes, The Little Ice Age as recorded in the stratigraphy of the tropical Quelccaya ice cap, Science, 234, 361-364, 1986.

Thorarinsson, S., The Oraefajokull eruption of 1362, Acta Naturalia Islandica, 2, 1-101, 1958.

Walker, G.P.L., The Taupo pumice: Product of the most powerful known (ultraplinian) eruption?, J. Volcanol. Geotherm. Res., 8, 69-94, 1980.

Walker, G.P.L., R.F. Heming, T.J. Sprod, and H.R. Walker, Latest major eruptions of Rabaul volcano, In·

Cooke-Ravian Volume of Volcanological Papers, R.W. Johnson (ed.), Geol. Surv. Papua New Guinea Mem., 10, 181-193, 1981.

Walker, G.P.L., S.Self, and L. Wilson, Tarawera 1886, New Zealand- a basaltic plinian fissure eruption, J. Volcanol. Geotherm. Res., 21, 61-78, 1984.

Williams, H., The great eruption of Coseguina, Nicaragua, in 1835, Univ. Calif. Publ. Geol. Sci., 29, 21-46, 1952.

Williams, S.N., The October 1902 eruption of Santa Maria volcano, Guatemala, MS Thesis, Dartmouth College, Hanover, N.H., 140 pp. (unpubl.), 1979.

Williams, S. N., and S. Self, The October 1902 plinian eruption of Santa Maria volcano, Guatemala, J. Volcanol. Geotherm. Res., 16, 33-56, 1983.

Wilson, C. J. N., N. N. Ambraseys, J. Bradley, and G. P. L. Walker, A new date for the Taupo eruption, New Zealand, Nature, 288, 252-253, 1980.

Wilson, C. J. N., and G. P. L Walker, The Taupo eruption, New Zealand, I. General aspects, Phil. Trans. R. Soc. Lond., 1985.

Wolff, J.A., S. Self, and M.R. Rampino, Basaltic fissure eruptions, fire fountains, and atmospheric aerosols, E.O.S., 65, 1148-1149.

Woods, D.C., R.L. Chuan and W. I. Rose, Science, 230, 170-172, 1984.

Zies, E.G., The Valley of Ten Thousand Smokes: I The fumarolic incrustations and the bearing ore deposition. II The acid gases contributed to the sea during volcanic activity, Nat. Geog.Soc. Contr. Tech. Pap., Katmai Ser., 4, 1-79, 1929.

# Section IV

# BIOGEOCHEMICAL CYCLES, LAND HYDROLOGY, LAND SURFACE PROCESSES AND CLIMATE

# UPTAKE BY THE ATLANTIC OCEAN OF EXCESS ATMOSPHERIC CARBON DIOXIDE AND RADIOCARBON

Bert Bolin and Anders Björkström

Department of Meteorology, University of Stockholm, Sweden

Berrien Moore

University of New Hampshire, Durham, New Hampshire

Abstract. Inverse methods have been used to deduce water circulation, spatial patterns of turbulent exchange and biological activity in the Atlantic Ocean, by using a set of stationary tracers and a condition of quasi-geostrophic flow. The solution yields a direct meridional circulation cell with descending motion in the northern Atlantic with an intensity of 20-25 Sverdrup, a reasonable distribution of vertical turbulent transfer in the uppermost ocean layers and comparatively large rates of detritus formation, about 4.5 Pg C $yr^{-1}$. - The solution is used to compute the invasion of tritium 1955-1983, and the uptake of excess radiocarbon and carbon dioxide during the period 1760-1983. A fair agreement between computed and observed changes of tritium and $^{14}C$ is obtained, but the period of observations is too short to serve as a conclusive test of the model. - The uptake of carbon dioxide during the 220 years period into the Atlantic Ocean is 33 ± 5 Pg and it is further found that significant variations of the uptake fraction of the $CO_2$ emissions may have occurred due to varying rates of emissions in course of time. The conclusion is tentatively drawn that the ocean and its carbonate system may not have been the only sink for anthropogenic emissions of carbon dioxide into the atmosphere. - Means for how to further improve the model and its capability to reproduce the ocean behaviour are discussed.

## Introduction

Burning of fossil fuels, deforestation and changing land use have changed the global carbon cycle very significantly during the last two centuries. We see this most clearly in the steadily increasing $CO_2$ concentration in the atmosphere, from a most likely value of about 280 ppm in the middle of the 18th century to 344 ppm in 1983 (cf. Siegenthaler and Oeschger 1987; Figure 1). This means an increase of the amount of carbon in the form of $CO_2$ in the atmosphere by 135 Pg ($Pg=10^{15}g$). The emissions, due to fossil fuel combustion, during this period of time have been about 180 Pg C (cf. Rotty, 1987), while estimates of the emissions due to man's interference with the terrestrial biota (vegetation and soils) are much more uncertain and range between 100-200 Pg C (cf. Bolin, 1986). The airborne fraction during this period thus has been between 36-48% (cf. Table 1). If we consider the two periods 1760-1957 and 1958-1985 separately we find that we are not able to tell if there has been a significant change of the airborne fraction during this period of time (cf. Table 1).

It has generally been accepted that a major part of the emissions of $CO_2$ into the atmosphere has been absorbed by the world oceans. Insufficient measurements are available that can quantitatively verify this view. It is rather primarily being based on simple models of the circulation of carbon in the sea (cf. Bolin, 1986). The most commonly used model is the one developed by Siegenthaler (1983), the so called box-diffusion model with polar outcrops. A simulation of the fossil fuel uptake since 1860 yields an airborne fraction of 60%. Siegenthaler (1983) argues, however, that the exchange between the ocean surface layers and the deep sea should be enhanced as compared with what is being obtained by determining the model parameters using steady state distribution of $^{14}C$. The modified model yields a value for the airborne fraction slightly above 50%. It is, however, questionable if such a modification of the box-diffusion model is justified. We need

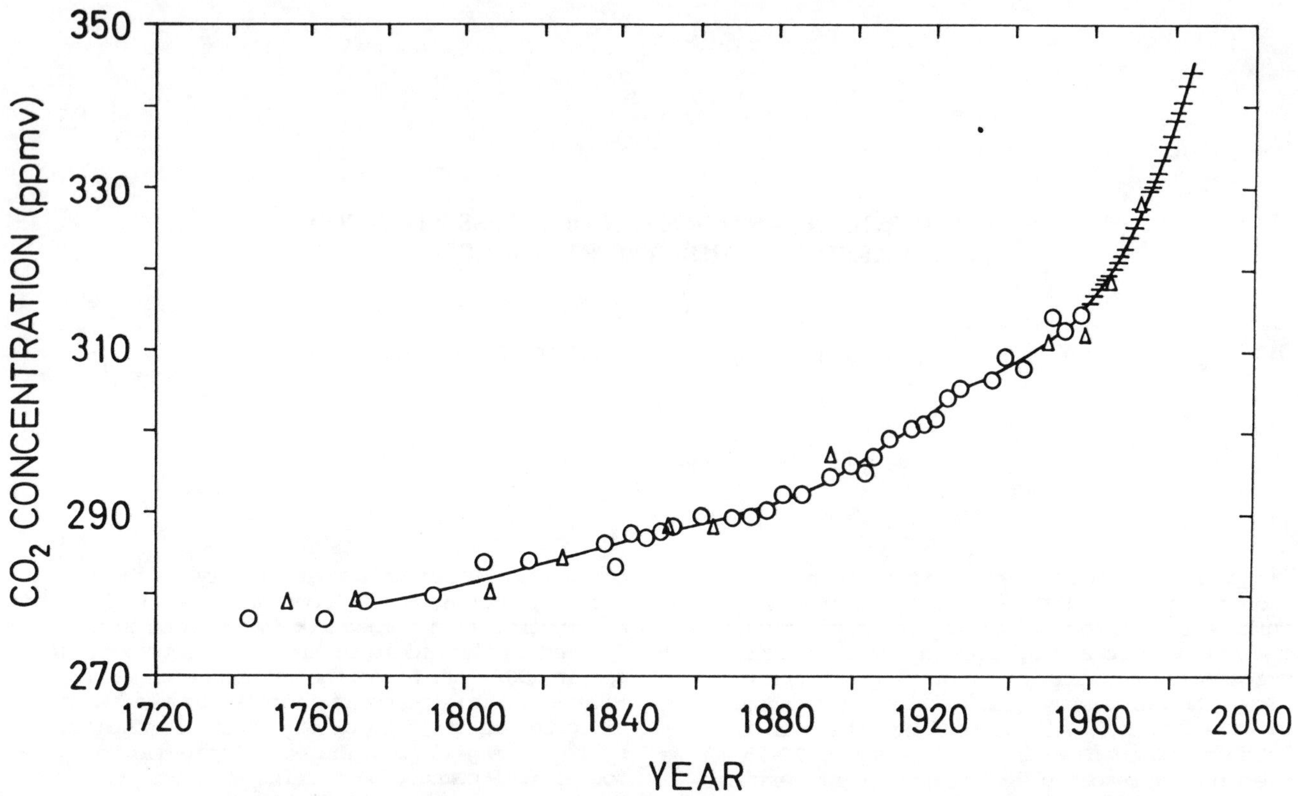


Fig. 1. Atmospheric $pCO_2$ 1760-1980 as determined by analysis of air bubbles in glacier ice and since 1957 by direct measurements (Compiled by Siegenthaler and Oeschger, 1987).

rather more detailed and realistic models to be able to determine more precisely the uptake capability of the oceans. For this purpose we should combine our knowledge about the circulation of the oceans as developed by the physical oceanographers and about chemical and biological processes that are of importance for understanding the role of the oceans in the global carbon cycle as is being studied by chemical oceanographers. If using General Ocean

TABLE 1. Estimated Changes of the Atmospheric Component of the Global Carbon Cycle for the Period 1760-1957 and 1958-1983.

| | | 1760-1957 | 1958-1983 | 1760-1983 |
|---|---|---|---|---|
| (1) | Increase of $CO_2$ ppm | 34 | 30 | 64 |
| | Increase of $CO_2$, PgC | 72 | 64 | 136 |
| (2) | Fossil fuel emissions, PgC | 80 | 102 | 182 |
| (3) | Terrestrial emissions, PgC (lower bound) | 80 | 20 | 100 |
| | Airborne fraction (upper bound) | 0.45 | 0.52 | 0.48 |
| (3) | Terrestrial emissions, PgC (upper bound) | 130 | 70 | 200 |
| | Airborne fraction (lower bound) | 0.34 | 0.37 | 0.36 |

(1) Measurements as reported by Keeling and colleagues (cf. Bolin, 1986) and Siegenthaler and Oeschger (1987).

(2) Rotty (1981); Rotty and Masters (1985).

(3) According to Bolin (1986). The lower estimate (100 PgC) implies very small emissions before1860, as direct assessments of releases since 1860 hardly are less than 100 PgC.The partitioning of the emissions to the two periods is approximate.

Circulation Models it is most important that the rates of ocean overturning be carefully validated by using tracer distributions, since direct current measurement are hardly adequate for this purpose. It is obvious that determination of gaseous uptake in which we are interested is crucially dependent on careful validation of model characteristics in this regard. Maier-Reimer and Hasselmann (1987) have recently presented studies of the circulation of carbon in the sea (both $^{12}C$ and $^{14}C$) by using general ocean circulation models. It is, however, obvious that a more detailed discussion of their results is needed in order to ascertain that the response characteristics of the model with regard to $CO_2$ uptake agree with those of the real ocean.

In the present paper we shall attempt an analysis of ocean uptake of $CO_2$ using an inverse methodology (cf. Bolin et al. 1987) in which case we shall employ the continuity equations for a set of key tracers (salinity, dissolved inorganic carbon, radiocarbon, alkalinity, phosphorus, and oxygen). It is, however, also important to make use of the extensive hydrographic data that have been collected for many years. We accordingly also impose a constraint of quasigeostrophic flow. Our aim is to develop a multiplebox model of the Atlantic Ocean that is carefully calibrated with the data referred to and that has sufficient spatial resolution so that the $CO_2$ uptake during the last few hundred years can be computed with reasonable accuracy.

The present analysis will be restricted to the Atlantic ocean because of data limitations and generalizations of our results to the world oceans are not yet possible. A few general observations will however be made questioning the view that the world oceans represent the only significant sink for excess atmospheric carbon. We shall also learn that the employment of inverse methods to develop boxmodels requires a careful and detailed analysis in order to show that the results are reliable. Our conclusions will thus not be final and firm, but will contribute to the fundamental problems we encounter when trying to assess the role of the oceans in the global carbon cycle.

## Model Development

### Derivation of a Basic Steady State Pattern of Circulation and Biochemical Processes

Bolin et al. (1987) have presented a general method for derivation of the steady state circulation pattern and biochemical processes by using inverse methods. We ask the question: What steady state patterns of water circulation, turbulent transfer, new primary production, decomposition of organic detrital matter and dissolution of biogenic carbonates are required to explain the observed quasi-steady distributions of temperature, salinity, total dissolved inorganic carbon (DIC), alkalinity, $^{14}C$, oxygen and phosphorus. Note that we do not consider the possible role of the formation of dissolved organic carbon in the decomposition process of organic matter. The following analysis is still of interest as a study of what the classical set of processes imply with regard to transfer patterns of carbon in the Atlantic Ocean.

We divide our domain into eight quasi-isopycnic layers (Figure 2) and twelve regions (Figure 3), based on our qualitative knowledge about the key features of water circulation. Altogether 84 water reservoirs, boxes, are defined in this way. An unknown vector x is specified having the 536 components:

- advective fluxes of water between all adjacent boxes (184 components)
- rates of turbulent exchange of matter between all adjacent boxes (184 components)
- production or decomposition of organic matter, one component for each box (84 components)
- carbonate production or dissolution, one component for each box (84 components).

Assuming that the steady state concentration of the seven tracers can be determined for the 84 boxes we are able to formulate 8 x 84 = 672 equations, one for each of the seven tracers and each box and one additional set of 84 equations expressing water continuity for each box. We have adopted the finite difference approximation that advection between two boxes carries the mean concentration of these in the direction of flow and that turbulent transfers matter in the direction of the gradient between them. It is further assumed that the change of the horizontal flow as a function of depth approximately agrees with the thermal wind computed from hydrographic data, which yields another set of 80 equations. This total set of 672 + 80 = 752 equations can be written in the form of a matrix equation (for details, see Bolin et al., 1987)

$$Ax = b \tag{1}$$

where A is a 752 x 536 matrix, determined by the concentration distributions in the sea, b quantifies the water or tracer flux across external boundaries and radioactive decay (in the case of $^{14}C$), and the prescribed vertical change of horizontal advective flow from one box to another in the equations for quasi-geostrophic flow. We determine the air sea exchange by assuming that there is no net flow of $CO_2$ to or from the Atlantic Ocean as a whole and that the inflow of $^{14}C$ is balanced by

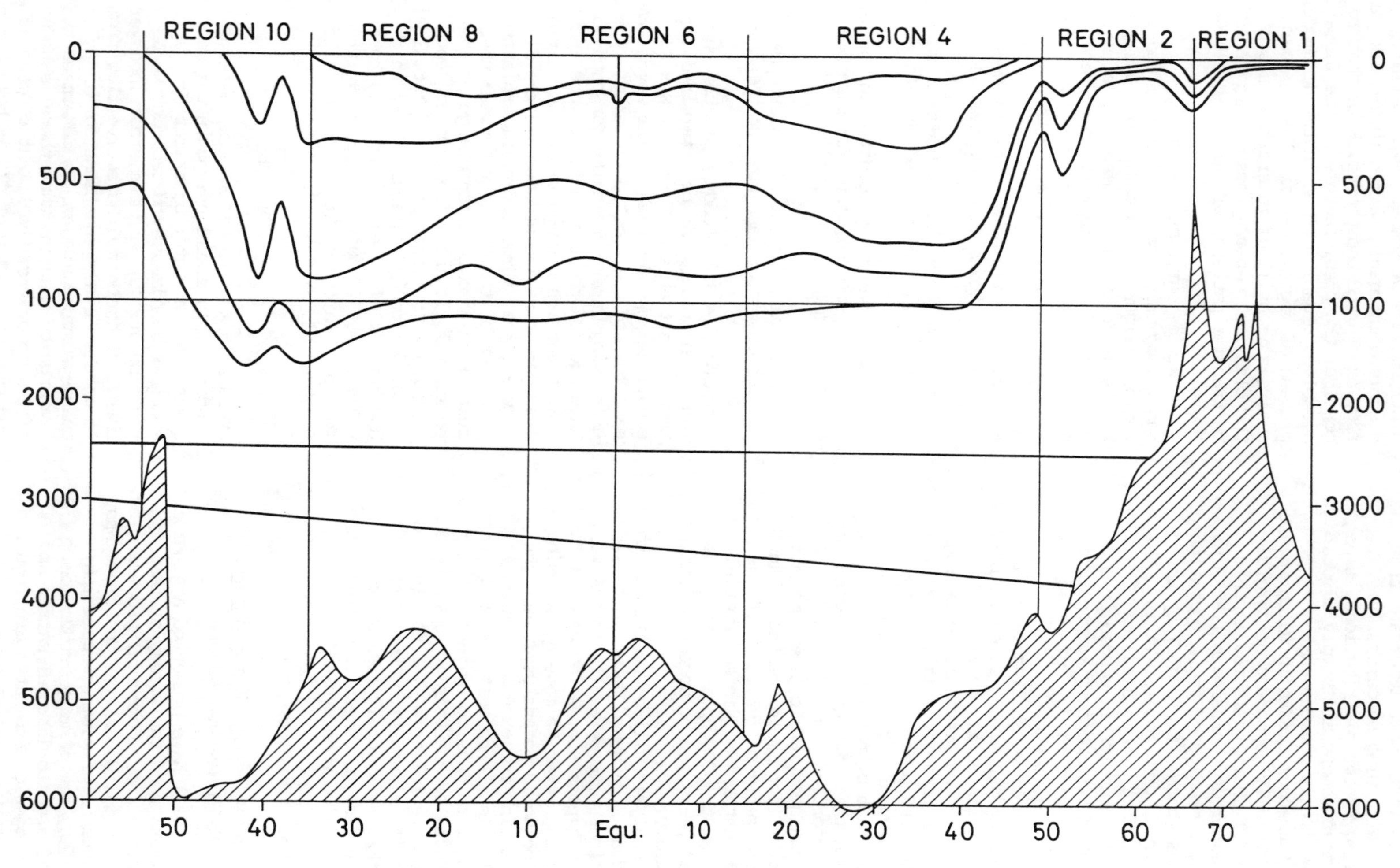


Fig. 2. Vertical structure of 84-box model (cf. Bolin et al., 1987).

Fig. 3. The division of the Atlantic Ocean into 12 regions for the definition of the 84-box model (cf. Bolin et al., 1987).

radioactive decay within the basin. In reality there may be a net through flow which implies an error in the way we determine rates of air-sea exchange. Such an error will affect the computed uptake of excess carbon and $^{14}C$ in the atmosphere. In addition the following inequality constraints are imposed

- turbulent flux of matter is always in the direction of the concentration gradient
- primary production only occurs in sunlit surface boxes and decomposition and dissolution of detrital matter take place at greater depth.

Since the model is approximate, equation system (1) is incompatible and a solution for the vector x is obtained by minimizing the norm in the least square sense. The result of such a procedure is of course dependent on the relative weighting of the equations in (1). Bolin et al. (1987) put equal weight on all tracer equations by normalizing the equations, but up-weighted the geostrophic equations in order to fulfill the geostrophic condition to about 15%. It is clear that the gross features of the circulation and biochemical processes were reasonably well reproduced as we know them from other studies, but resolution as well as specification of the key processes is not adequate to yield features such as the boundary currents or the coastal maxima of primary production. This is obviously a short-coming of the model, that has been discussed by Bolin et al. (1987).

Schlitzer (1988) in a similar attempt to use inverse methods to deduce the role of ocean circulation and biological processes for the quasi-steady distribution of tracers applied a more detailed and probably also more accurate method of analysis. It is, however, not possible in that way to arrive at a closed set of equations (1), which can be extended to a set of time dependent equations (1), that in turn can be used for study of transient changes, which is the prime objective of the present analysis.

Derivation of the Time Dependent Model

We shall use the model and the solution x derived in the previous section to determine how excess concentrations of $CO_2$, $^{14}C$ and tritium in the atmosphere penetrate into the ocean assuming that the water motions and biological processes, i.e. the vector x, remain unchanged. The injections of tritium and $^{14}C$ since 1954 are particularly interesting, because there are some data that show the invasion of these tracers into and within the Atlantic Ocean. If we are able to ascertain that the model is able to describe these processes adequately, we may also deduce with plausible reliability the likely uptake of carbon dioxide, although direct observations for verification are not available. It should be kept in mind, however, that the transient data for the last 20-30 years do not validate features of the solution that are important for studying the $CO_2$ uptake during the last centuries. We therefore implicitly still will be relying on the adequacy of the quasi-steady tracer distributions to derive the solution x, when deducing the response of the system for the time scales relevant for redistribution of $CO_2$ and $^{14}C$ in the Atlantic Ocean during the last 100-200 years.
The time dependent form of equation (1) can be written (Bolin et al. 1983)

$$dq/dt = - Ax + b \tag{2}$$

It is useful to rather transform this equation into the equivalent form

$$dq/dt = -Bq + b \tag{3}$$

where B is the matrix of coefficients that describes the way q will change and is defined by x. Equation (3) is a block diagonal matrix of the sets of tracer equations for the n tracers

$$dq^n/dt = -B^n q^n + b^n \tag{4}$$

where $q^n$ is the 84-component vector of concentrations of tracer n (tritium, Tr, carbon, C, or radiocarbon, *); $B^n$ is the (84 x 84) matrix which represents the subset of equations for tracer n and $b^n$ correspondingly the flux from or to the exterior and, for $^{14}C$, radioactive decay. Note that $B^n$ will not vary with time, while the vector $b^C$ for the DIC equations is dependent on the changes of the partial pressure of $CO_2$, p, in the atmosphere and the changing amount of DIC in the surface boxes caused by influx from the atmosphere and will accordingly vary with time. Similarly $b^*$ in the radiocarbon equation will be changing due to the transfer of bomb-produced $^{14}C$ from the atmosphere, while $b^{Tr}$ will be given as a function of time since return flow from the sea to the atmosphere can be neglected.

In order not to introduce artificial sources and sinks in the course of deriving a transient solution we must demand that the solution x is derived with the condition that the equations for the particular tracer (n) to be considered be well satisfied in steady state, i.e.

$$B^n q^n = A^n x = b^n \tag{5}$$

This condition was not forcefully imposed for any particular tracer in deriving the solutions given by Bolin et al. (1987), but the errors in the incompatible set of equations (1) were rather distributed equally between all tracer equations. We now rather derive a solution x, demanding that the subsets of equations referring to DIC and $^{14}C$ be well satisfied i.e. we upweight them markedly. To the extent small errors remain, these will be added to the right hand side of equation (4) when integrating with time making the implicit assumption that they are due to processes that have not been accounted for by the model and that these remain unchanged, i.e.

$$dq^n/dt = - B^n_1 q^n + b^n + e^n_1 \tag{6}$$

where $B^n_1$ is the matrix for the solution $x_1$ and $e^n_1$ is the error field in the set of equations (5) for the tracer n. This procedure is of course acceptable only if the solution $x_1$ is reasonably similar to the one previously derived with equal weighting for all tracers.

The transient solution for tritium can be obtained simply, since the water continuity equation is exactly satisfied and accordingly no artificial sources and sinks are introduced.

It should be remarked that the finite difference formulation of the advective terms as used may imply a potential numerical instability when employing the equation (4) for transient computations. The presence of turbulence, on the other hand, implies a stabilizing effect. Whether such an instability exists or not can easily be determined from the signs of the eigenvalues to the equation (4) and has only appeared in a few experiments and not in any of those described below.

The fact that the distributions of DIC and $^{14}C$ have been changing during the last few hundred years implies that we do not have accurate observations of steady state distributions of these tracers as required for the derivation of the vector x as outlined above. We assess, however, that the changes have been rather small ( $^{14}C$ data are available from the time before major injections occurred because of bomb-testing, see below) and we therefore adopt the following iterative procedure to derive a proper steady state solution, x, and the transient response of DIC and $^{14}C$ due to the perturbations induced by man.

We introduce the notations

$t = t_o = 1760$, preindustrial time, before which anthropogenic influences are assumed to have been negliable

$t = t_1 = 1957$, bomb testing had not yet significantly influenced the $^{14}C$ distribution

$t = t_2 = 1973$, time for the GEOSECS observations of tritium DIC and $^{14}C$ in the Atlantic Ocean

$t = t_3 = 1983$, the time to which transient computations were extended

We further use the notations $q^n_o$, $q^n_c$ and $q^n_e$ for observed, computed and estimated concentrations. The following iterative procedure is then applied:

1. A preliminary steady state solution x' is deduced by using the data sets $q^c_o\ (t_2)$ and $q^*_o\ (t_1)$ as was done by Bolin et al. (1987) although they do not represent a true steady state. It should be noted that the atmospheric $CO_2$ associated concentration $p_c\ (t_2)$ is computed in such a manner that no net air-sea exchange of $CO_2$ between the atmosphere and sea takes place. $p_c\ (t_2)$ differs little from $p_o\ (t_2)$.

2. Use the solution x' to derive transient solutions during the time period $t_o$ to $t_2$ assuming in this first iteration that the prescribed atmospheric carbon dioxide is given by

$$p_c(t) = p_c\ (t_2)+(p_o\ (t)-p_o\ (t_o)) \qquad (7)$$

using data for $p_o\ (t)$ as given by Siegenthaler and Oeschger (1987).

b) $\Delta^{14}C$ for atmospheric $CO_2$, is prescribed in accordance with data on the Suess effect during the time period $t_o < t < t_1$ and as due to release of bomb- produced $^{14}C$ for $t_1 < t < t_2$ (see further below).

3.a) The computed values of $q^c_c\ (t)$ during the period $t_o < t < t_2$ are used to derive a most plausible distribution of DIC at the time $t = t_o$, $q^c_e\ (t_o)$ , using the formula

$$q^c_e(t_o) = q^c_o(t_2) - (\ q^c_e(t_2) - q^c_o(t_2)\ ) \qquad (8)$$

i.e. we use the observed concentration at the time of GEOSECS minus a correction that is equal to the estimated increase between $t_o$ and $t_2$, as computed.

The value $p_c\ (t_o)$, corresponding to a state of no net flux of $CO_2$ between the atmosphere and the sea, is deduced for the field $q^c_e\ (t_o)$ thus derived.

b) Similarly the change of $\Delta\ ^{14}C$ in the sea during the period $t_o$ to $t_1$ is obtained and an estimated pre-industrial value

$q^*_e\ (t_o)$ is derived using the formula

$$q^*_e(t_o) = q^*_o(t_1) - (\ q^*_c(t_1) - q^*_o(t_1)\ ) \qquad (9)$$

4. The tracer distributions $q^c_e\ (t_o)$ and $q^*_e\ (t_o)$ together with the other sets of tracer distributions can be used to derive a second steady state solution x''

The procedure described above can of course be repeated but the two solutions x' and x'', as well as $p_e\ (t_o)$ and $p_o\ (t_o)$, are that similar that further iterations do not seem necessary.

The solutions x' or x'' are obtained by assuming that a steady state prevails. Bolin et al. (1987) ascertained that no net flux of any tracer occurs to or from the domain by assigning proper boundary conditions for exchange with the Pacific and Indian Oceans. In course of a transient computation this balance will be disturbed because of changing tracer concentrations in those reservoirs from which there is outflow of water. In order to assess in the best possible way the role of the Atlantic Ocean for uptake from the atmosphere it seems reasonable to change the tracer concentrations of the inflowing water in such a manner that the increased tracer outflow is balanced. It should be recognized, however, that the Atlantic Ocean may well be a pathway for uptake of tracers with excess atmospheric concentrations and that in reality a net flow from the Atlantic Ocean to adjacent oceans may take place. It is of course not possible to address such a question with a treatment of only the Atlantic Ocean as in the present case.

## Data on Tracer Distributions and Their Change

We shall use the data given by Bolin et al. (1987) for salinity, alkalinity, phosphorus and oxygen and assume that no changes have occurred during the time period considered. We describe below briefly the data for atmospheric $pCO_2$, DIC, $^{14}C$ and tritium that are used.

Atmospheric $CO_2$ and Dissolved Inorganic Carbon in the Sea

The changes of atmospheric $CO_2$ have been summarized by Siegenthaler and Oeschger (1987) and their data are shown in Figure 1. We notice that an increase by about 10 ppm occurred before significant emissions of $CO_2$ due to fossil fuel combustion began in the middle of last century.

The distribution of DIC in the Atlantic Ocean as determined during the GEOSECS program in 1973 supplemented by TTO data from 1981 in the northernmost parts of the sea have been used in the computations to be described (cf Bolin et al. 1987).

Radiocarbon, $^{14}C$

Broecker et al. (1960) have reported about 200 measurements in the Atlantic Ocean during 1957. These data and GEOSECS data from the deeper parts of the ocean have been used to deduce a most plausible pre-bomb $\Delta^{14}C$ distribution (cf Broecker and Peng, 1982; Bolin et al., 1987).

The $\Delta^{14}C$ distribution had, however, already been disturbed by human action at that time. The emissions of $^{14}C$-free carbon dioxide into the atmosphere since the middle of the 18th century led to a decrease of the $^{14}C/^{12}C$ isotope ratio in our environment, i.e. $\Delta^{14}C$ decreased. Stuiver and Quay (1981) conclude that atmospheric $\Delta^{14}C$ decreased by about 22‰ from an approximately constant value before the middle of last century to 1955, when bomb testing began, cf Figure 4. Druffel and Linick (1978) and Druffel and Suess (1983) have shown by the analysis of growth ring dated corals from the Florida Straits that $\Delta^{14}C$ for ocean surface water in this region decreased by about 12‰ during this same period.

Bomb-produced $^{14}C$ injections into the atmosphere began in 1952 but were hardly significant before 1955. Due to the major testing activities 1958-1962 $\Delta^{14}C$ for atmospheric $CO_2$ (lower troposphere) almost doubled, but has since then decreased to a value of about 300 ‰ above the pre-bomb value in 1980 (cf Figure 5; Bolin, 1986). A transfer into the ocean has also occurred, as is clear from direct observations. A $\Delta^{14}C$ concentration of about 100 ‰ was reached around 1970 and a slight decrease seems to have occurred since then. The increase of $\Delta^{14}C$ until 1973 within the Atlantic Ocean as observed during the GEOSECS expeditions has been evaluated for the 84 boxes and is given in Table 2.

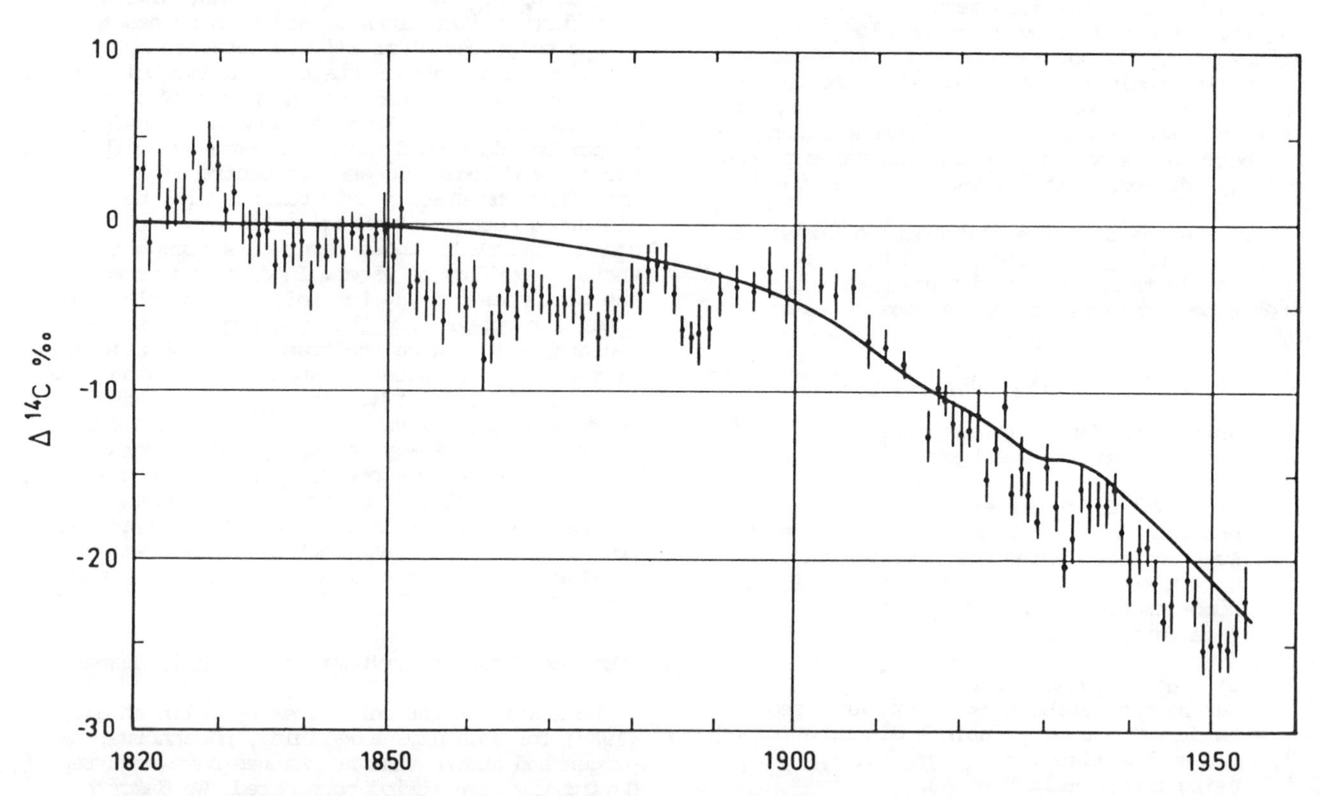


Fig. 4. Changes of atmospheric $\Delta^{14}C$ 1820-1955 as deduced from analysis of wood samples. (Stuiver and Quay, 1981).

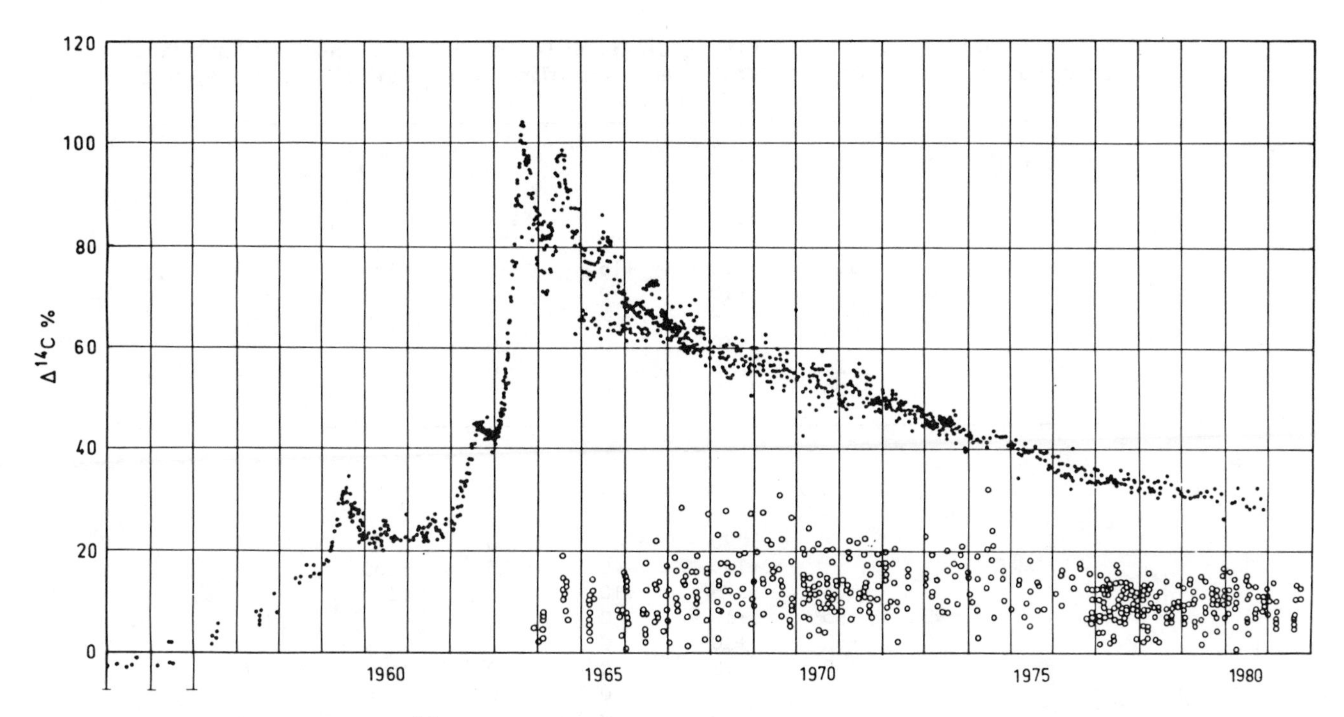


Fig. 5. Changes of $\Delta$ $^{14}C$ in atmospheric $CO_2$ (filled dots), and in dissolved inorganic carbon in ocean surface water (circles), 1954-1980 (Nydal and Lövseth, 1983).

Tritium

Observed changes in the atmosphere have been summarized by Weiss and Roether (1980). (Some misprints in tables of that article have been corrected after direct personal communication with Roether). Penetration of tritium into the sea, particularly the Atlantic Ocean has been described by Östlund et al. (1976, 1987) and the tritium penetration depth in the world oceans in 1973 has been determined (Broecker and Peng, 1982). We have used the GEOSECS data from 1973 to estimate the distribution in the 84 boxes that are defined by our model (see Table 3). We note that in comparison with concentrations observed in the sea 1957 we may safely assume that concentrations before bomb testing began were negligiable, i.e. $q^{Tr}$ $(t < t_1) = 0$. The boundary fluxes of tritium from the atmosphere, which determine $b^{Tr}$ in equation (2) (in addition to radioactive decay) have been deduced by Weiss and Roether (1980). Their method is to compute an input rate for each latitude and year, essentially as the product of a time factor with a latitude factor.

Results

Experimental Set-up

The discussion will primarily be based on three experiments, which are briefly described below, but several others have been performed, to which reference will occasionally be made.

I Weighting of the equations for $q^C$ and $q^*$ were assumed to be 50 and of the geostrophic equations 4; rate of air-sea exchange the same everywhere.

II Weighting of $q^C$, $q^*$ and geostrophic equations 50; air sea exchange the same everywhere.

III Weighting of $q^C$, $q^*$ and geostrophic equations 50; air-sea exchange increased by 70% in regions 1,2, 3 and 12 and reduced appropriately in other regions to maintain balance between $^{14}C$ inflow from the atmosphere and internal radioactive decay. There are reasons to believe that air-sea exchange is enhanced at higher latitudes as compared to tropical and subtropical latitudes because of stronger winds. The experiment is intended to test the sensitivity of the results caused by uncertainty due to prescribing a constant rate factor.

A few comments are of interest:

- The average rate of air-sea exchange of $CO_2$ deduced for I and II was 20.5 mol $m^{-2}yr^{-1}$ and for III 18.3 mol $m^{-2}yr^{-1}$ based on the initial data used. The corresponding values for the estimated (1760) air-sea exchange rates were 14.7 mol $m^{-2}yr^{-1}$ for I and II and 13.2 mol $m^{-2}yr^{-1}$ for III. The former values are of course not correct because the assumption of a steady

TABLE 2. Changes of $^{14}C$ (in o/oo) in Ocean Boxes 1957-1973 Based on Data From Broecker et al. (1960) and GEOSECS Data According to Östlund et al. (1976) and Östlund et al. (1987).

| Region Layer | 1 | 2 | 4 | 6 | 8 | 10 | 12 |
|---|---|---|---|---|---|---|---|
| | | | Western Basin | | | | |
| 1 | | | +210 | +160 | +175 | | |
| 2 | | | +165 | +70 | +130 | +145 | |
| 3 | | +100 | +65 | +15 | +25 | +75 | |
| 4 | +95 | +50 | +10 | 0 | 0 | +10 | +70 |
| 5 | +25 | 35 | +5 | 0 | 0 | 0 | +5 |
| 6 | | 35 | +10 | 0 | 0 | 0 | 0 |
| 7 | | 20 | +5 | 0 | 0 | 0 | 0 |
| 8 | | 20 | +5 | 0 | 0 | 0 | 0 |
| | | | Eastern Basin | | | | |
| | 1 | 3 | 5 | 7 | 9 | 11 | 12 |
| 1 | | | +220 | +140 | +120 | | |
| 2 | | | +190 | +40 | +115 | +135 | |
| 3 | | +105 | +175 | +10 | +20 | +105 | |
| 4 | +95 | +95 | +10 | 0 | 0 | | +70 |
| 5 | +25 | +55 | +10 | 0 | 0 | | +5 |
| 6 | | +30 | +5 | 0 | 0 | | 0 |
| 7 | | | +5 | 0 | 0 | 0 | 0 |
| 8 | | | 0 | 0 | 0 | 0 | 0 |

state is not a correct one. There is in reality a net inflow of $CO_2$ into the ocean, while an outflow of $^{14}C$ occurs to compensate for the decrease of $\Delta$ $^{14}C$ in the atmosphere because of fossil fuel combustion. An air-sea exchange rate of 14 mol $m^{-2}yr^{-1}$ therefore seems more appropriate than about 19 mol $m^{-2}yr^{-1}$ as is the outcome of the first inversion. There is of

TABLE 3. Concentrations of Tritium in 1973. Box Averages Based on GEOSECS Data (TU). (cf. Östlund, et al., 1987).

| Region Layer | 1 | 2 | 4 | 6 | 8 | 10 | 12 |
|---|---|---|---|---|---|---|---|
| | | | Western basin | | | | |
| 1 | | | 6.8 | 2.6 | 2.2 | | |
| 2 | | | 6.5 | 1.8 | 1.3 | 1.5 | |
| 3 | | 17.5 | 3.8 | 0.2 | 0.2 | 0.6 | |
| 4 | 11.0 | 11.3 | 1.0 | 0.1 | 0.1 | 0.2 | 0.7 |
| 5 | 3.7 | 7.3 | 0.8 | 0.0 | | 0.0 | 0.3 |
| 6 | | 3.9 | 0.6 | 0.0 | | 0.1 | 0.1 |
| 7 | | 1.5 | 0.1 | 0.1 | | 0.1 | 0.0 |
| 8 | | 2.9 | 0.2 | 0.1 | | 0.0 | 0.0 |
| | | | Eastern basin | | | | |
| Region Layer | 1 | 3 | 5 | 7 | 9 | 11 | 12 |
| 1 | | | | | | | |
| 2 | | | 7.2 | | | 1.6 | |
| 3 | | 7.5 | 5.0 | | | 0.6 | |
| 4 | 11.0 | 6.9 | 0.7 | | | 0.0 | 0.7 |
| 5 | 3.7 | 4.1 | 0.3 | | | 0.0 | 0.3 |
| 6 | | 2.1 | 0.2 | | | 0.0 | 0.1 |
| 7 | | | 0.1 | | | 0.0 | 0.0 |
| 8 | | | 0.0 | | | 0.0 | 0.0 |

course still considerable uncertainty about this value since $\Delta^{14}C$ for ocean surface waters before bombproduced $^{14}C$ was emitted into the atmosphere were not well determined. The value is considerably less than the value around 18 mol $m^{-2}yr^{-1}$ as given by Broecker and Peng (1982). The discrepancy may depend on the assumption of no net outflow of $^{14}C$ to adjacent oceans.

The assumption of a steady state distribution of DIC and $^{14}C$ in 1760 is not necessarily correct although more appropriate than in 1955. Figure 1 shows that atmospheric $CO_2$ concentrations rose by about 10 ppm between 1760 and 1860, although the emissions due to fossil fuel combustion were negligiable before the middle of the 19th century. Figure 4 shows that $^{14}C$ did not either start to decrease until about 1850. The increase of atmospheric $pCO_2$ must then either be due to a net input of carbon into the atmosphere due to deforestation and changing land use or to an unbalance associated with natural oscillations in the carbon cycle, possibly related to natural climatic variations, e.g. the little ice-age, during the preceding centuries. We shall also conduct a few experiments that reveal some response characteristics of the Atlantic Ocean of interest in this context.

Features of the Steady State Solution **x** .

A few statistical analyses are first of interest. The correlations between the first and second iterative solutions $\mathbf{x}'$ and $\mathbf{x}''$ and between the different solutions $\mathbf{x}''$ are given in Table 4.

We notice that the former are all above 0.9, i.e. the departure of the system from steady state before injection of bomb $^{14}C$ does not critically influence the solution **x**. We shall still use the more appropriate solution $\mathbf{x}''$.

A change of the weighting, on the other hand, influences the solution significantly. The correlation between the solutions in experiment I and II is about 0.7 and the same is true for anyone of these and the "base case" (Bolin et al., 1987), where all tracer equations had weight one, and the geostrophy equations had weight four.This is a genuine uncertainty due to incomplete or not representative data sets, coarse resolution, sensitivity of small scale features of the solution to errors in the data, as well as incomplete inclusion of the relevant processes that affect the tracer distributions (this matter will be further discussed below). We find that the intensity of advective flow patterns of inversions I and II are about 70% and 110% respectively of that of the "base case" presented by Bolin et al. (1987). The similarities between the solutions, particularly between II and the "base case", justify that we pursue the analysis, but we shall have to explore the uncertainties of our final results due to the uncertainties in the solution x introduced in these various ways.

The transient solutions primarily depend on the advective and turbulent terms, the $^{14}C$ equations to some degree also on the flux of detrital matter. The meridional circulation with sinking motion in the north Atlantic is of particular importance for the rate of transfer of matter injected into surface boxes to deeper layers. A comparison of the solutions I and II with geostrophic weighting of 4 and 50 respectively shows that the former is quite non-geostrophic because of the comparatively large weights for the tracer equations for DIC and $^{14}C$. The meridional circulation is also comparatively weak. Solution II, on the other hand, approximates the geostrophic constraints much better and is characterized by a meridional circulation with a return flow southward in the North Atlantic below 1000 m of about 22 Sv, of which 8 Sverdrup is below 2500 m (see Figure 6). This is similar to the base case presented by Bolin et al. (1987). The ocean circulations derived by Maier-Reimer and Hasselmann (1987),

TABLE 4. Correlations Between First and Second Iteration Solutions $\mathbf{x}'_i$ and $\mathbf{x}''_i$ for i = I, II, III and Between Second Iteration Solutions $\mathbf{x}''_i$ and $\mathbf{x}''_j$ for the Experiments i, j = I, II, III. Correlations for Subsets of Components of the Vectors $\mathbf{x}_i$ and $\mathbf{x}_j$, i.e. Advective, Turbulent and Organic Detritus Components Are Also Shown.

| | Total **x** | Adv. components | Turb. components | Organic detritus components |
|---|---|---|---|---|
| $\mathbf{x}'_I$ $\mathbf{x}''_I$ | 0.92 | 0.94 | 0.90 | 0.97 |
| $\mathbf{x}'_{II}$ $\mathbf{x}''_{II}$ | 0.97 | 0.99 | 1.00 | 0.97 |
| $\mathbf{x}'_{III}$ $\mathbf{x}''_{III}$ | 0.98 | 0.98 | 0.95 | 0.98 |
| $\mathbf{x}''_I$ $\mathbf{x}''_{II}$ | 0.73 | 0.69 | 0.78 | 0.94 |
| $\mathbf{x}''_I$ $\mathbf{x}''_{III}$ | 0.71 | 0.68 | 0.76 | 0.93 |
| $\mathbf{x}''_{II}$ $\mathbf{x}''_{III}$ | 0.99 | 1.00 | 0.98 | 1.00 |

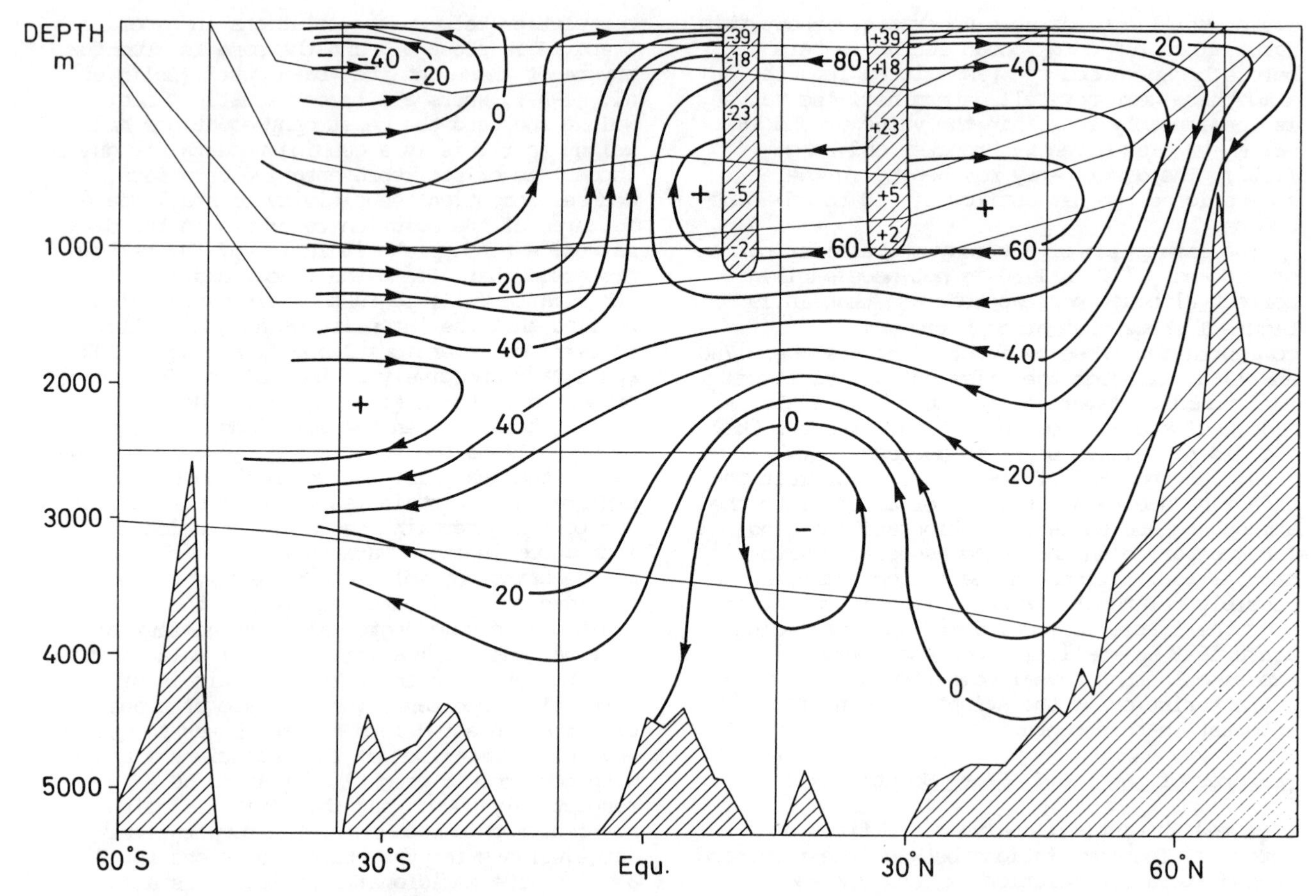


Fig. 6. Meridional water circulation in the Atlantic Ocean as determined from $x''$ for experiment II. Numbers on streamlines: flow in $10^{13}$ ton $yr^{-1}$ ( 3.2 Sverdrup). The analysis has not been extended southward of the boundary between boxes 8/9 and 10/11 (cf. Figure 3). The sink at latitude 10°N and the source at latitude 30°N due to the current through the Caribbean Sea and Mexican Gulf are shown as shaded regions in units of $10^{13}$ ton $yr^{-1}$. Note the expanded vertical scale above a depth of 1000 m.

on the other hand, shows significantly slower circulations and accordingly also less $^{14}C$ concentrations in the deep sea. In the light of these comparisons we shall use the solution II and the associated transient solution as our reference case.

The following features of the reference solution should be noticed (Figure 6):

1. The mean meridional circulation with sinking motion in regions 1-3, is about 0.70 $10^{15}$ ton $yr^{-1}$ ( 23 Sverdrup) in general agreement with the common view. A weak cell ( 0.10 $10^{15}$ ton $yr^{-1}$) in the opposite direction is found between the equator and 30°N below about 2.500 m.
2. Two meridional cells in the thermocline region with upwelling in equatorial regions and poleward flow in surface layers are superimposed on the general direct meridional cell as described in the previous paragraph. The northward flow of intermediate water in the Southern Hemisphere penetrates to about 10°N.
3. Upwelling takes place preferentially in the eastern basin and downwelling in the western one.
4. The northward flow in the surface and intermediate layers of the western basin outside the north American continent is about 0.7 $10^{15}$ tons $yr^{-1}$ ( 23 Sverdrup) and is obviously much weaker than the Gulf Stream. In reality some recirculation of Gulf Stream water takes place to the west of the mid-Atlantic ridge and can partly explain the discrepancy. Some of the tracer transfer by this major coastal current (as well as others elsewhere) is therefore implicitly accounted for by the horizontal turbulence.
5. The turbulent components are generally smaller than the advective ones, but do contribute to the solution. Vertical turbulent transfer plays a significant role only between

the uppermost boxes and those immediately below thus accounting for exchange between the surface mixed layer and the uppermost part of the thermocline.

6. Large horizontal turbulent components (along isopycnic surfaces) appear spotwise in our solutions. These are most likely due to the fact that tracer gradients often are comparatively small along isopycnic surfaces leaving the rate of turbulent exchange rather indeterminate. Accordingly they only contribute marginally to the errors. It should be remarked, however, that this would not necessarily be so, when computing tracer invasion into the sea that perferentially takes place along isopycnic surfaces. For this reason the rate of uptake that we will derive later may be an overestimate.

7. The detritus flux settling out of the surface boxes is 4.7 $10^{15}$ g C $yr^{-1}$ ( 50 g C $m^{-2}yr^{-1}$) which seems large as compared with the basic case presented by Bolin et al., (1987) and direct observations (cf Broecher and Peng, 1982). This may well be due to the comparatively large turbulent components derived (see above) which transfer DIC upwards from the layers of high concentrations into the mixed layer, for which a compensating detritus flux is derived.

8. The solution has been obtained without making use of the temperature data (cf Bolin et al, 1987). It is, however, possible to compute a posteriori the heat transfer that the steady-state solution implies. We find that the northward transfer (experiment II) from regions 4 and 5 into regions 2 and 3 is about 10 $10^{14}$ W compared to a range between 0 and 8 $10^{14}$ W as assessed by Wunsch (1984). Our larger value is primarily caused by turbulent transfer due to rather large horizontal turbulent components (cf 6 above). This implies that the $CO_2$ uptake that we compute below (see section 4.6) may also be somewhat too large as the penetration in the North Atlantic and horizontal transfer southward is a major pathway for $CO_2$ invasion into the Atlantic Ocean.

9. There are other details in the solution that deserve further study. It would obviously be of interest to relax the constraint due to the prescribed fixed values of the b-vector that the present method of the solution is based upon. The use of the simplex linear programming method for solution with given ranges of the components of the **b** vector is an interesting alternative (cf. Wunsch, 1984). Additional knowledge about the ocean circulation in terms of further constraints could also be added to the given set of inequality constraints (Bolin et al. 1987).

It is clear from the present solution that the use of inverse methods for deducing simultaneously the circulation and biochemistry of the ocean which is being considered, requires a careful and detailed data analysis in relation to the matrix inversion that provides our solution. As we shall see in the following sections the integral properties of the solution as revealed by accumulation of tracer material in course of transient invasions from e.g. the atmosphere are remarkably similar for different x solutions. The solution II is, however, plausibly one that overestimates the rate of tracer transfer within our ocean domain.

Invasion of Tritium into the Atlantic Ocean 1955-1973

Transient computations to determine the uptake of bomb produced tritium during the period 1955-1973 were done using the transfer vectors x'' from experiments I and II. The total computed uptake turned out to be 25-30% larger than the amount measured in the sea during the GEOSECS cruises in 1973, the difference between the two experiments being due to somewhat different flows to adjacent oceans. Sarmiento (1983) found a corresponding discrepancy between the given inflow according to Weiss and Roether (1980) and observed storage of about 20%. The correlations between observed and computed accumulation into the model boxes where uptake has occurred are 0.91 and 0.88 for the two transient runs I and II. Table 5 compares the uptake by region. We note the unrealistically large values in region 1, and also seemingly too large values in regions 6 and 8 (see further comments in section 4.5).

The Suess Effect

The emissions of $^{14}C$ free carbon dioxide due to fossil fuel burning have changed the distribution of $\Delta^{14}C$ both in the atmosphere and the sea (the Suess effect), already before injection of $^{14}C$ into the atmosphere due to bomb testing. We have used the data given by Stuiver and Quay (1981) for the changes of $\Delta^{14}C$ in the atmosphere since 1800 and assumed a constant

TABLE 5. Uptake of Tritium 1955-1973 by Region (in TU x $10^{15}$ ton), Observed According to GEOSECS and Computed in Transient Experiments I and II.

| Region | Obs. | I | II |
|---|---|---|---|
| 1 | 2 | 14 | 26 |
| 2 | 45 | 53 | 36 |
| 3 | 35 | 37 | 25 |
| 4 | 49 | 52 | 51 |
| 5 | 12 | 22 | 32 |
| 6 | 4 | 10 | 12 |
| 7 | | | |
| 8 | 5 | 10 | 7 |
| 9 | | | |
| 10 | 5 | 7 | 6 |
| 11 | 3 | 5 | 5 |
| 12 | 4 | 2 | 2 |
| Total | 164 | 212 | 202 |

TABLE 6. Decrease of $^{14}C$ in the Atlantic Ocean Due to Fossil Fuel Emissions Until 1955, Suess Effect, in o/oo.

| | Western basin | | | | | | |
|---|---|---|---|---|---|---|---|
| Region | 1 | 2 | 4 | 6 | 8 | 10 | 12 |
| Layer | | | | | | | |
| 1 | | | -8 | -5 | -7 | | |
| 2 | | | -5 | -4 | -5 | -5 | |
| 3 | | -5 | -4 | -2 | -2 | -4 | |
| 4 | -9 | -4 | -1 | 0 | -1 | -2 | -3 |
| 5 | -8 | -4 | -1 | 0 | -1 | 0 | 0 |
| 6 | -6 | -2 | -1 | 0 | 0 | 0 | 0 |
| 7 | 0 | 0 | 0 | 0 | 0 | 0 | 0 |
| 8 | 0 | 0 | 0 | 0 | 0 | 0 | 0 |
| | Eastern basin | | | | | | |
| | 1 | 3 | 5 | 7 | 9 | 11 | 12 |
| 1 | | | -5 | -5 | -7 | | |
| 2 | | | -4 | -4 | -5 | -7 | |
| 3 | | -5 | -3 | -1 | -1 | -3 | |
| 4 | -9 | -3 | -2 | 0 | 0 | -2 | -3 |
| 5 | -8 | -2 | 0 | 0 | 0 | 0 | 0 |
| 6 | -6 | -1 | 0 | 0 | 0 | 0 | 0 |
| 7 | 0 | 0 | 0 | 0 | 0 | 0 | 0 |
| 8 | 0 | 0 | 0 | 0 | 0 | 0 | 0 |

value before that time to determine the $\Delta\,^{14}C$ changes in the Atlantic Ocean during the period 1760-1955 with the aid of the solution II. A slow drift of less than 1 $^{o}/oo$ occurred before fossil fuel emissions began (about 1850). The computed Suess effect was accordingly derived as the $\Delta\,^{14}C$ change during the period 1850-1955. The result is shown in Table 6. The changes in the surface waters vary between -4 and -9 $^{o}/oo$, excluding the southernmost region for which the results may be questioned because of influence from the approximate treatment of exchange with adjacent seas. The average Suess effect in surface water deduced for 1955 is -5.5 $^{o}/oo$, compared with -22 $^{o}/oo$ for atmospheric carbon dioxide. In region 4 a value of -8 $^{o}/oo$ is deduced which is significantly less than the observed value of -12 $^{o}/oo$ as reported by Druffel and Suess (1983). The differences can hardly be explained by the deficiencies of the solution II as discussed in the previous section. The question is raised of how representative the measurements may be.

Uptake of Bombproduced $^{14}C$

Since 1955 large amounts of $^{14}C$ have been injected into the atmosphere and a transfer into the oceans has occurred. GEOSECS observations reveal the increase and Table 2 show the estimated changes during the period 1955-1973. The computed changes as determined with the reference solution II are given in Table 7. We compare observed and computed changes:

The area average increase of $\Delta\,^{14}C$ for all surface boxes is computed to have been 132 $^{o}/oo$ if using the reference solution II as compared with an observed change of 138 $^{o}/oo$. We note that the air sea exchange rate of $CO_2$ has been assumed to be 14.6 $Mm^{-2}yr^{-1}$ based on a $^{14}C$ balance between preindustrial inflow from the atmosphere and radioactive decay. As has already been pointed out this is less than the most commonly accepted value of about 18 $Mm^{-2}yr^{-1}$. (Broecker and Peng, 1982). The solution with enhanced air sea exchange in polar regions (III) yields almost the same value for $^{14}C$ uptake. In the case of a comparatively slow ocean circulation (I) the surface $\Delta\,^{14}C$ values increase more rapidly 163 $^{o}/oo$, presumably because of less effective transfer away from the ocean surface. The increase of surface concentrations depend more on the rate of transfer into deeper layers of the sea than on the precise value for the rate of air-sea exchange.

The $\Delta\,^{14}C$ changes in individual surface boxes show a somewhat more irregular pattern (Figure 7). The computed changes with time in surface box 4, layer 1, are compared with observed changes as deduced by analysis of corals in Florida Straits and Bermuda, Figure 8 (Druffel and Suess, 1983). The agreement between values computed with the reference solution (II)

TABLE 7. Changes of $^{14}C$ (o/oo) in Ocean Boxes 1955-1973 as Deduced in Experiment II.

| Western Basin | | | | | | | |
|---|---|---|---|---|---|---|---|
| Region<br>Layer | 1 | 2 | 4 | 6 | 8 | 10 | 12 |
| 1 | | | +219 | +126 | +137 | | |
| 2 | | | +107 | +97 | +96 | +118 | |
| 3 | | +105 | +71 | +28 | +58 | +85 | |
| 4 | +233 | +82 | +9 | +7 | +33 | +54 | +74 |
| 5 | +203 | +70 | +12 | +5 | +13 | +14 | +22 |
| 6 | +71 | +32 | +4 | +1 | +1 | 0 | 0 |
| 7 | +5 | +6 | 0 | 0 | 0 | 0 | 0 |
| 8 | +1 | +5 | 0 | 0 | 0 | 0 | 0 |
| **Eastern Basin** | | | | | | | |
| | 1 | 3 | 5 | 7 | 9 | 11 | 12 |
| 1 | | | +125 | +132 | +190 | | |
| 2 | | | +87 | +98 | +106 | +145 | |
| 3 | | +118 | +47 | +22 | +7 | +45 | |
| 4 | +233 | +64 | +25 | +2 | +5 | +46 | +74 |
| 5 | +203 | +13 | +7 | +5 | 0 | +7 | +22 |
| 6 | +71 | +2 | +1 | 0 | 0 | 0 | 0 |
| 7 | +5 | 0 | 0 | 0 | 0 | 0 | 0 |
| 8 | +1 | 0 | 0 | 0 | 0 | 0 | 0 |

and observations is excellent but probably somewhat fortuitous in the light of the irregularities between regions mentioned above. In case of a slower circulation (I) a substantially larger increase was obtained for the surface boxes, while the total uptake of $^{14}C$ by the sea was significantly less (see further below), again indicating that the less efficient transfer into the interior of the ocean lead to enhanced concentrations in the surface layer.

The penetration of $^{14}C$ into the ocean as computed (Table 7) exhibits both similarities and discrepancies compared with observations (Table 2). A short time elapsed between the major emissions into the atmosphere (1958-1961) and observations during GEOSECS cruises in 1973. We cannot expect that the interior ocean circulation has brought the invasion very far from the regions of entrance. There is, however, marked differences between uptake by different regions already after this short time, Table 8 shows the uptake integrated vertically for each region. Lack of observations in some regions (cf. Table 2) prevent us from a comparison of the total uptake but those that can be made between model computations and observations for boxes with data are of interest.

- A marked difference between uptake in middle latitudes and the tropical belt is clearly seen in both observed and computed fields. The model computations let the penetration proceed somewhat more quickly vertically in the tropics than observations indicate.
- The uptake by regions 2-5 is only about 70% of what observations show. Uncertainties of the observed pre-bomb concentrations and therefore of penetration could be the major cause for this discrepancy.
- The ratio of the uptake in the western basin to that in the eastern one, is considerably larger as deduced from the computations (48.0/28.2) than observed (46.6/38.7).
- The total observed uptake is about 10% larger than the one computed which difference hardly is significant with regard taken to both data and model uncertainties.
- A comparison between model computations and observed $^{14}C$ distribution in 1981/82 according to TTO observations could aid us further in our attempts to validate the results from the transient computations. This will be done when a more careful study of the sensitivity of the inverse solution to assumptions of weighting and the constraints being used has been completed (see further the concluding section of this paper).

In spite of discrepancies between observed and computed changes the overall fields show

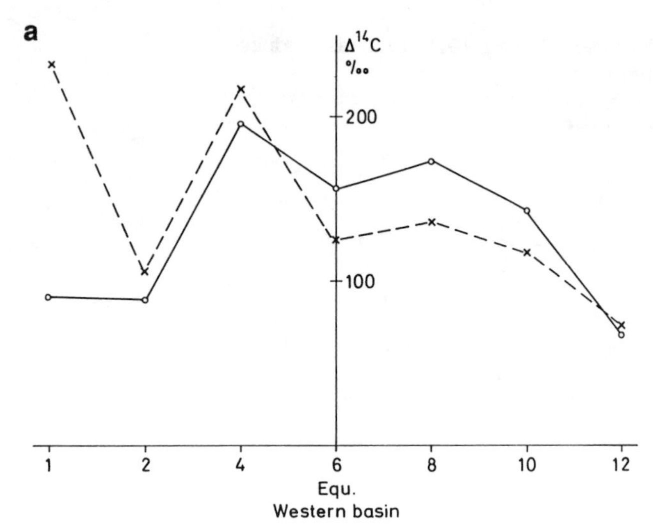


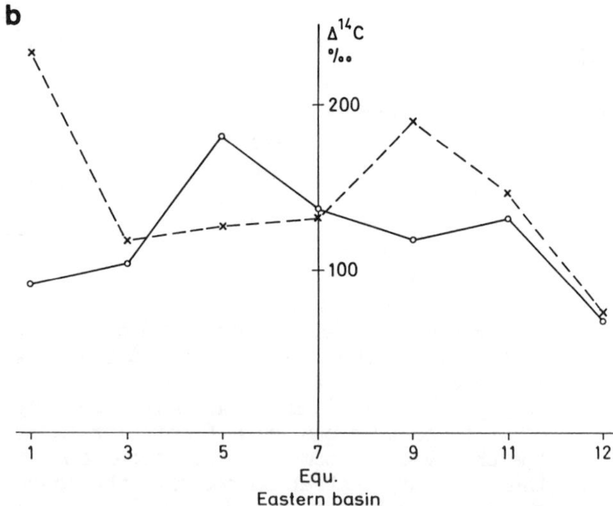


Figure 7. Observed (solid line) and computed (dashed line) changes of $\Delta^{14}C$ in surface boxes as computed with the aid of transient solution II during the period 1957-1973, for the western (a) and eastern (b) basins.

sufficient similarities to justify an analysis of the model uptake of $CO_2$ during an extended period of time using the solution II.

## Uptake of $CO_2$ During the Period 1760-1983

The annual rate of $CO_2$ uptake has been computed using the solutions I and II. In the light of the analysis in the previous sections we shall adopt the latter as one reference case. Before discussing the details of these results a few other features of the transient solutions are of interest.

Figure 9 is a scatter diagram showing the difference of the uptake by the different reservoirs in the experiments I and II. They are quite small but systematic. The correlation between the changes is about 0.9. The surface boxes, shown as circles, have systematically somewhat smaller concentrations in experiment II than in I which implies a larger $pCO_2$ difference between the atmospheric concentration and the mean $pCO_2$ of the ocean surface boxes. Accordingly the cumulative $CO_2$ flux from the atmosphere into the Atlantic Ocean was larger in experiment II (35.9 PgC in 1983) than in I (30.2 PgC in 1983). This is due to the more intense ocean circulation in II and accordingly more rapid transfer of $CO_2$ into the deeper strata of the ocean. We notice systematically higher concentrations in II in deeper layers than in experiment I. As a matter of fact the rate of air sea exchange is of less importance for the uptake than is the rate of ocean overturning as long as the air sea exchange is rapid enough not to let ocean surface $pCO_2$ lag too much behind the increasing atmospheric $pCO_2$ concentrations.

Figure 10 shows the vertical distribution of the uptake in a series of profiles for selected regions. The computed changes until 1860, 1955 and 1983 are reproduced. A few particular features of the penetration into the ocean (Figure 10) are worth noticing, even though careful improvement of the solution by using independent information may modify some of these (see further next section).

- Rapid penetration down to the bottom of the ocean occurs in region 2, while it does not reach layer 8 in region 3 because of upwelling from the deepest layer of the eastern basin (cf. also a similar feature of the water flow obtained by Schlitzer, 1987).
- The vertical distributions in regions 4 and 6 shows clearly the gradual invasion of carbon from the north into intermediate and deep layers due to the advective flow, while turbulent penetration from above is considerably less significant.
- The penetration of $CO_2$ into the intermediate and deep waters is remarkably similar when using the different vectors $x_{ij}$ that we have derived, in spite of significant differences between these, particularly with regard to the turbulent components. Clearly the small scale features of the solutions are not of prime importance for the integrated uptake.

Figure 11 shows some characteristic features of the transient solution. It should be emphasized that the integrations start from a state of no net transfer of $CO_2$ between the atmosphere and the sea i.e. the areal mean of $CO_2$ partial pressure in the sea is assumed to be equal to the atmospheric $pCO_2$ in 1760. Figure 11 shows how a difference between these develop in course of time and accordingly the uptake increases with time.

The transient computations imply that we assume a net emission of $CO_2$ into the atmosphere due to man, causing an increase of the

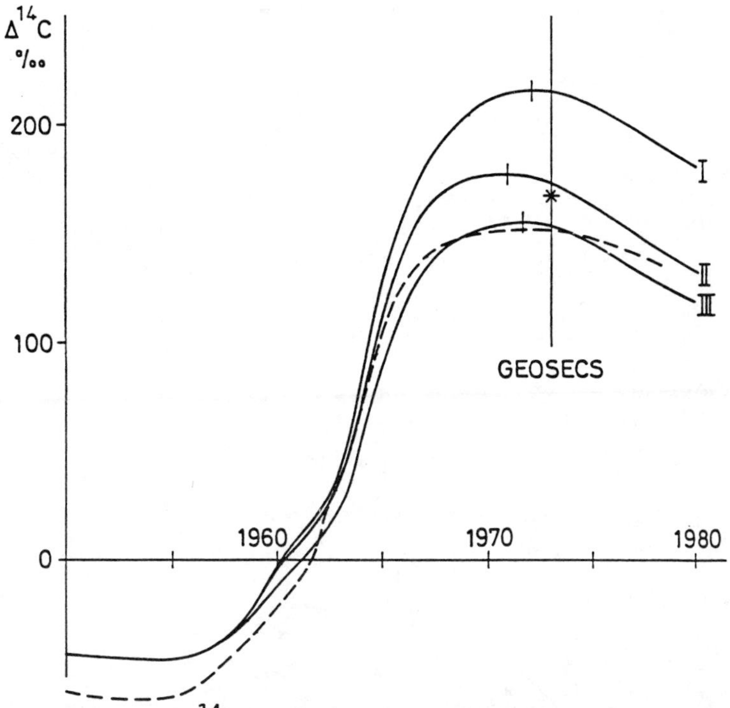


Fig. 8. Computed changes of $\Delta^{14}C$ in the surface layer of region 4 as deduced with the transient solutions I, II and III (solid lines) compared with measured changes as determined from analysis of corals in Florida Straits (dashed line; Druffel and Suess, 1983). The average value in 1973 as determined during the GEOSECS project is also shown (*).

atmospheric $pCO_2$ and a response of the Atlantic Ocean in terms of a net uptake. We can perhaps better visualize the characteristics of what happens by assuming hypothetically that the rest of the world oceans would behave in an analogous way as described by our model of the Atlantic Ocean. Not knowing the total emissions we explore how cases in which we assume that the

TABLE 8. Uptake of Excess Atmospheric $^{14}C$ by Region During the Period 1955-1973. Observations are from the GEOSECS Expeditions. Computations are According to the Transient Experiments II and III.

| Region | Total inventory $10^{26}$ atoms | | | Inventory $10^{14}$ atoms/m$^2$ | | |
|---|---|---|---|---|---|---|
| | Obs. | II | III | Obs. | II | III |
| 1 | 1.0 | 2.8 | 3.7 | 0.22 | 0.62 | 0.83 |
| 2 | 5.8 | 5.1 | 5.6 | 1.20 | 1.07 | 1.16 |
| 3 | 7.2 | 4.1 | 4.1 | 1.56 | 0.88 | 0.89 |
| 4 | 16.9 | 11.8 | 9.0 | 1.59 | 1.11 | 0.85 |
| 5 | 11.5 | 5.1 | 5.0 | 1.57 | 0.70 | 0.68 |
| 6 | 2.8 | 3.5 | 3.3 | 0.43 | 0.54 | 0.50 |
| 7 | 2.9 | 4.6 | 3.9 | 0.30 | 0.46 | 0.39 |
| 8 | 6.9 | 9.0 | 8.4 | 0.86 | 1.11 | 1.05 |
| 9 | 4.1 | 3.9 | 3.2 | 0.63 | 0.60 | 0.49 |
| 10 | 4.5 | 7.2 | 7.3 | 0.66 | 1.07 | 1.08 |
| 11 | 7.4 | 3.9 | 5.0 | 1.17 | 0.61 | 0.80 |
| 12 | 1.7 | 1.9 | 2.6 | 0.15 | 0.17 | 0.23 |
| Total | 72.7 | 62.9 | 61.1 | | | |

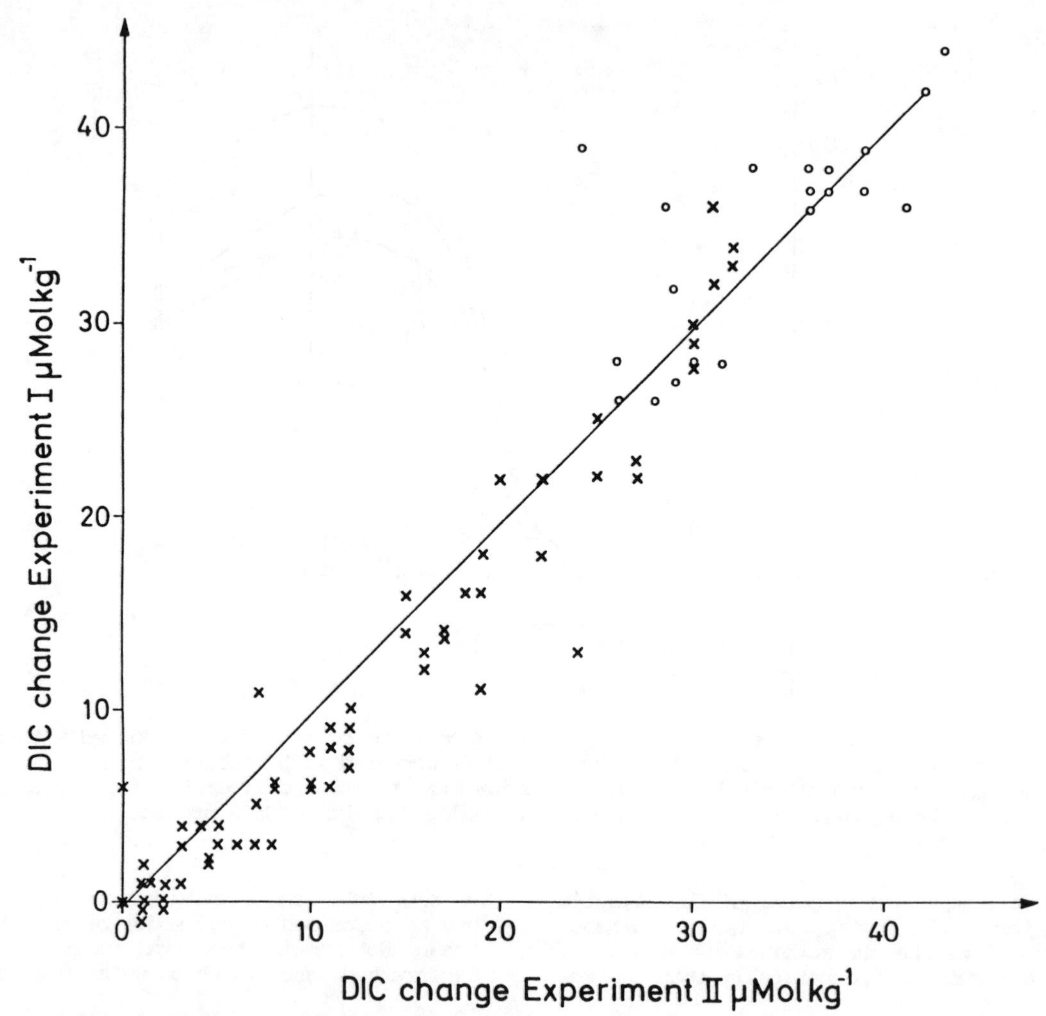


Fig. 9. Scatter diagram of computed concentration changes of dissolved inorganic carbon, $q^C$, for the 84 model boxes 1760-1983 as deduced with transient solutions I and II. Surface boxes are shown as circles, other boxes with crosses.

average air-borne fraction during the period 1760-1983 were 0.50 (case A) and 0.45 (case B). This implies for this period

| | | |
|---|---|---|
| Total emission (Pg) | 278 | 310 |
| Fossil fuel emissions (Pg) | 179 | 179 |
| Biosphere emissions (Pg) | 99 | 131 |
| Ratio of the total uptake to the Atlantic Ocean | 3.9 | 4.8 |

We note that the ratio of the total ocean surface to that of the Atlantic Ocean is 4.0 or 4.5 dependent on whether region 12 is included or not. Accordingly the uptake by the Atlantic Ocean per unit area is larger than that of the rest of the oceans in case A and vice versa in case B.

The curves a in figure 11 shows how the airborne fraction ($\alpha$) would have varied with time in cases A and B. During the first 120 years $\alpha$ decreases from an initial value of 100% to 35-40% in 1870 at which time the atmospheric concentrations began to increase more rapidly, but slowed down after the turn of the century. Because of the accelerating increase since 1950, $\alpha$ increased significantly during the last 35 years. Although we cannot trust the precise values as shown in figure 11, it seems likely that this general feature of our model is one which might well also characterize the world oceans (cf also Oeschger and Siegenthaler, 1985). Because of the fact that irregular variations in reality are superimposed on such slow trends, we usually cannot discover these until a decade or two have gone by.

Although the calculations described above are based on an analysis of the Atlantic Ocean it seems plausible that the airborne fraction of

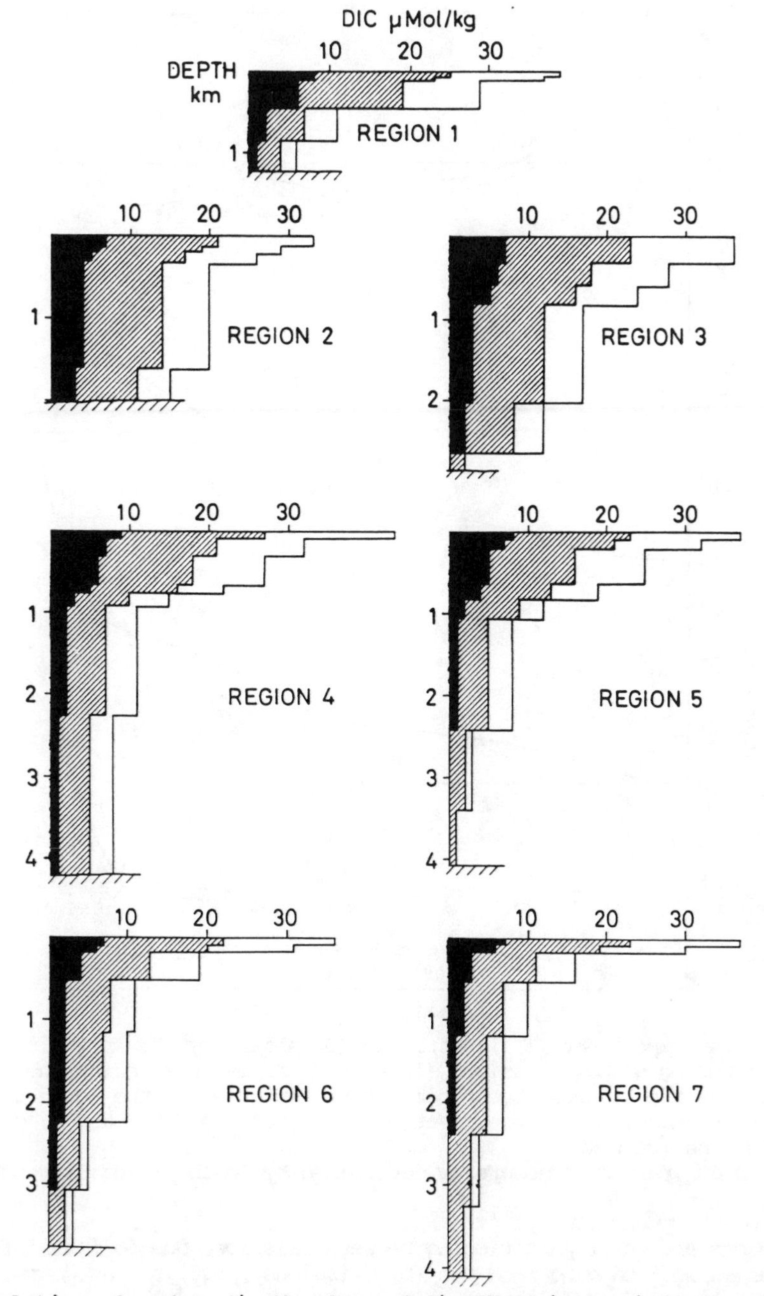


Fig. 10. Accumulation of carbon (in the form of dissolved inorganic carbon in the Atlantic Ocean) as a result of increasing atmospheric $CO_2$ concentrations as shown in Figure 1. Vertical profiles of concentrations are shown for selected regions for 1880, 1955 and 1983 as deduced with the aid of the transient solution II.

man's emissions of $CO_2$ into the atmosphere has increased from 1950 to 1973 and then decreased, when the annual increase of $CO_2$ emissions by 4.5% per year due to fossil fuel combustion in 1973 decreased to only about 1% per year, which situation has prevailed until recently.

$CO_2$ emissions to the atmosphere before 1850 due to fossil fuel burning were small probably 0.05 PgC $yr^{-1}$ or less and insignificant before 1830. The total emissions until 1860 can hardly have been more than 4 PgC (cf Rotty, 1981). The increase of the atmospheric $CO_2$ content

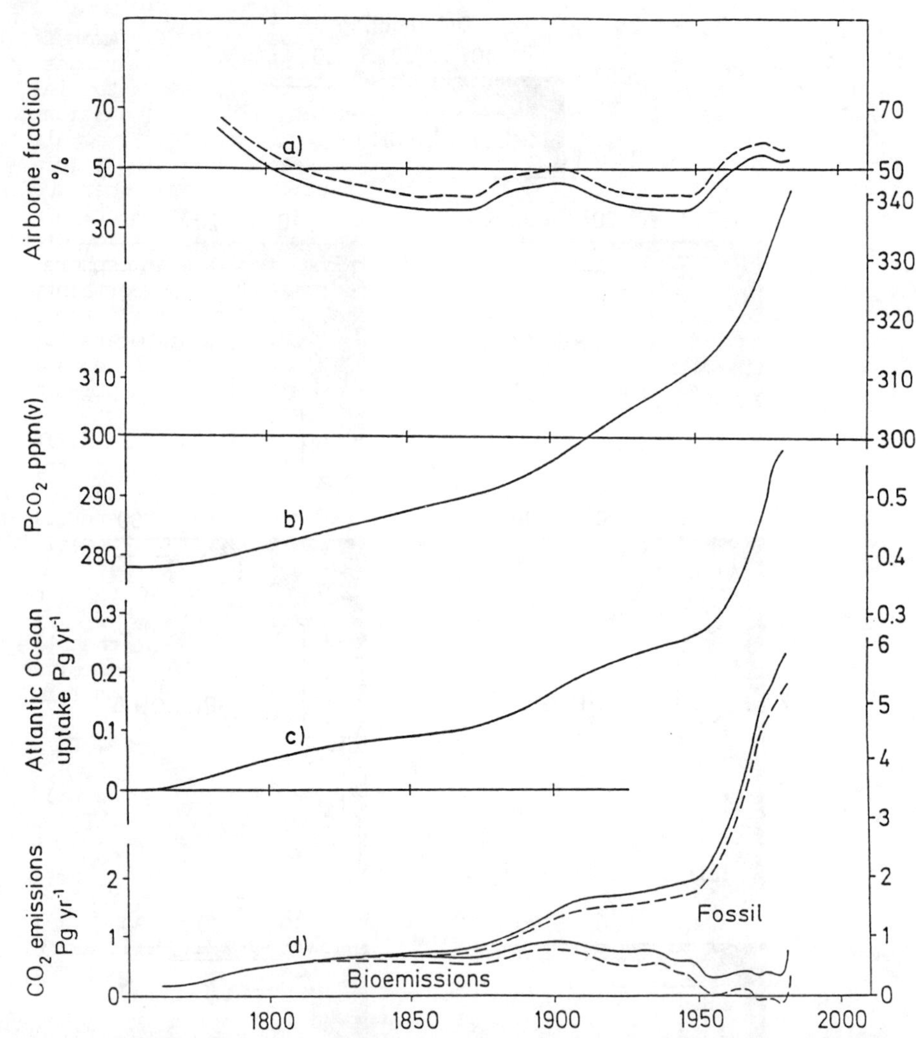


Fig. 11. Transient experiment II illustrating $CO_2$ uptake by the Atlantic Ocean.
a) Airborne fraction of $CO_2$ emissions (in %) deduced by considering the response characteristics of the Atlantic Ocean being representative of the world oceans as a whole and assuming that the mean airborne fraction 1760-1983 was 50% (case A; dashed line), 45% (case B, solid-line (cf text).
b) Atmospheric $CO_2$ partial pressure as determined by Oeschger and Siegenthaler (1987), cf Figure 1.
c) Atlantic Ocean uptake (PgC $yr^{-1}$).
d) $CO_2$ emissions and their partitioning between emissions due to fossil fuel combustion and deforestation and changing land use (bioemissions), PgC $yr^{-1}$ as computed for the two cases A (dashed line), B (solid line).

1760-1880 was about 20 PgC, and the uptake by the Atlantic Ocean about 6 PgC according to our computations. We assess that the total oceans will have taken up 20-25 Pg. Thus, if the equilibrium between the atmosphere and the sea was not markedly disturbed already before 1760, the total emission to the atmosphere during the period 1760-1880 were above 35 PgC, of which accordingly only about 10% was due to fossil fuel combustion. Emissions due to deforestation and expanding agriculture may accordingly on the average have been about 0.3-0.4 PgC $yr^{-1}$ during this period.

Figure 11d finally shows the emissions that we deduce corresponding to a total airborne fraction of 45% and 50% (solid and dashed lines respectively). The values until 1950 seem plausible yielding a total emission due to deforestation and changing land use (bioemissions) of 115 and 97 PgC respectively, deduced as the difference between the computed total emissions and those due to fossil fuel combustion (Rotty, 1981). During the last 25 years, however, the bioemissions become quite small. This result leads to a suspicion that other sinks than the oceans may have been present to absorb the larger bioemissions that seems plausible.

Let us finally return to the discussion in the introduction as summarized in Table 1. We have found that about 35 PgC may have been emitted into the atmosphere before 1880 due to deforestation and changning land use. In the light of the assessments by Bolin (1986) 100 PgC seems minimum of these emissions for the period 1860-1980. Thus the total emissions from deforestation and changing land use for the total period is likely to have been well above 100 PgC, most likely 135 PgC. Total emissions, i.e. including those due to fossil fuel emissions therefore probably have been larger than 300 PgC, i.e. Case B is applicable rather than A. On the other hand it does not seem likely that the uptake per unit area by the Atlantic Ocean is less than that of the rest of the ocean. Other sinks for $CO_2$ may be of importance than those considered here.

## Summary of Results and Conclusions

1. Inverse methods (Bolin et al, 1987) have been used to deduce ocean circulations, a spatial pattern of turbulent exchange and biological activity that best satisfy the steady state destribution of a set of ocean tracers.
2. An analysis of a series of such solutions and comparison with some independant data show fair agreement for the large-scale pattern (cf Bolin et al, 1987) but also significant uncertainties on the smallest scale of resolution particularly with regard to the pattern of turbulent flux.
3. A prominent feature of the most plausible solution is characterized by a direct mean meridional circulation cell with an intensity of 0.7 $10^{15}$ ton $yr^{-1}$ ( 23 Sverdrup) with sinking motion in the north and two coupled circulation cells in the termocline region with 0.7 $10^{15}$ ton $yr^{-1}$ ( 23 Sverdrup), upwelling in the equatorial regions. Turbulent and detrital transfer seem too high and implies probably somewhat too rapid tracer transfer in transient computations.
4. The errors in satisfying the set of tracer equations are due to the incompatibility because of crude spatial resolution, approximate description of chemical and biological processes and above all inaccurate data fields. It would be desirable to obtain solutions in which both the b-vector (primarily determined by the prescribed boundary conditions) and the coefficients in the transfer terms could be given values within ranges as determined by the errors in the data field. Some of the tracer continuity equations in A x = b could rather be given as inequality constraints in the set G x > h. The Simplex method of solution offers another such possibility.
5. The steady state solution $x''$ has been used to compute the invasion of tritium, $^{14}C$ and carbon from the atmosphere.
6. In spite of the uncertainty of the basic transfer solution $x''$ the integrated uptake of $^{14}C$ during the years 1955-1973 is quite stable and agrees reasonably well with observations during GEOSECS.
7. The more detailed pattern of $^{14}C$-penetration as computed show some differences compared with observations. A further analysis of these descrepancies might yield a more reliable solution. Generally great care must be exercised in the use of available data and their uncertainties in order to improve sucessively the solutions of the basic matrix equation.
8. The Suess effect in the Atlantic Ocean in 1955 being due to the emission of $^{14}C$-free carbon dioxide by fossil fuel combustion in the past is obtained with the aid of a transient computation. The computed values, the average of which is 5.5 $^{o}$/oo, are considerably less than the observed being of the order of 10 $^{o}$/oo. The discrepancy can hardly be due to inadequacies of the transient solution and the result warrants further analysis.
9. The total uptake of carbon dioxide by the Atlantic Ocean during 1760-1983 is computed to have been $33 \pm 5$ Pg C, of which about 6 Pg occurred before 1860.
10. Considerable temporal changes of the uptake in terms of a percentage of the atmospheric increase (equivalent to a varying air-borne fraction) are found. It seems plausible that the air-borne fraction of the emissions was considerably greater in the 1950' and 1960's than during the earlier part of this century and it may have decreased since 1973.

## References

Bolin, B., 1986. How much $CO_2$ will remain in the atmosphere? In Bolin B., Döös, B. Warrick, R. and Jäger, J. (Eds) The greenhouse effect, climatic change and ecosystems. SCOPE Report 29, pp 93-155. J. Wiley, Chichester, England.

Bolin, B., Björkström, A., Holmén, K. and Moore, B., 1987. On inverse methods for combining chemical and physical oceanographic data: A steady-state analysis of the Atlantic Ocean. Report CM-71, Dept. of Meteorology, Univ. of Stockholm, 220 p.

Broecker, W.S., Gerard, R., Ewing, M. and Heezen, B.C., 1960. Natural radiocarbon in the Atlantic Ocean. J. Geophys. Res., 65, 2903-2931.

Broecker, W.S. and Peng, T.H., 1982. Tracers in the sea. Lamont-Doherty Geological Observatory, Columbia Univ. N.Y.

Druffel, E. and Linick, T., 1978. Radiocarbon in annual coral rings of Florida. Geophys. Res. Lett. 5, 913-916.

Druffel, E. and Suess, H., 1983. On the radiocarbon record in banded corals: Exchange parameters and net transport of $^{14}CO_2$ between atmosphere and surface ocean. J. Geosphys. Res. 88 C2, 1271-1280.

Maier-Reimer, E. and Hasselman, K., 1987. Transport and storage of $CO_2$ in the ocean - an inorganic ocean-circulation carbon cycle model. Climate Dynamics 2, 63-90.

Nydal, R., and Lövseth, K. 1983. Tracing bomb $^{14}C$ in the atmosphere (1962-1980). J. Geophys. Res., 88, 3621-3642.

Östlund, H.G., Dorsey, H.G. and Brescher, R. 1976. GEOSECS Atlantic radiocarbon and tritium results (Miami). Tritium Laboratory Data Report 5, Rosenstiel School of Marine and Atmospheric Science, Univ. of Miami.

Östlund, H.G., Craig, H., Broecker, W.S. and Spencer, D. 1987. GEOSECS Atlantic, Pacific and Indian Ocean expeditions. Volume 7: Shorebased data and graphics. Superintendent of Documents, U.S. Government Printing Office, Washington D.C. 20402.

Rotty, R.M. 1981. Data for global $CO_2$ production from fossil fuels and cement. In: Bolin, B. (Ed.) Carbon cycle modelling. J. Wiley and Sons, Chichester, 121-125.

Rotty, R.M. and Masters, C.D. 1985. Carbon dioxide from fossil fuel combustion: Trends, resources, and technological implications. In: Trabalka, J.R. (Ed.) Atmospheric carbon dioxide and the global carbon cycle. U.S. Department of Energy, DOE/ER-0239, 63-80.

Sarmiento, J.L., 1983. A simulation of bomb tritium entry into the Atlantic Ocean. J. Physical Oceanogr. 13, 1924-1939.

Schlitzer, R. 1987. Renewal rates of east Atlantic deep water estimated by inversion of $^{14}C$ data. J. Geophys. Res. 92 C3, 2953-2969.

Schlitzer, R., 1988. Modeling the Nutrient and Carbon Cycles of the North Atlantic. Part I: Circulation, Mixing Coefficients, and Heat Fluxes. J. Geophys. Res. To be published.

Siegenthaler, U. 1983. Uptake of excess $CO_2$ by an outcrop-diffussion model of the ocean. J. Geophys. Res. 88, C6, 3599-3608.

Siegenthaler, U. and Oeschger, H. 1987. Biospheric $CO_2$ emissions during the past 200 years reconstructed by deconvolution of ice core data. Tellus 39B, 140-154.

Stuiver, M. and Quay, P.D., 1981. Atmospheric $^{14}C$ changes resulting from fossil fuel $CO_2$ release and cosmic ray flux variability. Earth and Planetary Science Letters 53, 349-362.

Weiss, W. and Roether, W. 1980. The rates of tritium input to the world oceans. Earth and Planetary Science Letters 49, 435-446.

Wunsch, C. 1984. An eclectic model of the Atlantic Circulation - Part I. The meridional flux of heat. J. Phys. Oceanogr. 14,1712-1733.

# AFRICAN DROUGHT: CHARACTERISTICS, CAUSAL THEORIES AND GLOBAL TELECONNECTIONS

Sharon E. Nicholson

Department of Meteorology, Florida State University, Tallahassee, FL 32306

## Introduction

Three quarters of the African continent is predominantly semi-arid drought-prone land. This fact has been dramatically observed during the last few years, during which only a small strip of equatorial rain forest and the hyperarid core of the Sahara desert have not experienced severe drought. In 1983, one of the worst years, rainfall was below normal over virtually the whole continent (Fig. 1), a sizable portion of the global land surface. In the semi-arid Sahel along the Sahara's southern border, dry conditions have persisted since the late 1960s; annual rainfall in recent years has been about half that of the 1950s (Tab. 1).

By some estimates, famine in the Sahel claimed over 100,000 lives in 1973 alone. Thus, African drought is a geophysical problem of tremendous importance. Unfortunately, the phenomenon is not well understood and the causes of drought in most of Africa have not been established. On the basis of the available evidence, the vast majority of meteorologists concur that the factors which reduce rainfall can be traced to anomalous large-scale patterns of atmospheric circulation, which may or may not be a result of sea-surface temperature (SST) variations. This point is important, as it has often been said that drought in the Sahel has resulted from human-induced environmental changes.

The latter viewpoint, now fostered mostly by non-scientists, is usually based on the supposition that droughts of this extent and severity are a recent phenomenon. This concept was introduced into the meteorological community by Charney (1975). His numerical modelling efforts suggested that land surface changes, if extreme, could initiate drought in the Sahel. The physical characteristics of the surface, such as albedo and soil moisture, can be altered by such anthropogenic factors as deforestation, irrigation, cultivation, and overgrazing. However, a reduction of rainfall by 50% in a semi-arid region such as the Sahel will have a similar effect on the landscape. Currently, a widely accepted version of the land-surface feedback hypothesis is that in the Sahel, modification of the land surface might have sufficient impact on atmospheric conditions to intensify and perpetuate an existing drought, but not enough to initiate one. Here the word "hypothesis" must be emphasized.

Given the complexities of the phenomenon, African drought is truly an interdisciplinary problem, an understanding of which requires a contribution of the geophysical and biological sciences. Clearly, there are many unanswered questions. Foremost among these are the causes of drought, whether it can be predicted, whether it will continue and, for the Sahel, what the human role has been. In this paper, I will show the extent to which we can currently answer these questions by examining and summarizing the results of numerous investigations dealing with long-term climatic change in Africa, temporal and spatial characteristics of recent droughts, atmospheric circulation anomalies of wet and dry years, tropical teleconnections to African rainfall, and land-surface processes.

## Paleo- and Historical Climates of Africa

Paleo-climates and paleo-environments of Africa have been studied extensively. Much of the research is summarized in reviews and overviews, such as those of Butzer et al., 1972; Nicholson and Flohn, 1980; Street and Grove, 1979; Williams and Faure, 1980; Street-Perrott and Roberts, 1983; Street-Perrott and Harrison, 1985; Cockcroft et al., 1987; and Deacon and Lancaster, 1988. The changes which have occurred since the late Pleistocene are schematically generalized in Fig. 2.

Toward the end of the late Pleistocene, from about 20,000 to 12,000 B.P., the desert expanded to cover most of the continent (Fig. 2). Humid climates all but disappeared and tropical forests were confined to a few highland refuges. Conditions changed abruptly about 12,000 B.P. and deep lakes formed in many formerly arid locations. The first humid period climaxed about 9,000 years ago and

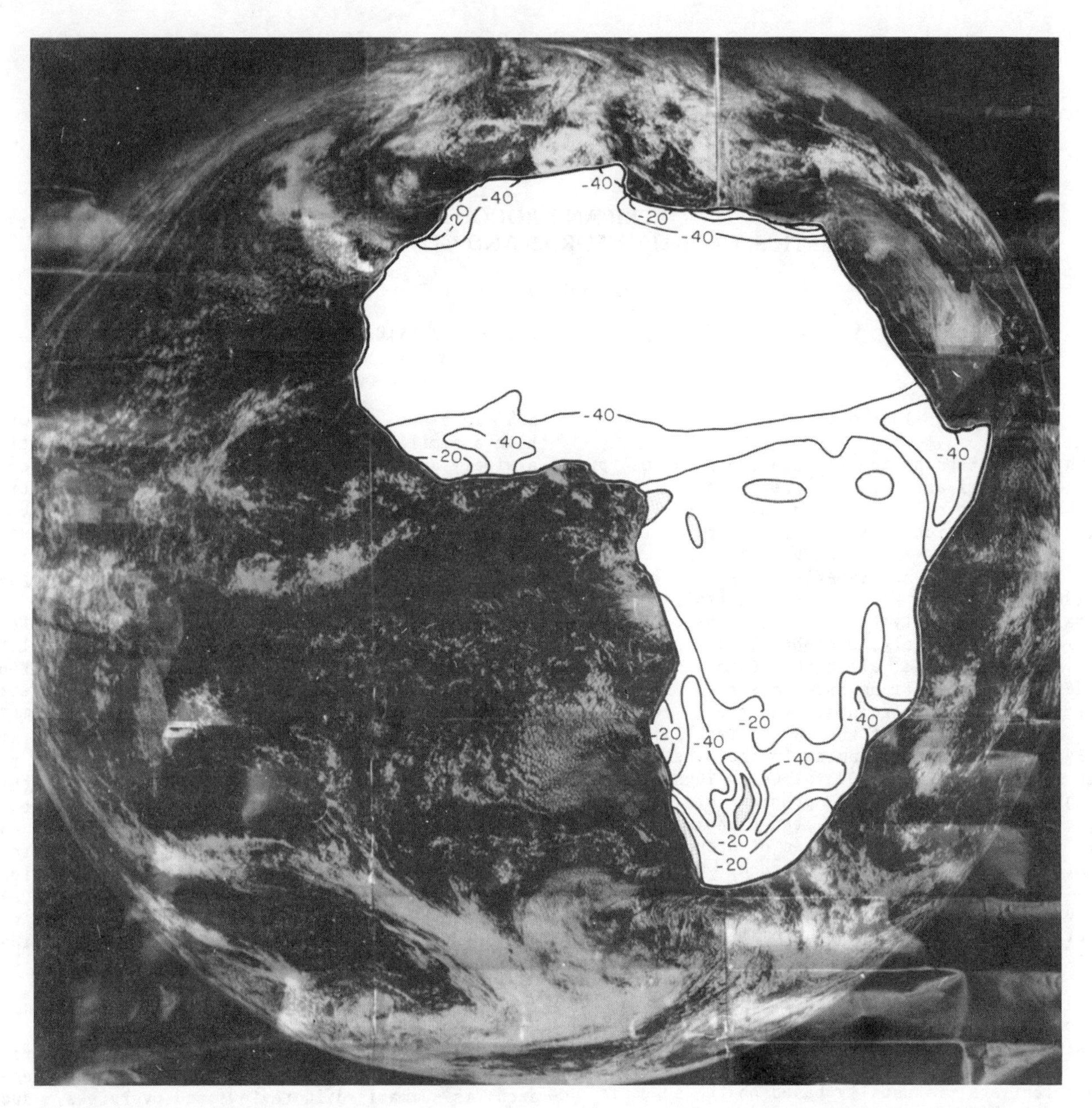


**Fig. 1. Schematic of rainfall departures (% above or below normal) for 1983 superimposed on Meteosat view.**

apparently encompassed the equatorial regions and the subtropics of both hemispheres. A second humid period prevailed from about 6,000 to 4,000 B.P. The desert had all but vanished; lowland marshes formed in the northwestern Sahara while Neolithic man herded cattle in the central Sahara, surrounded by fauna grazing on what was apparently a savanna landscape. Lakes dotted now arid regions of Mauritania; fish hooks uncovered in archaeological sites there attest to human occupation and a livelihood that is no longer possible. Rift Valley lakes were several hundred feet deeper than at present and expanded to form a huge hydrologic system. Lake Chad, almost entirely desiccated in recent years, was then over a hundred meters deep and expanded to ten times its normal twentieth century size.

The onset of more arid conditions occurred about 3,000 years ago and since that time conditions generally resembling those of the current century

TABLE 1. Mean Annual Rainfall (mm)

| | 1950-59 | 1970-84 |
|---|---|---|
| BILMA | 20 | 9 |
| ATBARA | 92 | 54 |
| NOUAKCHOTT | 172 | 51 |
| KHARTOUM | 178 | 116 |
| AGADEZ | 210 | 97 |
| TIMBUKTOO | 241 | 147 |
| NEMA | 381 | 210 |
| DAKAR | 609 | 308 |
| BANJUL | 1409 | 791 |

have prevailed. However, extensive periods of significantly more humid or more arid climate have occurred. These are best illustrated by the last two centuries, since a greater availability of data allows for more reliable historical reconstructions than for earlier periods. Conditions for three periods commencing in the 1820s are summarized in Fig. 3; details can be found in Nicholson, 1978, 1979a, 1980a, and 1981a.

A general desiccation occurred throughout much of Africa toward 1800; in the subtropics and equatorial regions subnormal rainfall or drought prevailed for most of two decades in the 1820s and 1830s (Fig. 3). The situation was particularly severe in the Sahel, which had experienced similar persistent droughts in the 1680s and in the 1730s through 1750s. Lake Chad was almost totally desiccated, as it has been during recent drought years. After the 1830s, rainfall began to increase and within decades, conditions considerably wetter than those of the current century again prevailed.

This relatively humid period persisted from the 1870s to the mid-1890s. Lake levels from Chad in the north to Ngami in the south rose dramatically, by several meters in many cases. The high stands which persisted for tens of years were achieved at most briefly, and generally not at all, during the twentieth century. Near Timbuktoo, wheat production thrived to such an extent from the continually good Niger floods that grain was exported to surrounding areas. Numerous measurements of rainfall and river flow confirm the wetter conditions and suggest that rainfall in subtropical regions was then 25-35% greater than the 20th century normals. Conditions again changed abruptly around 1895, and a continent-wide decrease in rainfall culminated in a long period of severe droughts in the 1910s. The steady decline is evidenced by a variety of indicators, including rainfall, river flow, lakes, agriculture and harvests.

## Temporal and Spatial Variability of Rainfall during the Present Century

Rainfall fluctuations in Africa show preferred spatial and temporal patterns of occurrence. The dominant spatial modes of variability have been analyzed by Klaus (1978), Gregory (1982) and Motha et al. (1980) for West Africa, by Dyer (1975) for South Africa, and on a continental scale by Nicholson (1980b, 1986a, 1986b). The most important characteristics are the large-scale spatial coherence and the relatively small number of typically occurring patterns of rainfall anomalies. The most common mode is drought or below-average rainfall throughout the subtropics of Africa but increased rainfall in equatorial latitudes; this occurs during 13 years between 1910 and 1973. Another common pattern, typified by 1983 (Fig. 1), is drought or below-average rainfall continent-wide, a situation occurring during the two historical dry periods illustrated in Fig. 3. The remaining major patterns are the inverse of these: continentally "wet" conditions or abnormally high rainfall in the subtropics and rainfall deficits in equatorial latitudes. This clearly implies that large-scale circulation changes, and not regional-scale phenomena, impose the dominant control on rainfall variability.

The large-scale spatial coherence also implies that a few key regional time series can serve to describe rainfall variability over much of the continent. Four such series are shown in Fig. 4. That for the Sahel is generally representative of the area from 10°N to 25°N and extending across most of the east-west expanse of Africa. That for East Africa is broadly representative of the equatorial zone; and the northern and southern Kalahari series are averages of the subtropical latitudes of southern Africa. The droughts of recent years and the wet period in the subtropics in the 1950s are clearly apparent.

These time series also illustrate an important contrast between the Sahel region and other parts of Africa. Rainfall variability in the Sahel is characterized by long-term persistence of anomalies, e.g., the long series of above-average rainfall in the 1950s and early 1960s and the continual sequence of dry years since 1968. In comparison, the remaining series show only high frequency variability, the largest continually wet or dry period being five years. The persistence in the Sahel has been pointed out by numerous authors (Bunting et al., 1976; Walker and Rowntree, 1977; Kraus, 1977a; Nicholson, 1979b, 1982, 1983, 1985; Lamb, 1982; and Dennett et al., 1985). This characteristic and its absence in southern Africa, a region where teleconnections with Sahel rainfall are evident, may be indicative of a more local positive feedback mechanism acting to reinforce rainfall anomalies initiated by large scale forcing. Thus,

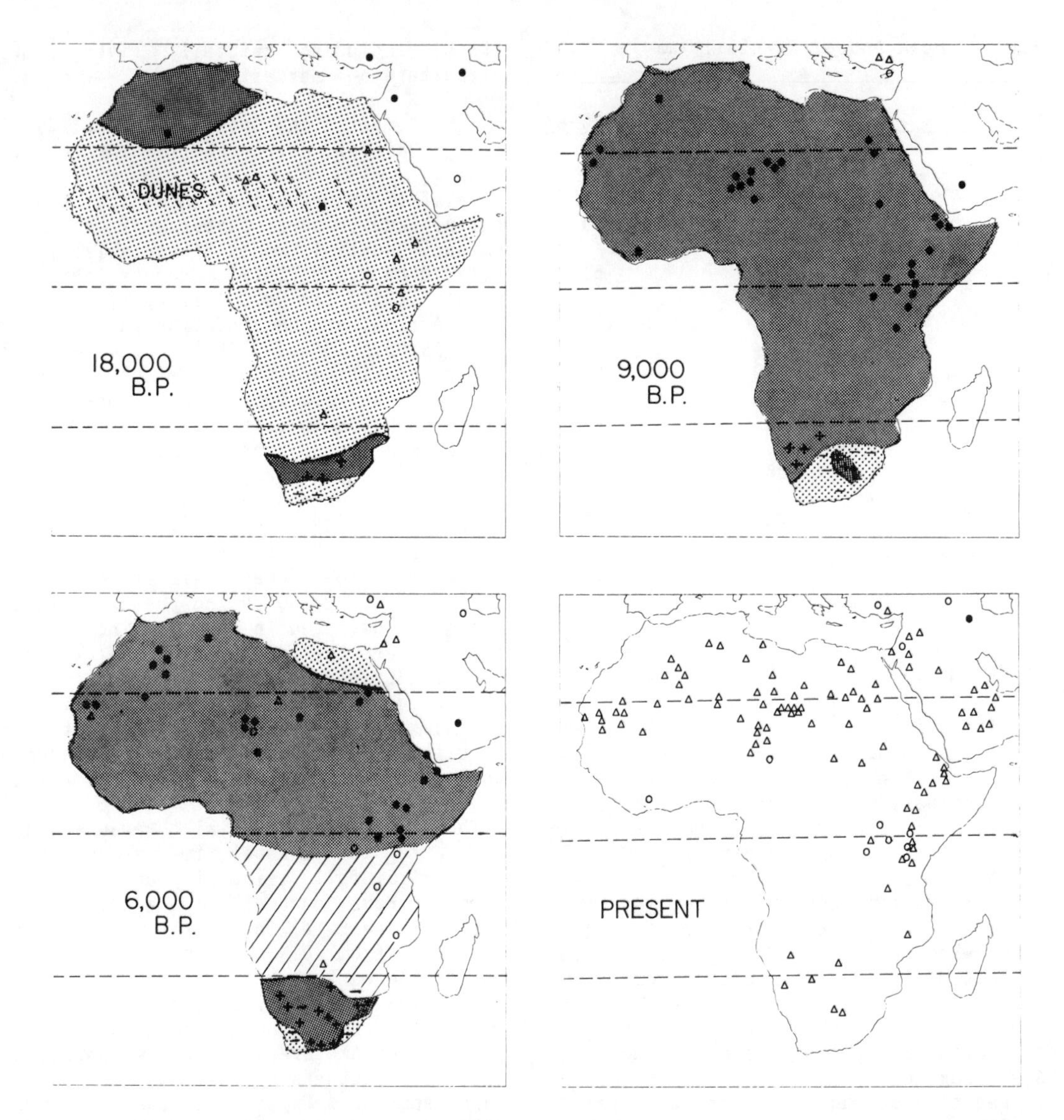


Fig. 2. African paleoclimates c. 18,000 B.P., 9,000 B.P. and 6,000 B.P. compared to present conditions (rainfall: shading or plus signs, generally greater than at present; dots or minus signs, less than at present; hatching, decreasing rainfall; lakes: ● = high stands, O = intermediate stands, △ = low stands; based on Street-Perrott et al., Nicholson and Flohn, Cockcroft et al., and others).

the persistence can be used as a point of evidence in support of the land-surface/atmosphere feedback hypothesis previously mentioned and described in more detail in Section 6.

The temporal characteristics suggested by the time series in Fig. 4 are quantified using spectral analysis (Fig. 5). The spectra for East Africa and the Kalahari regions each show three significant peaks in the range 2.2-2.4, 3.3-3.8 and 5-6 years (Nicholson and Entekhabi, 1986). In contrast, that for the Sahel shows "red noise", which is indicative of persistence, and little power at higher frequencies. Quasi-periodic fluctuations on these time scales are characteristic of rainfall throughout equatorial and southern Africa (see also Tyson et al., 1975; Rodhe and Virje, 1976; Dyer and Marker, 1978; Tyson, 1980), but are notably absent elsewhere on the continent (Fig. 6). They also characterize numerous other atmospheric and oceanic phenomena in the tropics. All three are evident in the Southern Oscillation; the 5-6 year time scale is the dominant one for SST variability in the Atlantic and Indian Oceans (Fig. 7); and the Quasi-Biennial Oscillation is a well-known phenomenon.

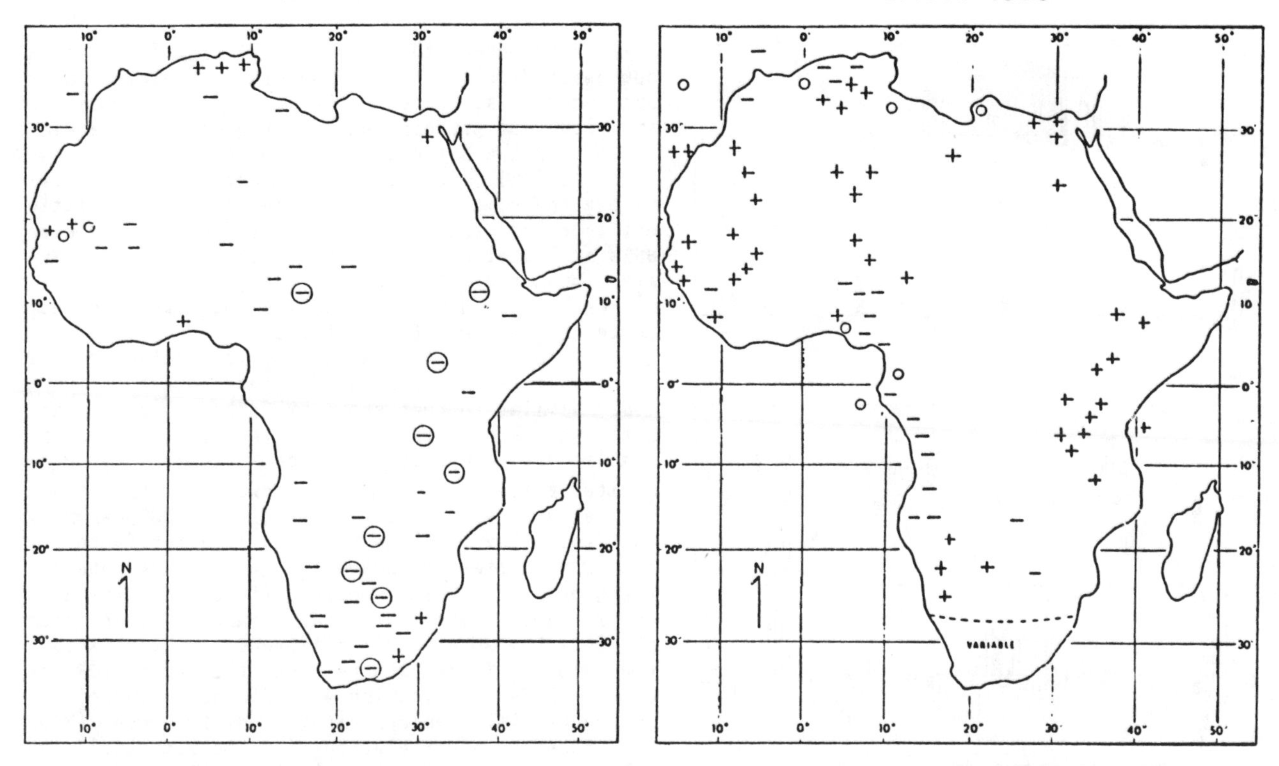


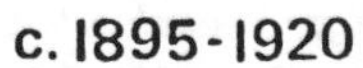


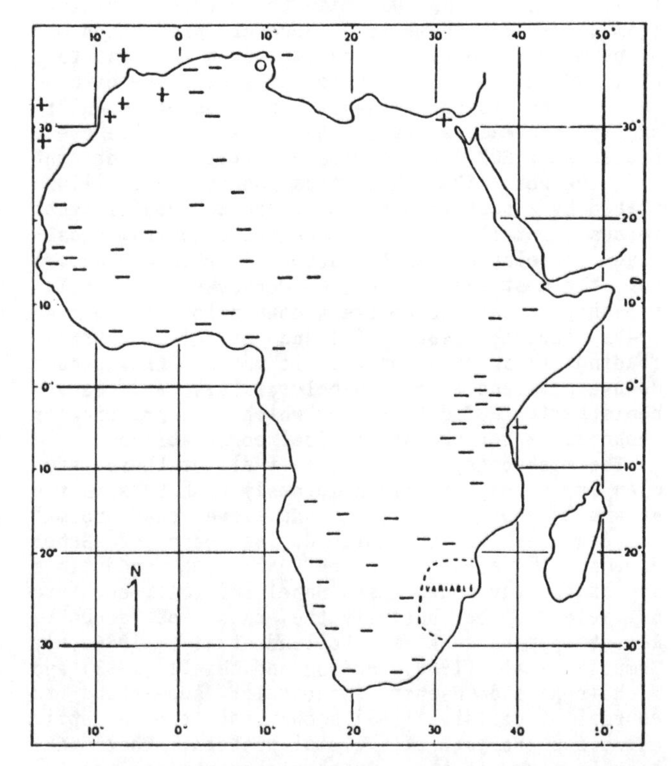


Fig. 3. African rainfall anomalies for three historical periods (minus signs denote evidence of drier conditions; plus signs denote evidence of above-average rainfall; zeroes, near normal conditions; circled symbols denote regional integrators, such as lakes or rivers).

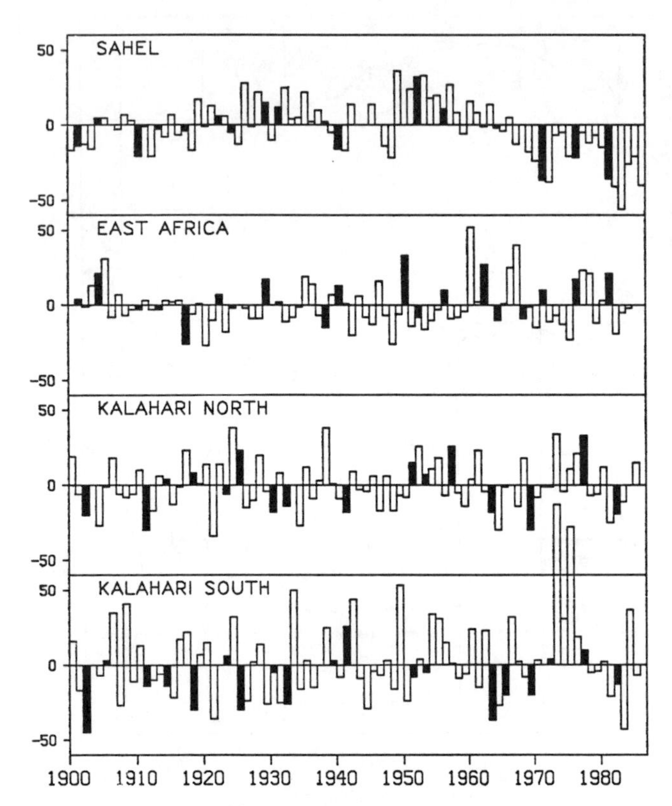


Fig. 4. Rainfall fluctuations in four sectors of Africa 1901-1984, expressed as a percent departure from the long-term mean (ENSO years are shaded for Sahel and East Africa, ENSO +1 is shaded for the two Kalahari series).

## Tropical Teleconnections to African Rainfall

The correspondence between the dominant time scales of rainfall variability and those of other tropical parameters suggests possible forcing mechanisms of rainfall variability. Here, several of these are examined, beginning with sea-surface temperatures. Most studies relating rainfall to SSTs have been concerned only with the Sahel and a number of viewpoints have emerged. Only a handful of studies have dealt with other locations. In many cases, weak or strong statistical associations have been documented but the dynamic mechanisms accounting for these associations have not been established.

One of the first studies to examine the relationship between Sahel rainfall and sea-surface temperatures was that of Lamb (1978a,b). Using a Sahel rainfall index heavily weighted toward the western Sahel (west of 0°W), he concluded that Sahelian wet years are associated with abnormally high temperatures throughout most of the tropical Atlantic, but below average SSTs to the north and west of the convergence zone over the Atlantic. Lamb attributed the lower Sahel rainfall to a southward displacement of the wind convergence zone and equatorial trough, thermally forced by the anomalous SST pattern, but other atmospheric anomalies over the Atlantic are also evident (Hastenrath, 1984a). Lough (1986), examining approximately the same area and time period, found little evidence of the consistent relationship between SSTs and Sahel rainfall which Lamb suggested. She also found that SSTs appear to lag Sahel rainfall, a result which implies that SST patterns do not force the rainfall fluctuations.

Folland et al. (1986a,b) found an Atlantic SST anomaly pattern for Sahel dry years which is markedly similar to that shown by Lamb; this is significant, since only one of the five years in Lamb's dry composite was also in their dry composite. Folland et al. also showed that the Atlantic anomalies are part of a global SST anomaly pattern. They conclude that the relevant factor is the SST difference between the two hemispheres; the global inter-hemispheric difference correlates considerably better with Sahel rainfall than does the interhemispheric SST difference for the Atlantic alone. These authors use the third EOF of global SSTs (explaining 4.7% of the variance) to represent interhemispheric SST differences (Fig. 8). Based on a correlation of .6 between EOF 3 and Sahel rainfall (significant at the 99% confidence level), an attempt was made to predict rainfall from global SST anomalies. However, prediction may be premature since SST changes often lag rainfall (Fig. 8) and there is little correspondence between the two series during several periods of significantly abnormal rainfall. As an example, the coefficients of EOF 3 are similar for the 1910s drought and the wet 1950s. This situation is well illustrated by a scatter diagram of the EOF coefficients versus rainfall (Fig. 9); despite a general positive correlation, much scatter is evident except for the most extreme years. Moreover, the relationship appears to be break down prior to the mid-1940s (Lough, 1986; Folland et al., 1986a), leading to speculation that it may be the steady downward trend in both Sahel rainfall and inter-hemispheric SST differences which is principally responsible for the statistical correlations.

The most consistent result of all of these studies is a pattern of anomalously high SSTs in the Atlantic south of c. 10°N and lower than normal temperatures to the north during years of Sahel drought. The authors agree that such conditions appear to favor decreased Sahel rainfall, but are unlikely to be uniquely the cause of drought. Recent papers such as Folland et al. (1986a,b), Semazzi et al. (1988), Hsiung and Newell (1983) and Nicholson and Nyenzi (manuscript submitted to Journal of Climate, 1989) show that the Atlantic anomalies are part of a global pattern. That pattern in turn, or the atmospheric anomalies producing or responding to it, may be the critical factor in Sahel rainfall variability. Numerical simulations (Palmer, 1986; Semazzi, personal communica-

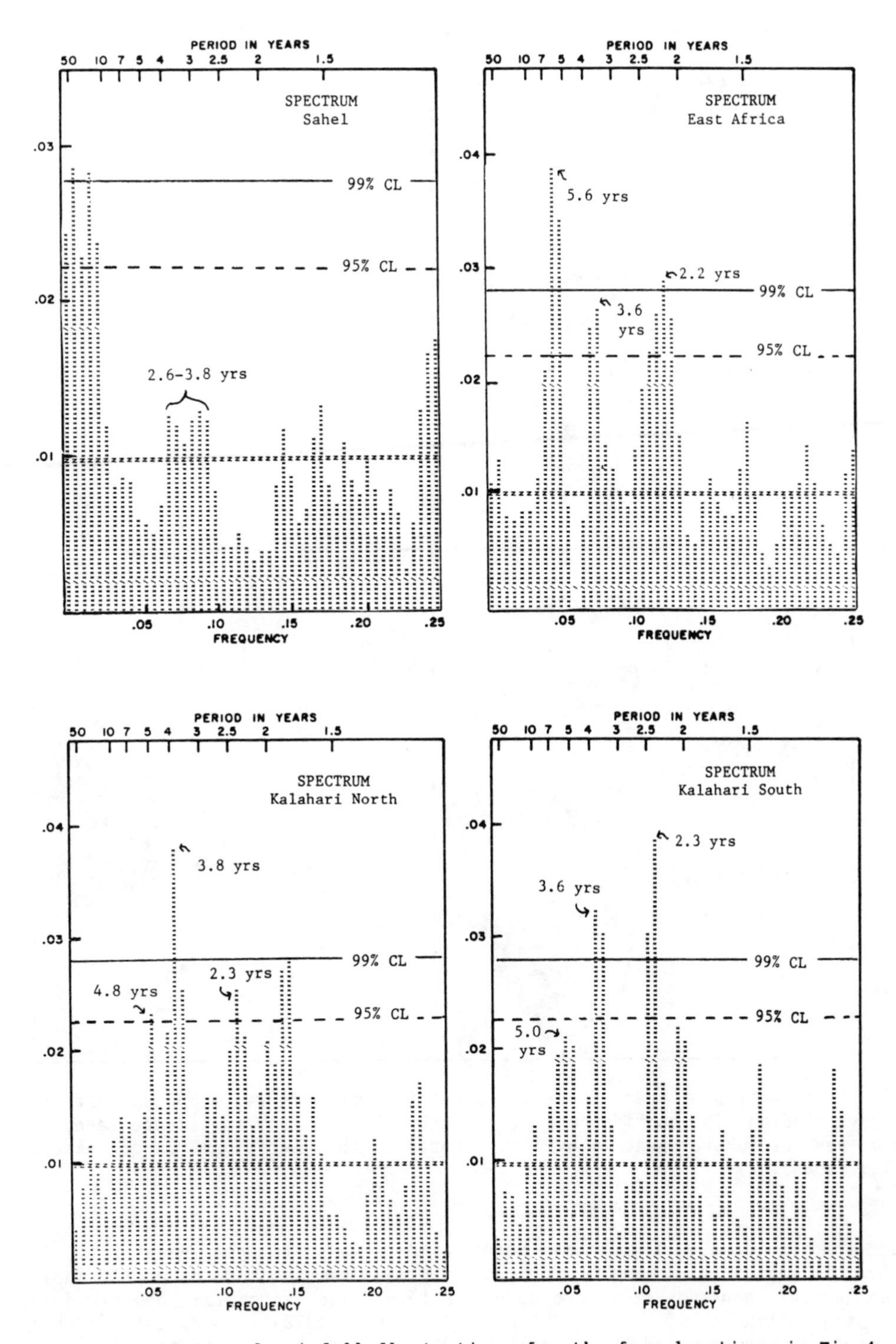


Fig. 5. Spectra of rainfall fluctuations for the four locations in Fig.4.

tion) likewise suggest that global, Pacific Ocean, and Indian Ocean SSTs all have a greater influence on the Sahel than Atlantic SSTs do. This finding is more compatible with Sahelian climatology, especially the relationship with the Indian Ocean, since the systems which "fail" in the western Sahel in dry years develop from pressure perturbations in the eastern Sahel and traverse most of the east-

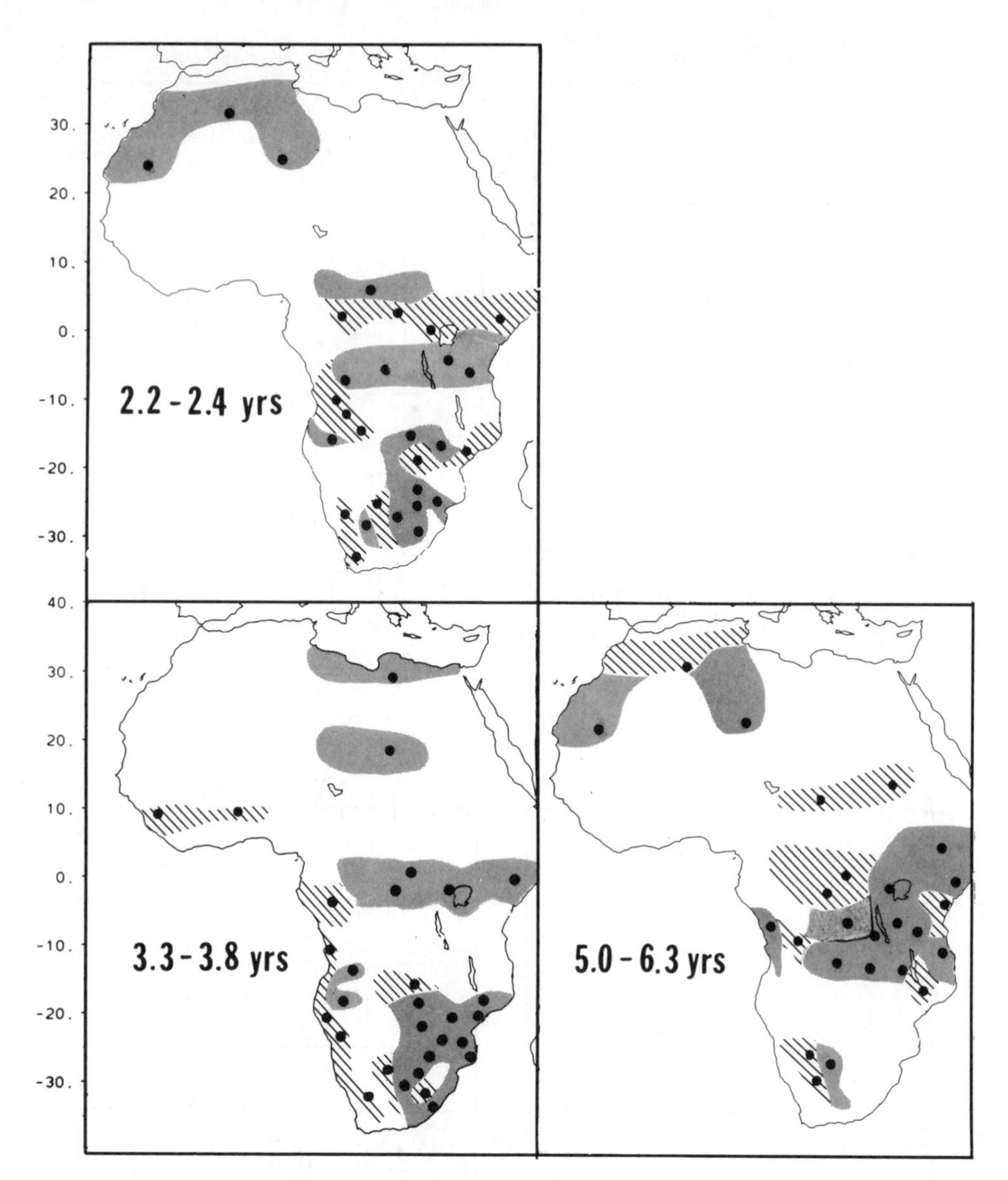


Fig. 6. Distribution of spectral peaks in rainfall at 2.2-2.4, 3.3-3.8, and 5.0-6.3 years (dark shading, 95% confidence level; light shading, 90% confidence level; dots indicate regions where peak is evident; based on the 84 regions in Entekhabi and Nicholson, 1986).

west extent of the continent. These are more likely influenced by factors "upstream" in the Indian Ocean than several thousand kilometers downstream in the tropical Atlantic.

A handful of studies have examined the relationship between sea-surface temperatures and rainfall variability elsewhere in Africa. Hirst and Hastenrath (1983a, b) showed that increased rainfall during the March/April rainy season along the Angolan coast and in the western equatorial region (Zaire basin) is associated with a weakening of the high pressure over the Atlantic and reduced strength of the southeasterly trades and westward wind stress in the equatorial Atlantic. This implies remote forcing of SST anomalies via the Kelvin wave mechanism described by Adamec and O'Brien (1978). Surprisingly, they found that SST, pressure and wind anomalies accompanying wet conditions during the second rainy season of October and November were dissimilar to those of wet March/April seasons and in some ways (e.g., high pressure and strong southeast trades) resembled the March/April dry composites.

A strong relationship between SSTs and rainfall along the Benguela coast, which includes the Angolan sector studies by Hirst and Hastenrath, was

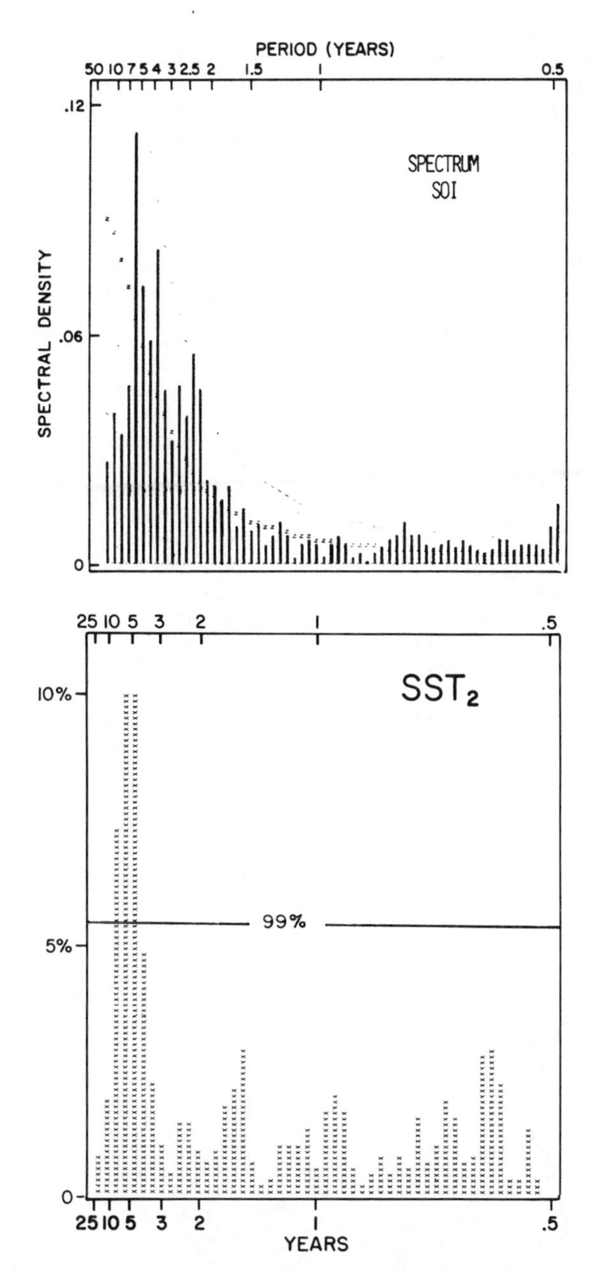


Fig. 7a. Spectra of SST fluctuations along the Benguela coast of the southeastern Atlantic (10°S - 20°S, 1948-72) and the Tahiti-Darwin pressure index of the Southern Oscillation (1935-73).

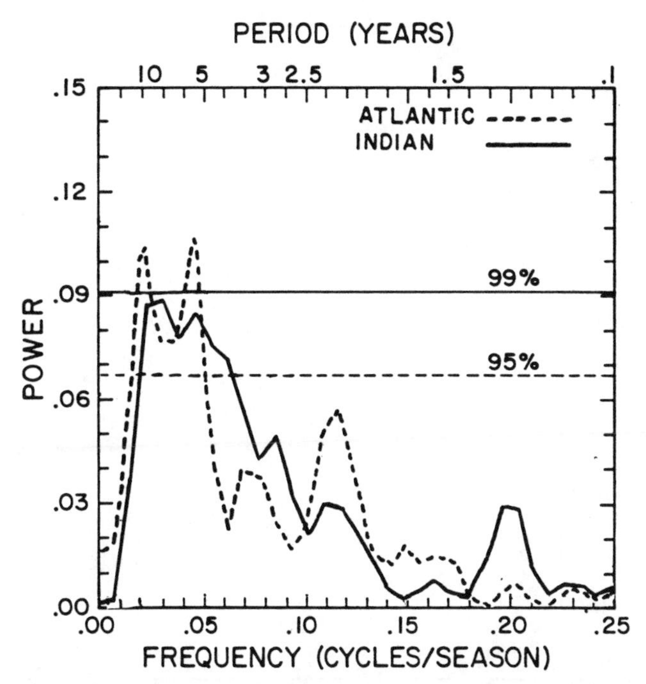


Fig. 7b. Spectra of the first principal component of sea-surface temperatures in the tropical Atlantic and tropical Indian Oceans (from Nicholson and Nyenzi, 1988, manuscript submitted to the Journal of Climate).

demonstrated by Nicholson and Entekhabi (1987). They further showed that the variance of both SSTs and rainfall is concentrated on time scales of 4-6 years (Figs. 6 and 7). As described in section 3, the variance of rainfall throughout the equatorial zone and in much of southern Africa is similarly dominated by 4-6 year quasi-periodicities and is strongly coherent with SST fluctuations on this time scale (Fig. 10). Cadet (1987, personal communication) also shows that a relationship exists between East African rainfall and the Indian Ocean. During unusually wet periods of the September-to-December season, pressure is below normal, cloudiness and SSTs above normal and trade-wind convergence is reduced in the western tropical-Indian Ocean while the ITCZ is abnormally strong. The

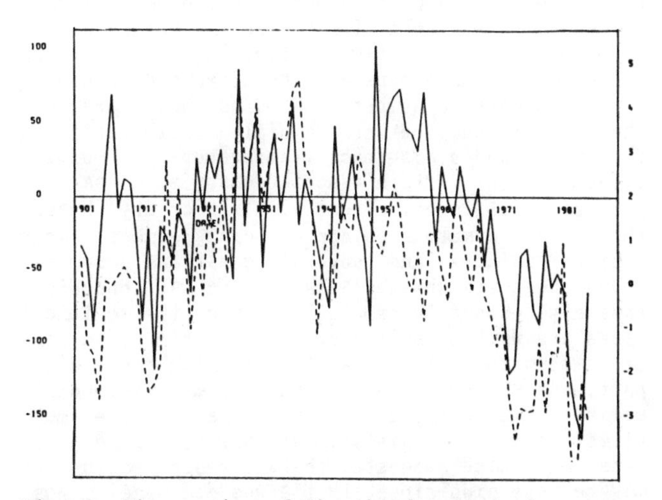

Fig. 8. Time series of the third principal component of global sea-surface temperatures and Sahel rainfall, 1901-1984 (from Folland et al., 1986a).

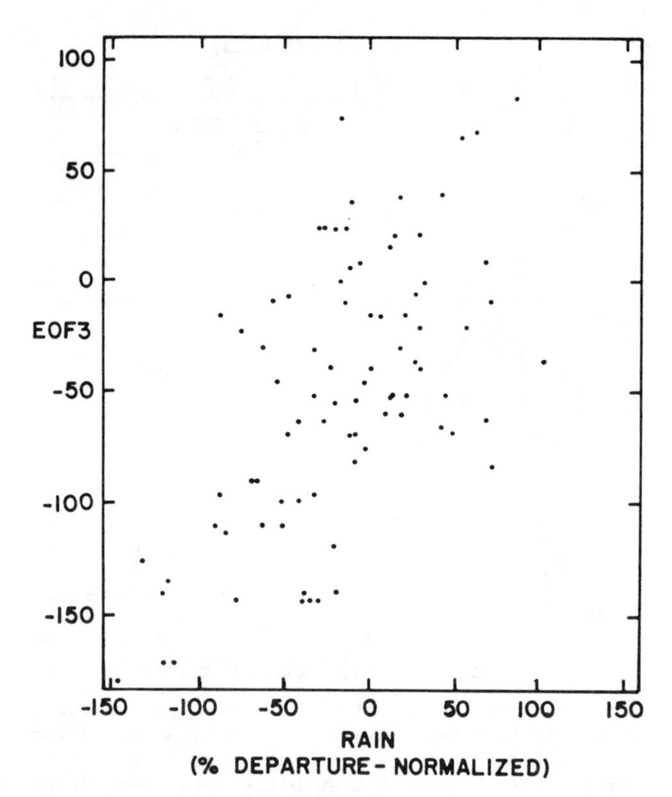


Fig. 9. Scatter diagram of the data in Fig. 8, comparing component scores of EOF3 and Sahel rainfall.

opposite patterns occur in dry years and in the eastern Indian Ocean. Although Cadet considers only East Africa, the strong teleconnections in rainfall between East and southern Africa (Nicholson, 1986b) and the coherent SST variability between in the Indian and Atlantic Oceans suggest that the Indian Ocean would influence rainfall elsewhere in southern Africa. Again, the relationships may well be part of a global anomaly pattern influencing both oceans and the African continent.

In summary, a number of studies have examined the relationship between rainfall variability in the Sahel and sea-surface temperatures. Despite agreement on certain associations, such as the pattern of SSTs coincident with wet or dry years, there are numerous points of disagreement among various authors and several questions remain. Among these are whether SSTs may lag and therefore possibly respond to rainfall, whether the critical links are with oceanic sectors in proximity to the Sahel or with a global pattern of which these are a part, and whether rainfall and SSTs may respond to common atmospheric forcing which may be the more direct factor in rainfall variability. Some studies have also suggested that oceanic influences may be more significant in the western Sahel, possibly providing an explanation for the often differential behavior of western and eastern sectors (Nicholson, 1980b). Likewise, SSTs may be more important as high frequency forcing of the interannual variability than as a mechanism producing the longer-term rainfall trends in the Sahel.

Evidence for a global teleconnection derives directly from studies relating rainfall to tropical phenomena such as the Southern Oscillation/El Nino (ENSO) and the 30-60 day oscillation. Only a few studies, however, have evaluated African rainfall in a global text. The two most comprehensive are those of Nicholson and Entekhabi (1986) and Ropelewski and Halpert (1987). These studies used basically the same data set but different analysis techniques, the former dealing primarily in the frequency domain and the latter solely in the time domain. Both studies show that the regions with the strongest statistical associations between rainfall and the Southern Oscillation are equatorial East Africa and southern Africa and that no relationship is evident in the Sahel. As shown in Fig. 4, East Africa generally receives above normal rainfall in the ENSO year, usually during the "short rains" of October and November, while droughts tend to occur in southern Africa early in the year following ENSO. The association with droughts for South and southeastern Africa was also shown by Stoeckenius (1981), Rasmusson and Wallace (1983), Dyer (1979), Schulze (1983), and Lindesay et al., 1986.

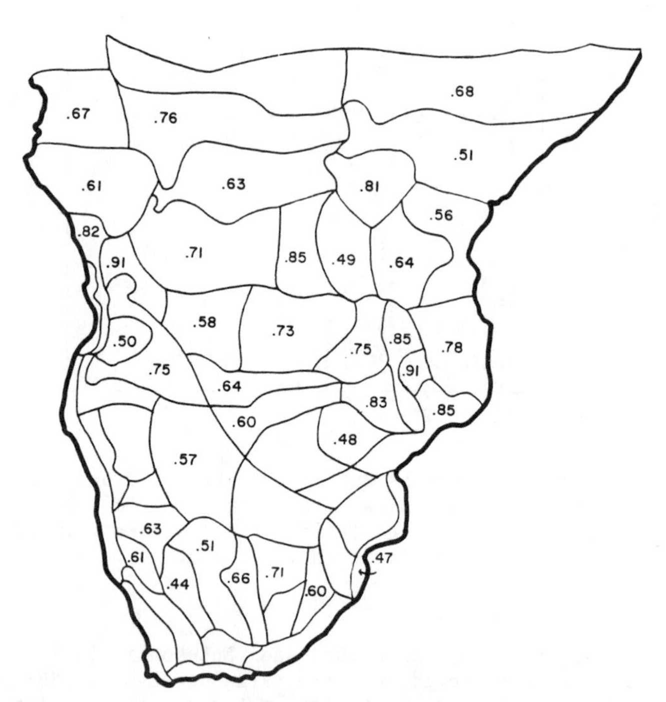


Fig. 10a. Coherence square between sea-surface temperatures along the Benguela coast (10°-20°S) in the 5.0-6.3 year spectral band for all rainfall regions with discrete spectral peaks in this band (95% confidence level = .58, 99% confidence level = .63; from Nicholson and Entekhabi, 1987).

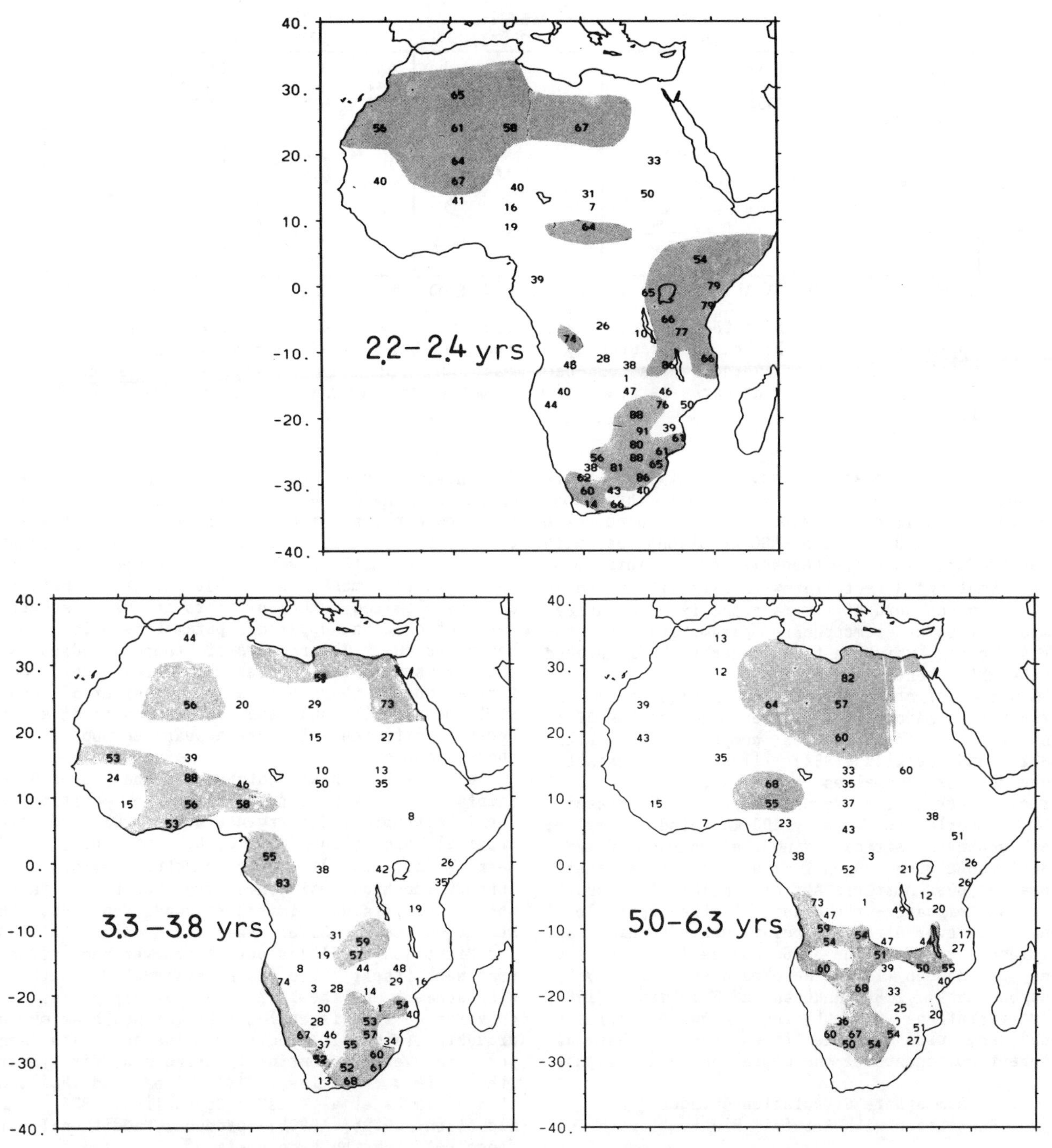


Fig. 10b. Coherence-square between rainfall and the Southern Oscillation in three spectral bands for all regions with significant spectral peaks in rainfall in these bands (shading denotes values exceeding the 95% confidence level, from Nicholson and Entekhabi, 1986).

Nicholson and Entekhabi (1986) provide more detail on the relationship between ENSO and African rainfall. The three spectral peaks consistently evident in rainfall series (~2.3, ~3.5 and 5-6 years) (Fig. 5) also characterize the Tahiti-Darwin surface pressure index of the Southern Oscillation (Fig. 7). The coherence-square (Fig. 10) indicates that the relationship between the two is strongest

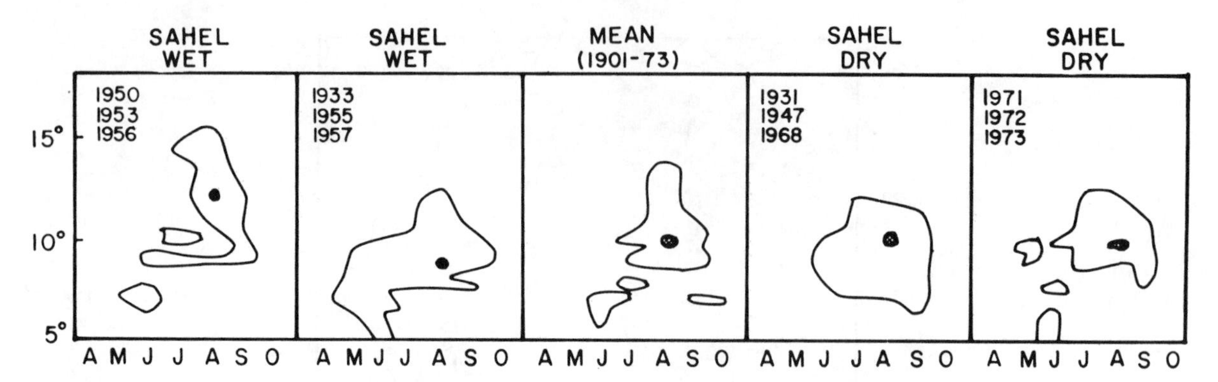


Fig. 11. Location of the ITCZ over West Africa during wet years and dry years, as approximated by the "center of gravity" of the rain belt for the months of April to October (three-year averages plus long-term mean; circled areas are rainfall exceeding 200 mm/month; shading marks latitude of maximum rainfall in August; from Nicholson, 1981b).

at 5-6 years in equatorial regions, at ~3.5 years in southern Africa, and at ~2.3 years throughout the eastern half of the subcontinent. In no case is the relationship with ENSO as strong as with Atlantic SSTs along the Benguela coast. This suggests that the direct link with rainfall in equatorial and southern Africa is probably sea-surface temperatures or atmospheric parameters in the marine sector, which in turn appear to be modulated by the ENSO events.

A number of other global studies illustrate the influence of major tropical phenomenon on the African sector. The ENSO wind composites of Arkin (1982) and Selkirk (1984) illustrate consistent sequences of anomalies over certain parts of Africa. Arkin, for example, shows a progression from easterly to westerly 200-mb wind anomalies over southern Africa during the course of ENSO years, and a progression from westerly to southerly anomalies over northern Africa. Selkirk's results likewise emphasize the change of the upper-level zonal flow over Africa in relationship to high and low index periods. The 30-60 day oscillation studies of Madden (1987), Weickmann et al. (1985), Knutson et al. 1986, Knutson and Weickmann (1987) show variations in outgoing longwave radiation (OLR) and wind over the African sector that are coherent components of the global 30-60 day signal.

## Atmosphere Circulation Changes Associated with Rainfall Fluctuations

Although numerous statistical analyses and model simulations have linked African rainfall to sea-surface temperature fluctuations and tropical atmospheric phenomena, few studies have dealt directly with drought mechanisms in terms of local atmospheric factors (e.g., circulation changes, atmospheric moisture, stability). This lack may be due to the dearth of upper air data over Africa. Sufficient radiosonde and pibal data exist for only a few years, most prior to the recent severe droughts. Modern satellite-derived data sets, which are hard to compare with conventionally observed data, are available only for recent years; for the Sahel, this includes only drought years and no true "normal" or wet year for comparison. The result is that most studies have either considered a very limited number of years or were severely restricted in the upper air parameters which could be assessed. The results of these studies are briefly summarized. Three areas, the Sahel, East Africa and southern Africa, are considered separately because the analyses have tended to do so and because different factors appear to operate in these locations.

An early hypothesis interpreted changes of Sahel rainfall mainly as a function of the position of the Intertropical Convergence Zone (ITCZ) and subtropical high (Kraus, 1977a, b; Greenhut, 1977; Beer et al., 1977). The only actual evidence to support the hypothesis came from Lamb (1978a,b), whose analysis was limited to conditions over the Atlantic Ocean. Later, Newell and Kidson (1984) and Nicholson (1981b) showed that over the African continent itself there was no systematic southward displacement of the ITCZ. In many dry years, the convergence zone is at least as far south as during droughts (Fig. 11) and it remains near its mean position over the continent during most dry years. Others (e.g., Namias, 1974; Miles and Folland, 1974; Tanaka et al., 1975; Schupelius, 1976; and Nicholson, 1979b, 1980b) produced additional evidence refuting the hypothesis.

A number of more plausible explanations for drought in the Sahel have since been proposed. Various investigations have shown that, compared with wet years, drought years in the sub-Saharan region are characterized by a weaker 200 mb tropical easterly jet (TEJ), a stronger mid-tropospheric (700 mb) African easterly jet (AEJ), weaker shear in the 700 mb easterly jet south of the equator, enhanced Hadley-type overturnings (and weaker Walker-type overturnings), increased geopotential of the 700 mb surface, increased vertical shear over West Africa (thus, enhanced horizontal temperature gradients), and the virtual disappearance of the 850 mb trough over West Africa (Kidson, 1977; Tanaka et al., 1975; Kanamitsu and Krishnamurti, 1978; Newell and Kidson, 1979, 1984; and Dennett et al., 1985). Surprisingly, there appears to be no systematic difference in the depth of the monsoon layer, its humidity or advective moisture flux over West Africa between wet and dry years in the Sahel (Lamb, 1983). In many cases, linkages with other tropical weather anomalies are evident (Kanamitsu and Krishnamurti, 1978; Krueger and Winston, 1975; Kraus, 1977a; Fleer, 1981; and Hastenrath and Kaczmarczyk, 1981).

Even fewer systematic studies of wet and dry years have been carried out for other parts of Africa. In East Africa, local forecasters generally associate westerly wind anomalies in the mid- or lower troposphere with abnormally high rainfall but that is not always the case (Kiangi and Temu, 1984). The two monsoon flows prevailing during most of the year are generally divergent and a large-scale diffluence of the NE monsoon over East Africa is readily observable (Flohn, 1964; Anyamba, 1984). The suggested causes of rainfall variability are quite varied, as the factors controlling rainfall are complex and diverse in various parts of the region. Moreover, most investigations of rainfall variability have been confined to case studies of one or two years (e.g., Minja, 1985; Anyamba, 1983, 1984; Agumba, 1984; Anyamba and Ogallo, 1986; Ogallo and Anyamba, 1986; and Flohn, 1988). Therefore, few generalizations can be made. Factors which have been associated with abnormally wet conditions, such as those of 1961/62 or 1977/78, include strong low-level westerlies from the Atlantic and simultaneous intensification of the Mascarene and Arabian Highs. This is consistent with Cadet's finding of a strengthening of ITCZ over the western Indian Ocean during wet years in East Africa. During droughts, such as 1972 or 1983/84, low or mid-level easterly anomalies often appear in association with a Mascarene High which is unusually intense or more zonally oriented. There is also evidence that East Africa is influenced by fluctuations in the tropical divergent circulations (Kanamitsu and Krishnamurti, 1978; Chen and van Loon, 1987). Anomalous zonal divergent circulations (Walker-type overturnings) may account for the apparent inverse relationship between the intensity of East African rainfall and the Indian monsoon, as suggested by studies of Stoeckenius (1981), Rasmusson and Carpenter (1982), and Ropelewski and Halpert (1987).

Although the synoptic systems of southern Africa are fairly well described (Bhalotra, 1973; Acharya and Bhaskara Rao, 1981), in most cases our understanding of mechanisms of interannual rainfall variability in the region can be reduced to a few generalizations. The importance of easterly flow in the upper-troposphere in producing high rainfall has been established (Taljaard, 1981a,b and Tyson, 1984), but a dynamical explanation for this association is not readily apparent. The interaction between tropical and extra-tropical systems (Kumar, 1978; Riehl, 1979) is likewise a significant determinant of the character of the rainy season. As an example, the coincidence of an eastward-moving westerly (extratropical) trough with a westward-propagating trough in the tropical easterlies can rapidly transform a stable tropical air mass into an intense disturbance. Knowledge of the relationship between such systems and interannual variability of rainfall is woefully inadequate. A better picture, however, is starting to emerge. Several recent papers have shown the importance of diagonal cloud bands (also a system manifested by tropical/extra-tropical interactions) in determining the seasonal character of rainfall (Harrison, 1983, 1984; Harangozo and Harrison, 1983). These systems influence most of the southern subcontinent and are clearly linked to large-scale patterns of tropical flow, as evidenced by their relationship to the ENSO phenomenon.

## Land Surface Processes

In the early 1970s it became apparent that human activities were dramatically altering the natural global landscape. Semi-arid lands, the predominant African environment, were shown in many cases to be undergoing a process of "desertification", the result of overcultivation, deforestation, overgrazing, improper irrigation, slash/burn agriculture and other forms of land mismanagement (United Nations, 1977). The desertified landscape, relatively barren of vegetation with bare soil exposed and subsequently eroded, was illustrated by several satellite photos showing a contrast between high albedo in altered sectors and low albedo in adjacent protected areas (e.g., Glantz, 1977). Other physical changes expected to accompany the removal of vegetation include reduced soil moisture, evapotranspiration and surface roughness, higher surface temperatures, and increased soil erosion and dust generation (Anthes, 1984). A drought would likely produce similar land surface changes.

The potential feedbacks on the atmosphere include changes of latent and sensible heat transfer, thermal stability and convergence. The hypothesized feedback was first modelled by Charney (1975). His initial study involved an albedo change over the Sahara from 35% to 14%; the result was a significant reduction in rainfall over the Sahel and a southward displacement of the rain zone in the high albedo case. In subsequent GCM simulations (Charney et al., 1975, 1977) the concept of reduced evapotranspiration was added and a similar reduction in rainfall resulted (Fig. 12). A large number of studies have similarly examined land-surface feedback mechanisms, simulating not only albedo and evapotranspiration or soil moisture but also surface roughness. These are beyond the scope of this paper but summarized in review articles such as Mintz (1984) or Nicholson (1988). Despite

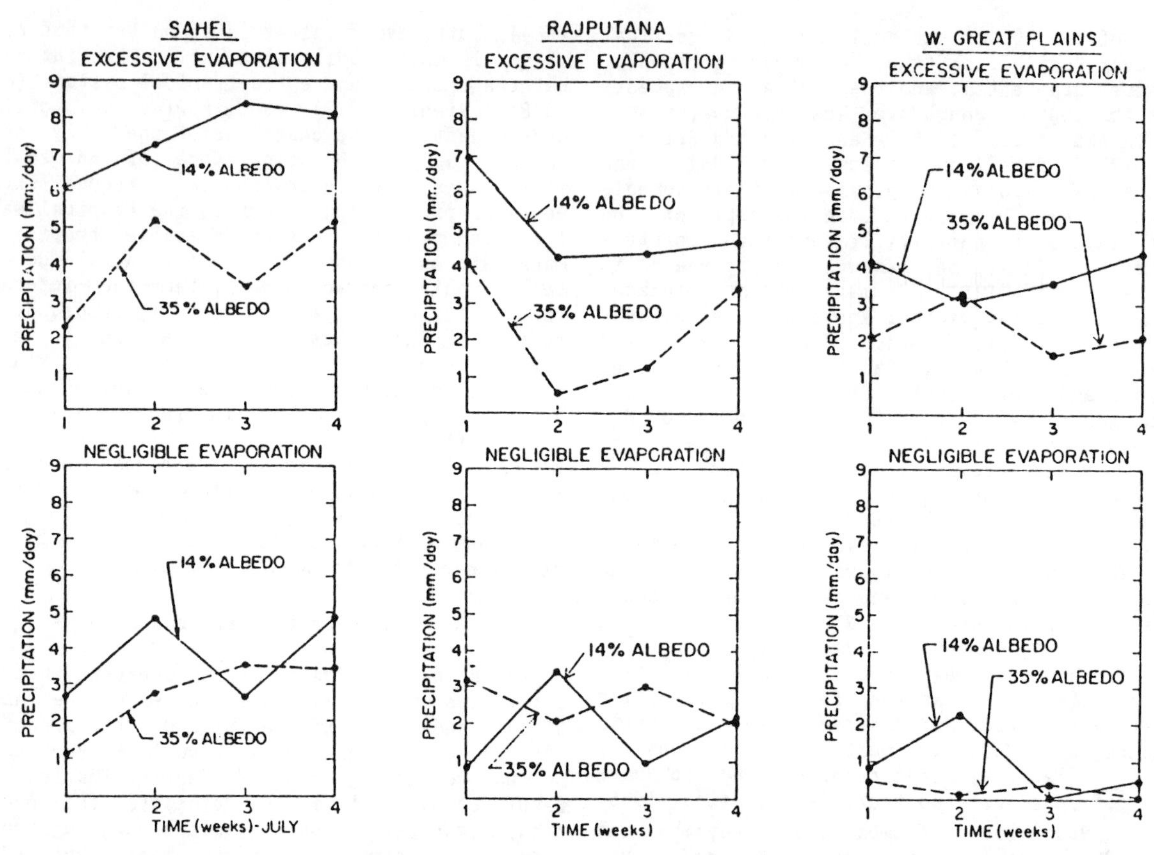


Fig. 12. Weekly average rainfall rates (mm/day) for July in three semi-arid regions from model simulations using surface albedos of 14% and 35% in the test area (from Charney et al., 1977).

considerable differences in mechanisms, locations considered or nature of the models used, nearly all lead to the conclusion that large-scale land surface modification can alter climate and that the Sahel is one of the regions most sensitive to such effects.

Although there is general agreement between the diverse numerical simulations, strong empirical evidence of land/atmosphere feedback is lacking. A fair amount exists for regional impact (see Anthes, 1984) and one statistical study (Walsh et al., 1985) suggests large-scale influence of soil moisture on atmospheric variables. The multi-year persistence of rainfall anomalies in the Sahel, as described in Section 3, could well be a manifestation of such feedback (Nicholson, 1986c). Even so, a number of additional questions must be answered in order to determine whether or not land-surface change can influence rainfall in the Sahel (or elsewhere) and to what extent human factors contribute. These include the extent of environmental change, the sensitivity of the atmosphere to surface forcing, and the quantitative physical impact of landscape changes.

An important question concerning sensitivity is the source of atmospheric moisture in the Sahel region. Current model results suggest that changes of soil moisture and/or evapotranspiration would have a greater impact than albedo. Unfortunately, studies of atmospheric moisture over West Africa have reached different conclusions concerning its sources (e.g., low-level southwest monsoon flow, Indian Ocean, the Mediterranean) and little is known about the contribution of local evaporation to the supply (Cadet and Houston, 1984; Cadet and Nnoli, 1987).

Despite the belief in widespread "desertification" over West Africa (United Nations, 1977), there have been few systematic studies of land surface change and there has been virtually no attempt to distinguish between human and natural factors in any observed changes. For West Africa, rainfall decreased by a factor of two during the time period over which desertification has been assessed. The reduced rainfall, rather than human impact on the land, is the likely cause of the suggested "advancement" of the Sahara. Moreover, in most cases, broad generalizations have been made

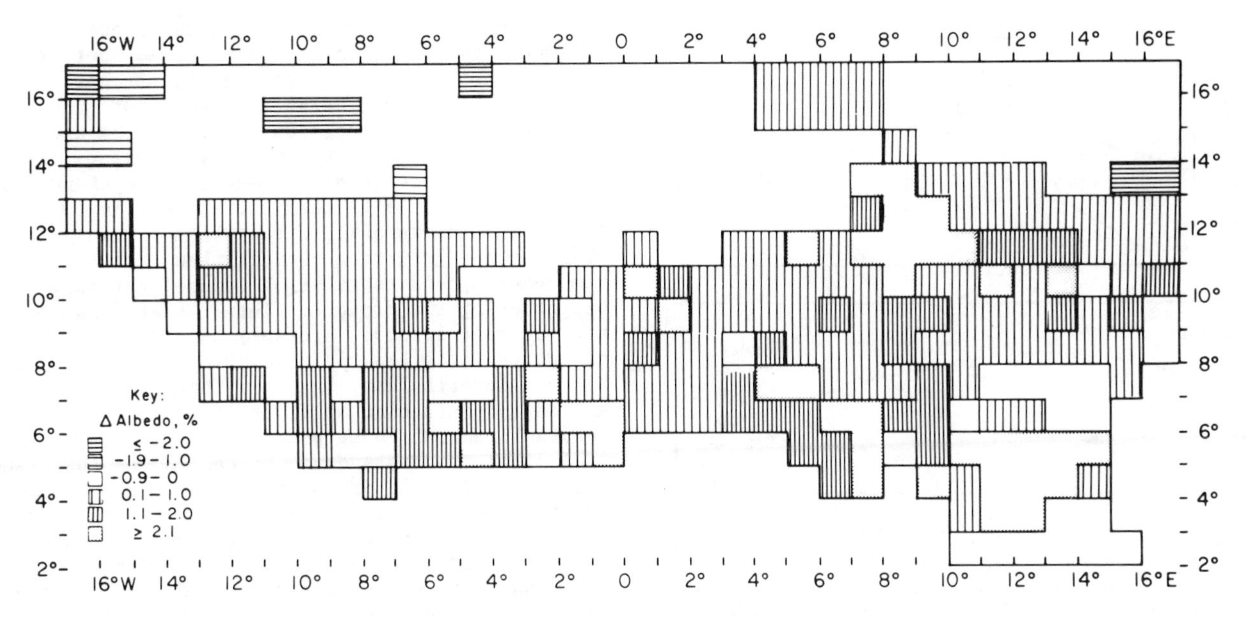


Fig. 13. Map of albedo changes over West Africa, pre-agriculture to present (from Gornitz, 1985).

from a few local studies (e.g., Ibrahim, 1978; Wendler and Eaton, 1983). Recent papers which have attempted to verify the claims of anthropogenic desertification (Hellden, 1984; Olsson, 1983, 1985) could not do so and have also demonstrated the importance of rainfall fluctuations in altering surface characteristics. Similarly, Gornitz (1985), in a monumental study of anthropogenic vegetation change over West Africa during the last century, has shown that only small changes of surface albedo have accompanied it (Fig. 13).

Questions concerning the physical impact of surface change also remain. For example, the assumed increase of Sahelian albedo accompanying drought or a denuded land surface (Charney, 1975) cannot be firmly established. The albedo of "protected" patches in Niger and two areas of southern Tunisia (.34, .36 and .26 respectively) is considerably higher than in surrounding overgrazed areas (.42, 1983). Courel et al. (1984), combining their own albedo measurements with those of Norton et al. (1979), have shown that albedo increased significantly as the drought progressed from 1967 to 1973, going from ~.31 to ~.38 in the more northern Sahel (Fig. 14). However, as the drought continued in the late 1970s, albedo decreased to values well below those prior to the drought. Surface temperature changes associated with decreased vegetation are likewise complex and controversial. Otterman (1974) reported lower radiative temperatures over the overgrazed Sinai than in the Negev, where grazing is more controlled. This result was challenged by Jackson and Idso (1975), among others, as being a manifestation of the lower emissivity of the more denuded surface. A more recent study of Otterman and Tucker (1985) provides some confirmation of the earlier result.

One of the shortcomings of the feedback hypothesis is that the proposed feedback mechanisms do not adequately provide a "memory" from one year to the next. If the land surface acts to promote multiyear drought, the surface characteristics responsible for the feedback must be related to rainfall in the previous year. This does not seem to be true for soil moisture, since the soil is dry throughout the the dry season. Nor does albedo show the anticipated relationship with rainfall of the previous year.

The surface-related parameter which appears to change most consistently in response to Sahel rainfall is dust production. Prospero and Nees (1977, 1986) have shown that African dust, which is transported across the Atlantic and Caribbean, appears at Barbados (West Indies) in dramatically increased concentration during years of Sahel drought (Fig. 15). The correlation with Sahel rainfall in the previous year is exceedingly high. The dust influences atmospheric stability, its presence contributing to the formation of inversion layers above the Sahel/Sahara (Prospero and Carlson, 1972; Carlson and Prospero, 1972). Unfortunately, the impact of dust on atmospheric dynamics over West Africa has not been extensively modelled but is potentially quite significant (see Coakley and Cess, 1985). This phenomenon has received more attention in recent years, in part because of the French ECLATS experiment, and a number of studies of dust mobilization and transport and its radiative effects have appeared (Carlson and Benjamin, 1980; ben Mohamed and Frangi, 1983, 1986; Druil-

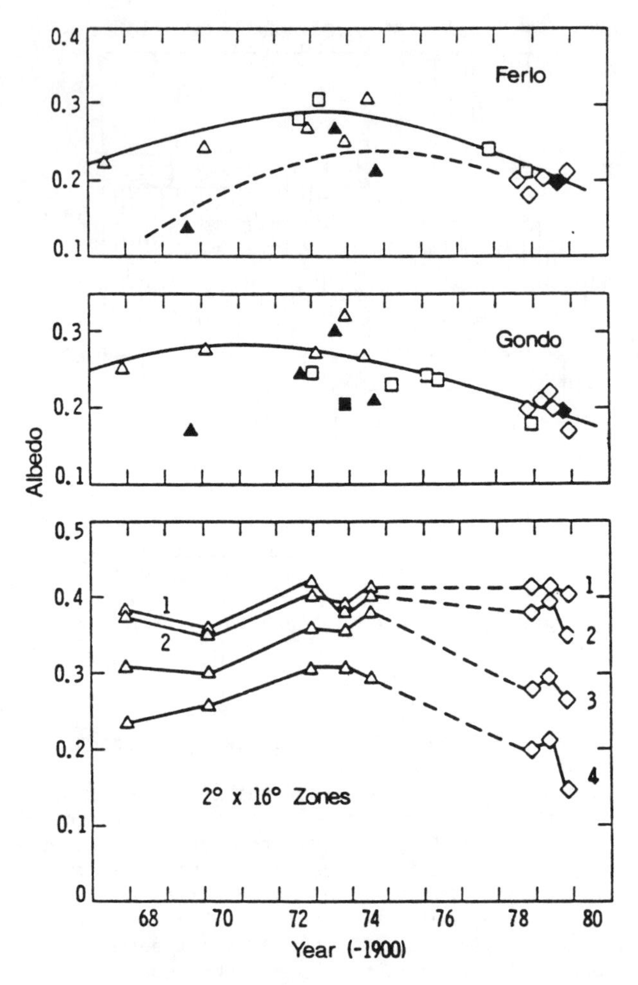


Fig. 14. Albedo history of the western Sahara and Sahel between 1967 and 1979. Curves 1-4 are integrated albedo values for four strips 2° wide in latitude, from 16-0° W: 1 = 20-22°N, 2 = 18-20°N, 3 = 16-18°N, 4 = 14-16°N (from Courel et al., 1984).



het et al., 1982; Druilhet and Durand, 1984; Guedalia et al., 1984; Reiff et al., 1986; d'Almeida, 1986, 1987; Fouquart et al., 1987a, b; Westphal et al., 1987; and Helgren and Prospero, 1987).

Currently, few scientists accept the original hypothesis that human-induced land surface changes caused the recent drought in the Sahel. Several points of evidence contradict such a suggestion. For one, strong atmospheric teleconnections between the Sahel and other African locations are evident and droughts often occur more or less synchronously throughout the continent. Secondly, the Sahel droughts are simply too large in scale to result from regional factors. Also, such droughts have occurred in the historical past, long before the human impact on the land became significant. However, there is considerable support for the idea that land surface processes can intensify and prolong a Sahel drought initiated by anomalous large-scale atmospheric forcing. The surface changes are likely a response to the initial decline in rainfall (as much as 50%), rather than human activities. These surface changes then reinforce the atmospheric conditions which reduced rainfall, thus perpetuating the drought. The 20-year duration of the current episode, a highly unusual occurrence elsewhere, is often cited as evidence of such feedback. Nevertheless, the idea that Sahel droughts are self-reinforcing as a result of land-atmosphere feedback remains to be proven.

A real understanding of the role the land-surface plays in promoting drought in the Sahel requires a quantified description of changes of land-surface properties accompanying drought or human-induced surface modification. The most relevant parameters are surface albedo, temperature, vegetation cover and soil moisture (Bollé and Rasool, 1985), and the related fluxes of latent and sensible heat. Establishing these on a regional or continental scale would be nearly impossible with conventional observational techniques, but remote-sensing methodologies are currently being developed which will soon permit such large-scale monitoring of the land surface. These are reviewed by Nicholson (1989). Recent papers applying these to Africa include studies of albedo (Rockwood and Cox, 1978; Norton et al., 1979; Courel et al., 1984: Pinty and Szejwach, 1985; Pinty et al., 1985; and Pinty and Tanré, 1986), surface temperature, thermal inertia, outgoing radiation and evapotranspiration (Duvel and Kandel, 1985; Abdellaoui et al., 1986), vegetation cover (Tucker et al., 1983, 1985a,b;

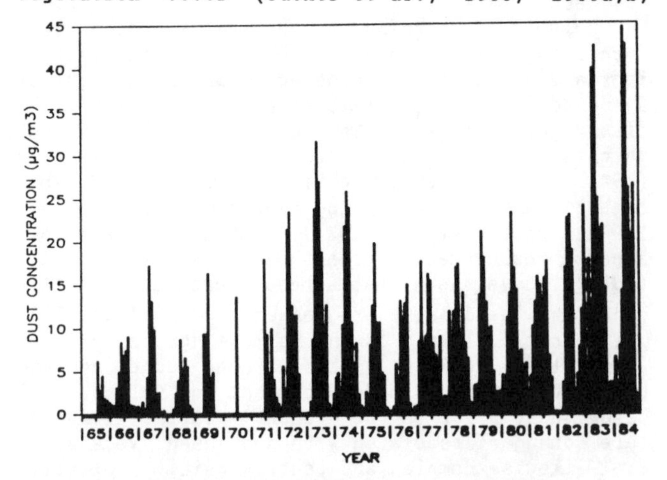


**Fig. 15. Monthly mean trade wind mineral aerosol concentration at Barbados (Units $10^{-6}$ g/m$^3$; from Prospero and Nees, 1986, compare with Figure 4).**

McGinnis and Tarpley, 1985), and hydrology (Schneider et al., 1985; McCauley et al., 1982).

## Implications for Forecasting and Future Research

To improve our ability to forecast droughts in Africa, several lines of investigation must be pursued. First of all, the links between synoptic or regional scale phenomenon and planetary-scale circulation must be better established. This is particularly true for the Sahel and for East Africa, where the factors which modify the rain-bearing disturbances from year-to-year are poorly understood. This precludes relying solely on numerical models to evaluate interannual variability of rainfall. A related need is further study of the characteristics of atmospheric circulation and weather systems over the African continent during wet years and dry years. Requisite to such study is increased acquisition and improved dissemination of rainfall and upper air data. Finally, the teleconnections with the global tropics must be further explored, particularly with quasi-global phenomena like ENSO and with the tropical Atlantic and Indian Oceans. Pending the results of such investigations, forecasting schemes might be developed. These may soon be feasible for equatorial and southern Africa.

For the Sahel, however, reliable methods to forecast drought cannot be developed until the question of land surface feedback is resolved. If it is assumed that large-scale factors initiate a drought and that the resultant land surface changes reinforce the initial atmospheric forcing, the character of a given rainy season represents the interplay of these two types of forcing. Presumably, then, a drought would continue until an atmospheric perturbation favoring rainfall occurs and is sufficiently strong to overcome the surface-imposed drought-promoting feedback. To forecast drought in the Sahel, it must first be definitively established whether the land surface does play such a role and, if so, through what mechanism. So far, this has been attempted primarily through numerical models and observational evidence of feedback is lacking.

A major weakness of the theory is that the proposed feedback mechanisms, e.g. soil moisture or vegetation, do not adequately provide the requisite "memory" from one year to the next. In order to determine the appropriate variables and resolve the question of "memory", observational studies of such non-atmospheric factors as soil erosion and structure or efficiency of plant growth must be carried out. The best dependence on antecedent years so far appears to be atmospheric dust (Prospero and Nees, 1986). A number of studies are assessing the impact of dust on the atmosphere, but the missing link to rainfall variability is an understanding of the atmospheric factors (e.g., static or dynamic stability) governing such parameters as the number, size or intensity of rain-bearing disturbances and how these are influenced by aerosols. Thus, forecasting drought in the Sahel is considerably more complex than elsewhere in Africa since two very different types of forcing must be evaluated. It will be quite some time before these issues are resolved.

## Summary and Conclusions

During the last two decades, most of the African continent has experienced extensive, severe and prolonged droughts. The area most affected has been the semi-arid subtropical region of the northern hemisphere. While recent droughts have been the worst of the century, such events are an inherent part of the long-term climatic history of Africa and have occurred in past centuries. Despite numerous claims that the Sahel drought was anthropogenically induced via human impact on the land surface, at present no evidence exists to support such a hypothesis. A vast number of numerical simulations do, however, suggest that land surface changes, if sufficiently large and extensive, can influence large-scale climate, especially in the Sahel. Currently these are viewed as potential factors in perpetuating rather than initiating drought, the ultimate trigger being some change in the large-scale atmospheric dynamics.

For the Sahel, little is known about the actual causes of drought. Elsewhere in Africa, the synoptic situations associated with dry years are fairly well known, but larger scale factors controlling these are not thoroughly understood. Research has centered on sea-surface temperature fluctuations and atmospheric circulation and/or stability changes. The latter might be caused by sea-surface temperature fluctuations but could also occur independently.

In the absence of comprehensive knowledge of drought-producing mechanisms over Africa, it is presently impossible to forecast rainfall fluctuations. Thus, the question as to whether the current drought situation will persist cannot be answered. In view of the linkages with tropical phenomenon like the Southern Oscillation, a potential exists for short- or long-term forecasting in some regions if the physical mechanisms producing the statistically established associations can be determined. In such a case, forecasting schemes might be developed, such as those of Nicholls (1983), Shukla and Paolino (1983), Shukla and Mooley (1987), and Hastenrath (1987) for monsoon rainfall and Hastenrath (1984b) for Northeast Brazil.

Acknowledgments. This paper was presented at the IUGG in Vancouver, August 1987. I would like to acknowledge the contributions of the FSU Foundation and the National Science Foundation (Grant ATM 86 14208) in providing travel funds to attend the meeting. Laura Easter, Andrew Lare and Ada Malo assisted in the preparation of this manuscript. Figures were drafted by Dewey Rudd.

References


Abdellaoui, A., F. Becker and E. Olory-Hechinger, 1986: Use of Meteosat for mapping thermal inertia and evapotranspiration over a limited region of Mali. Journal of Climate and Applied Meteorology, 25, 1489-1506.

Acharya, U.R. and N.S. Bhaskara Rao, 1981: Meteorology of Zambia, Part II. Government Printers, Lusaka, 49 pp.

Adamec, D. and J.J. O'Brien, 1978: The seasonal upwelling in the Gulf of Guinea due to remote forcing. Journal of Physical Oceanography, 8, 1050-1060.

Agumba, F., 1984: Climatological fluctuations of spring-rains in Kenya in relation to large-scale circulations. Extended Abstracts of Papers Presented at the WMO Regional Scientific Conference on GATE, WAMEX and Tropical Meteorology, World Meteorological Organization Tropical Meteorology Programme Report Series No. 16, Geneva, 258-262.

Anthes, R.A., 1984: Enhancement of convective precipitation by mesoscale variations in vegetative covering in semiarid regions. Journal of Climate and Applied Meteorology, 23, 541-554.

Anyamba, E.K., 1983: On the monthly mean tropospheric circulation and the anomalous circulation during the 1961/62 floods in East Africa. M.S. Thesis, Department of Meteorology, University of Nairobi.

Anyamba, E.K., 1984: Some aspects of the origin of rainfall deficiency in East Africa. Extended Abstracts of Papers Presented at the WMO Regional Scientific Conference on GATE, WAMEX and Tropical Meteorology, World Meteorological Organization Tropical Meteorology Programme Report Series No. 16, Geneva, 110-113.

Anyamba, E.K. and L.J. Ogallo, 1986: Anomalies in the wind field over Africa during the East African rainy season of 1983/84. Long-Range Forecasting Research. Report Series No. 6, 1, World Meteorological Organization, Geneva, 128-131.

Arkin, P.A., 1982: The relationship between interannual variability in the 200-mb tropical wind field and the Southern Oscillation. Monthly Weather Review, 110, 1393-1404.

Bhalotra, Y.P.R., 1973: Disturbances of the summer season affecting Zambia. Technical Memorandum No. 3, Zambian Meteorological Department.

Bhalotra, Y.P.R., 1973: Disturbances of the summer season affecting Zambia. Technical Memorandum No. 3, Zambian Meteorological Department, 7 pp.

Beer, T., G.K. Greenhut and S.E. Tandoh, 1977: Relations between the Z criterion for the subtropical high, Hadley Cell parameters and rainfall in northern Ghana. Monthly Weather Review, 105, 849-855.

**ben Mohamed, A., and J.-P. Frangi, 1983: Humidity and turbidity parameters in Sahel: A case study for Niamey (Niger). Journal of Climate and Applied Meteorology, 22, 1820-1823.**

ben Mohamed, A. and J.-P. Frangi, 1986: Results from ground-based monitoring of spectral aerosol optical thickness and horizontal extinction: some specific characteristics of dusty Sahelian atmospheres. Journal of Climate and Applied Meteorology, 25, 1807-1815.

Bolle, H.J. and S.I. Rasool, 1985: Development of the implementation plan for the International Satellite Land - Surface Climatology Project (ISLSCP). Phase I: World Meteorological Organization, WCP-34, Geneva, 56 pp.

Bunting, A.H., M.D. Dennett, J. Elston, and J.R. Milford, 1976: Rainfall trends in the West African Sahel. Quarterly Journal of the Royal Meteorological Society, 102, 59-64.

Butzer, K.W., G.L. Isaac, J.L. Richardson, and C. Washbourn-Kamau, 1972: Radiocarbon dating of East African lake levels. Science, 175, 1069-1076.

Cadet, D.L. and S.H. Houston, 1984: Precipitable water over Africa and the Eastern/Central Atlantic Ocean during the 1979 Summer. Journal of the Meteorological Society of Japan, 62, 761-774.

Cadet, D.L. and N.O. Nnoli, 1987: Water vapour transport over Africa and the Atlantic Ocean during Summer 1979. Quarterly Journal of the Royal Meteorological Society, 113, 581-602.

Carlson, T.N. and J.M. Prospero, 1972: The large scale movement of Saharan air outbreaks over the northern equatorial Atlantic. Journal of Applied Meteorology, 11, 283-297.

Carlson, T.N. and S.G. Benjamin, 1980: Radiative heating rates for Saharan dust. Journal of the Atmospheric Sciences, 37, 193-213.

Charney, J.G., 1975: Dynamics of deserts and drought in the Sahel. Quarterly Journal of the Royal Meteorological Society, 101, 193-202.

Charney, J.G., P.H. Stone and W.J. Quirk, 1975: Drought in the Sahara: A biogeophysical feedback mechanism. Science, 187, 434-435.

Charney, J.G., W.J. Quirk, S.H. Chow and J. Kornfield, 1977: A comparative study of the effects of albedo change on drought in semiarid regions. Journal of the Atmospheric Sciences, 34, 1366-1385.

Chen, T.-C., and H. Van Loon, 1987: Interannual variation of the Tropical Easterly Jet. Monthly Weather Review, 1739-1759.

Coakley, J.A. and R.D. Cess, 1985: Response of the NCAR community climate model to the radiative forcing by the naturally occurring tropospheric aerosol. Journal of the Atmospheric Sciences, 42, 1677-1692.

Cockcroft, M.J., M.J. Wilkinson and P.D.Tyson, 1987: The application of a present-day climatic model to the Late Quaternary in Southern Africa. Climatic Change, 10, 161-181.

Courel, M.F., R.S. Kandel and S.I. Rasool, 1984: Surface albedo and the Sahel drought. Nature, 307, 528-531.

d'Almeida, G., 1986: A Model for Saharan dust transport. Journal of Climate and Applied Meteorology, 25, 903-916.

d'Almeida, G., 1987: On the variability of desert aerosol radiative characteristics. Journal of Geophysical Research, 92, 3017-3026.

Deacon, J. and N. Lancaster, 1988: Late Quaternary Palaeoenvironments of Southern Africa. Clarendon Press, Oxford, 225 pp.

Dennett, M.D., J. Elston and J.A. Rodgers, 1985: A reappraisal of rainfall trends in the Sahel. Journal of Climatology, 5, 353-362.

Druilhet, A. and P. Durand, 1984: Etude de la couche limite convective Sahélienne en presence de brumes seches/Experience (ECLATS). La Météorologie, 51-77.

Druilhet, A., J.P. Frangi and P. Durand, 1982: Donneés statistiques sur le bilan d'energie de la couche de surface Sahélienne. La Météorologie, 29-30, 227-237.

Duvel, J.P. and R.S. Kandel, 1985: Regional-scale diurnal variations of outgoing infrared radiation observed by METEOSAT. Journal of Climate and Applied Meteorology, 24, 335-349.

Dyer, T.G.J., 1975: The assignment of rainfall stations into homogeneous groups: An application of principal component analysis. Quarterly Journal of the Royal Meteorological Society, 101, 1005-1013.

Dyer, T.G.J., 1979: Rainfall along the east coast of southern Africa, the Southern Oscillation and the latitude of the subtropical high pressure belt. Quarterly Journal of the Royal Meteorological Society, 105, 445-452.

Dyer, T.G.J. and M.E. Marker, 1978: On the variation of rainfall over southwest Africa. The South African Geographical Journal, 60, 143-149.

Fleer, H., 1981: Large-scale tropical rainfall anomalies. Bonner Meteorologische Abhandlungen, 26, 114 pp.

Flohn, H. 1964: Investigations on the Tropical Easterly Jet. Bonner Meteorologische Abhandlungen, 26, 83pp.

Flohn, H., 1988: East African rains of 1961/2 and the abrupt change of the White Nile discharge. Palaeoecology of Africa, 18, in press.

Folland, C.K., D.E. Parker, M.N. Ward and A.W. Colman, 1986a: Sahel rainfall, Northern Hemisphere circulation anomalies and worldwide sea temperature changes. Long-Range Forecasting and Climate Research, 7, Meteorological Office, Bracknell.

Folland, C.K., T.N. Palmer, and D.E. Parker, 1986b: Sahel rainfall and world-wide sea temperatures 1901-85; Observational, modelling, and simulation studies. Nature, 320, 602-607.

Fouquart, Y., B. Bonnel, G. Brogniez, J.C. Buriez, L. Smith, J.J. Morcrette and A. Cerf, 1987a: Observations of Saharan aerosols: Results of ECLATS field experiment. Part I: Optical thicknesses and aerosol size distributions. Journal of Climate and Applied Meteorology, 26, 28-37.

Fouquart, Y., B. Bonnel, M. Chaoui Roquai and R. Santer, 1987b: Observations of Saharan aerosols: Results of ECLATS field experiment Part II: Broadband radiative characteristics of the aerosols and vertical radiative flux divergence. Journal of Climate and Applied Meteorology, 26, 38-52.

Glantz, M.H., ed., 1977: Desertification: Environmental Degradation in and around Arid Lands. Westview Press, Boulder, 346 pp.

Gornitz, V., 1985: A survey of anthropogenic vegetation changes in West Africa during the last century--climatic implications. Climatic Change, 7, 285-326.

Greenhut, G.K., 1977: A new criterion for locating the subtropical high in West Africa. Journal of Applied Meteorology, 16, 727,734.

Gregory, S., 1982: Spatial patterns of Sahelian annual rainfall. Archives fuer Meteorologie, Geophysik und Bioklimatologie, Ser. B., 31, 273-286.

Guedalia, D., C. Estournel, and R. Vehil, 1984: Effects of Sahel dust layers upon nocturnal cooling of the atmosphere (ECLATS) experiment. Journal of Climate and Applied Meteorology, 23, 644-650.

Harangozo, S. and Harrison, M.S.J., 1983: On the use of synoptic data indicating the presence of cloud bands over southern Africa. South African Journal of Science, 79, 413-414.

Harrison, M.S.J., 1983: The Southern Oscillation, zonal equatorial circulation cells and South African rainfall. Preprints of the 1st International Conference on Southern Hemisphere Meteorology, Americn Meteorological Society, Boston, 302-305.

Harrison, M.S.J., 1984: The annual rainfall cycle over the central interior of Africa. South African Journal of Science, 66, 47-64.

Hastenrath, S.L. and E.B. Kaczmarczyk, 1981: On the spectra and coherence of tropical climate anomalies. Tellus, 33, 453-462.

Hastenrath, S.L., 1984a: Interannual variability and annual cycle: mechanisms of circulation and climate in the tropical Atlantic sector. Monthly Weather Review, 112, 1097-1107.

Hastenrath, S., 1984b: Predictability of north-east Brazil droughts. Nature, 307, 531-533.

Hastenrath, S., 1987: Predictability of Java monsoon rainfall anomalies: A case study. Journal of Climate and Applied Meteorology, 26, 133-141.

Helgren, D.M. and J.M. Prospero, 1987: Wind velocities associated with dust deflation events in the western Sahara. Journal of Climate and Applied Meteorology, 26, 1147-1151.

Hellden, U., 1984: Drought impact monitoring: a remote sensing study of desertification in Kordofan, Sudan. Lunds Universitets Naturgeografiska Institution, 61, 61 pp.

Hirst, A.C. and S. Hastenrath, 1983a: Atmosphere-ocean mechanisms of climate anomalies in the Angola-tropical Atlantic sector. Journal of Physical Oceanography, 13, 1146-1157.

Hirst, A.C. and S. Hastenrath, 1983b: Diagnostics of hydrometeorological anomalies in the Zaire (Congo) basin. Quarterly Journal of the Royal Meteorological Society, 109, 881-892.

**Hsiung, J. and R. Newell, 1983: The principal non-seasonal modes of variation of global sea-surface temperature.** Journal of Physical Oceanography, 13, 1952-1967.

Ibrahim, F., 1978: Anthropogenic causes of desertification in Western Sudan. Geographical Journal, 23, 243-254.

Jackson, R.D. and S.B. Idson, 1975: Surface albedo and desertification. Science, 189, 1012-1013.

Kanamitsu, M. and T.N. Krishnamurti, 1978: Northern summer tropical circulations during drought and normal rainfall months. Monthly Weather Review, 106, 331-347.

Kiangi, P.M.R. and J.J. Temu, 1984: Equatorial westerlies in Kenya. Are they all rain laden? Extended abstracts of papers presented at the WMO Regional Scientific Conference on GATE, WAMEX and Tropical meteorology in Africa, World Meteorological Organization Tropical Meteorology Programme Report Series No. 16, Geneva, 144-146.

Kidson, J.W., 1977: African rainfall and its relation to the upper air circulation. Quarterly Journal of the Royal Meteorological Society, 103, 441-456.

Klaus, D., 1978: Spatial distribution and periodicity of mean annual precipitation south of the Sahara. Archives fuer Meteorologie, Geophysik und Bioklimatologie Ser. B., 26, 17-27.

Knutson, T.R., and K.M. Weickmann, 1987: 30-60 day atmospheric oscillations: composite life cycles of convection and circulation anomalies. Monthly Weather Review, 115, 1407-1436.

Knutson, T.R., K.M. Weickmann and J.E. Kutzbach, 1986: Global-scale intraseasonal oscillations of outgoing longwave radiation and 250 mb zonal wind during northern hemisphere summer. Monthly Weather Review, 114.

Kraus, E.B., 1977a: Subtropical droughts and cross-equatorial energy transports. Monthly Weather Review, 105, 1009-1018.

Kraus, E.B., 1977b: The seasonal excursion of the intertropical convergence zone. Monthly Weather Review, 105, 1009-1018.

Krueger, A. and J.S. Winston, 1975: Large-scale circulation anomalies over the tropics during 1971-72. Monthly Weather Review, 103, 465-473.

Kumar, S., 1978: Interaction of upper westerly waves with intertropical convergence zone and their effect on the weather over Zambia during the rainy season. Government Printers, Lusaka, 36 pp.

Lamb, P.J., 1978a: Case studies of tropical Atlantic surface circulation patterns during recent sub-Saharan weather anomalies: 1967 and 1968. Monthly Weather Review, 106, 482-491.

Lamb, P.J., 1978b: Large-scale tropical Atlantic surface circulation patterns associated with Subsaharan weather anomalies. Tellus, 30, 240-251.

Lamb, P.J., 1982: Persistence of Sub - saharan drought. Nature, 299, 46-47.

Lamb, P.J., 1983: West African water vapor variations between recent contrasting Subsaharan rainy seasons. Tellus, 35A, 198-212.

Lindesay, J.A., M.S.J. Harrison and M.P. Haffner, 1986: The Southern Oscillation and South African rainfall. South African Journal of Science, 82, 196-198.

Lough, J.M., 1986: Tropical Atlantic sea-surface temperatures and rainfall variations in Sub-saharan Africa. Monthly Weather Review, 114, 561-570.

Madden, R.A., 1987: Seasonal variations of the 40-50 day oscillation in the tropics. Journal of the Atmospheric Sciences, 43, 3138-3158.

McCauley, J.F., G.G. Schaber, C.S. Breed, M.J. Grolier, C.V. Haynes, B. Issawi, C. Elachi, and R. Blom, 1982: Subsurface valleys and geoarchaeology of the eastern Sahara revealed in shuttle radar. Science, 218, 1004-1017.

McGinnis, D.F. and J.D. Tarpley, 1985: Vegetation cover mapping from NOAA/AVHRR. Advances in Space Research, 5, 359-369.

Miles, M.K. and C.K. Folland, 1974: Changes in the latitude of the climatic zones of the Northern Hemisphere. Nature, 252, 616.

Minja, W.E.S., 1985: A comparative investigation of weather anomalies over West Africa. M.S. Thesis, Department of Meteorology, University of Nairobi.

Mintz, Y., 1984: The sensitivity of numerically simulated climates to land-surface boundary conditions. The Global Climate, J.T. Houghton, Ed., Cambridge University Press, 70-105.

Motha, R.P., et al., 1980: Precipitation patterns in West Africa. Monthly Weather Review, 108, 1567-1578.

Namias, J., 1974: Suggestions for research leading to long-range precipitation forecasting for the tropics. Preprints of the International Tropical Meteorology Meeting, Nairobi, American Meteorological Society, 141-144.

Newell, R.E. and J.W. Kidson, 1979: The tropospheric circulation over Africa. Saharan Dust (Scope Report 14), C. Morales, Ed., Wiley and Sons, 133-170.

Newell, R.E. and J.W. Kidson, 1984: African mean wind changes in Sahelian wet and dry periods. Journal of Climatology, 4, 1-7.

Nicholls, N., 1983: Predicting Indian monsoon rainfall from sea-surface temperature in the Indonesia-North Australia area. Nature, 306, 576-577.

Nicholson, S.E., 1978: Climatic variations in the Sahel and other African regions during the past five centuries. Journal of Arid Environments, 1, 3-24.

Nicholson, S.E., 1979a: The methodology of historical climate reconstruction and its application to Africa. Journal of African History, 20, No. 1, 31-49.

Nicholson, S.E., 1979b: Revised rainfall series for the West African subtropics. Monthly Weather Review, 107, 620-623.

Nicholson, S.E., 1980a: Saharan climates in historic times. The Sahara and the Nile, M.A.J. Williams and H. Faure, Eds., A.A. Balkema, Rotterdam, 173-200.

Nicholson, S.E., 1980b: The nature of rainfall

fluctuations in sub-tropical West Africa. Monthly Weather Review, 108, 473-487.

Nicholson, S.E., 1981a: The historical climatology of Africa. Climate and History, T.M.L. Wigley, M.J. Ingram and G. Farmer, Eds., Cambridge University Press, Cambridge, 249-270.

Nicholson, S.E., 1981b: Rainfall and atmospheric circulation patterns during drought periods and wetter years in West Africa. Monthly Weather Review, 109, 2191-2208.

Nicholson, S.E., 1982: The Sahel: A Climatic Perspective. Club du Sahel, Paris, 80 pp.

Nicholson, S.E., 1983: Sub-Saharan rainfall in the years 1976-80: evidence of continued drought. Monthly Weather Review, 111, 1646-1654.

Nicholson, S.E., 1985: Sub-Saharan rainfall 1981-1984. Journal of Climate and Applied Meteorology, 24, 1388-1391.

Nicholson, S.E., 1986a: The spatial coherence of African rainfall anomalies: interhemispheric teleconnections. Journal of Climate and Applied Meteorology, 25, 1365-1381.

Nicholson, S.E., 1986b: The nature of rainfall variability in Africa south of the equator. Journal of Climatology, 6, 515-530.

Nicholson, S.E., 1986c: African drought: an example of the influence of land-surface properties on climate? ISLSCP: Parameterization of Land-Surface Characteristics; Use of Satellite Data in Climate Studies; First Results of ISLSCP. European Space Agency, Brussels, SP-248, 405-410.

Nicholson, S.E., 1988: Land surface-atmosphere interaction: physical processes and surface changes and their impact. Progress in Physical Geography, 12, 36-65.

Nicholson, S.E., 1989: Remote sensing of land surface parameters of relevance to climate studies. Progress in Physical Geography, in press.

Nicholson, S.E. and D. Entekhabi, 1986: The quasi-periodic behavior of rainfall variability in Africa and its relationship to the Southern Oscillation. Archives for Meteorology, Geophysics and Bioclimatology, Ser. A., 34, 311-348.

Nicholson, S.E. and D. Entekhabi, 1987: Rainfall variability in equatorial and Southern Africa: relationships with sea-surface temperatures along the southwestern coast of Africa. Journal of Climate and Applied Meteorology, 26, 561-578.

Nicholson, S.E. and H. Flohn, 1980: African environmental and climatic changes and the general atmospheric circulation in late Pleistocene and Holocene. Climatic Change, 2, 313-348.

Norton, C.C., F.R. Mosher and B. Hinton, 1979: An investigation of surface albedo variations during the recent Sahel drought. Journal of Applied Meteorology, 18, 1252-1262.

Ogallo, L.J. and E.K. Anyamba, 1986: Droughts of tropical central and eastern Africa. Long-Range Forecasting Research Report Series No. 6, 1, World Meteorological Organization, Geneva, 67-72.

Olsson, L., 1983: Desertification or climate. Lund Studies in Geography, Ser. A., Physical Geography, No. 60, The University of Lund, 36 pp.

Olsson, L., 1985: An integrated study of desertification, Lunds Universitets Geografiska Institution Avhandlingar, XCVIII, 170 pp.

Otterman, J., 1974: Baring high-albedo soils by overgrazing: a hypothesized desertification mechanisms. Science, 186, 531-533.

Otterman, J., 1981: Satellite and field studies of man's impact on the surface in arid regions. Tellus, 33, 68-77.

Otterman, J. and C. J. Tucker, 1985: Satellite measurements of surface albedo and temperatures in semi-desert. Journal of Climate and Applied Meteorology, 24, 228-234.

Palmer, T.N., 1986: The influence of the Atlantic, Pacific, and Indian Oceans on Sahel rainfall. Nature, 322, 251-253.

Pinty, B., G. Szejwach, and J. Stum, 1985: Surface albedo over the Sahel from METEOSAT radiances. Journal of Climate and Applied Meteorology, 24, 108-113.

Pinty, B. and G. Szejwach, 1985: A new technique for inferring surface albedo from satellite observations. Journal of Climate and Applied Meteorology, 24, 741-750.

Pinty, B. and D. Tanré, 1987: The relationship between incident double-way transmittances: an application for the estimate of surface albedo from satellites over the African Sahel. Journal of Climate and Applied Meteorology, 26, 892-896.

Prospero, J.M. and T.N. Carlson, 1972: Vertical and areal distribution of Saharan dust over the west equatorial North Atlantic Ocean. Journal of Geophysical Research, 77, 5255-5265.

Prospero, J.M. and R.T. Nees, 1977: Dust concentrations in the atmosphere of the of the equatorial North Atlantic: possible relationship to the Sahelian drought. Science, 196, 1196-1198.

Prospero, J.M. and R.T. Nees, 1986: Impact of the North African drought and El Nino on mineral dust in the Barbados trade winds. Nature, 320, 735-738.

Rasmusson, E.M. and T.H. Carpenter, 1982: Variations in tropical sea surface temperature and surface wind fields associated with the Southern Oscillation/El Niño. Monthly Weather Review, 110, 354-384.

Rasmusson, and J.M. Wallace, 1983: Meteorological aspects of the El Nino/Southern Oscillation. Science, 222, 1195-1202.

Riehl, H., 1979: Climate and Weather in the Tropics. Academic Press, New York, 611 pp.

Reiff, J., G.S. Forbes, F.T.M. Spieksma, and J.J. Reynders, 1986: African dust researching north-western Europe: a case study to verify trajectory calculations. Journal of Climate and Applied Meteorology, 25, 1543-1567.

Rockwood, A.A. and S.K. Cox, 1978: Satellite infrared surface albedo over north-western Africa. Journal of the Atmospheric Sciences, 35, 513-522.

Rodhe, H. and H. Virji, 1976: Trends and periodicities in East African rainfall data. Monthly Weather Review, 104, 307-315.

Ropelewski, C.F. and M.S. Halpert, 1987: Global and regional scale precipitation and temperature patterns associated with El Nino/Southern Oscillation. Monthly Weather Review, 115, 1606-1626.

Schneider, S.R., D.F. McGinnis, Jr. and G. Stephens, 1985: Monitoring Africa's Lake Chad basin with LANDSAT and NOAA satellite data. International Journal of Remote Sensing, 6, 59-73.

Schupelius, G.D., 1976: Monsoon rains over West Africa. Tellus, 28, 533-537.

Schultze, G.C., 1983: 'N Moontlike verband tussen die Suidelike Ossilasie/El Nino verskynsels en droogtes oor die somerreenvalstreke van Suid-Afrika. 'N Voorlopige studie. South African Weather Bureau Newsleter, 410, 79-84.

Selkirk, R., 1984: Seasonally stratified correlations of the 200 mb tropical wind field to the Southern Oscillation. Journal of Climatology, 4, 365-382.

Semazzi, F.H.M., V. Mehta and Y.C. Sud, 1988: An investigation of the relationship between sub-Saharan rainfall and global sea surface temperatures. Atmosphere-Ocean, 26, 118-138.

Shukla, J. and D.A. Mooley, 1987: Empirical prediction of the summer monsoon rainfall over India. Monthly Weather Review, 115, 695-703.

Shukla, J. and D.A. Paolino, 1983: The Southern Oscillation and long-range forecasting of the summer monsoon rainfall over India. Monthly Weather Review, 111, 1830-1837.

Stoeckenius, T., 1981: Interannual variations of tropical precipitation patterns. Monthly Weather Review, 109, 1233-1247.

Street, F.A. and A.T. Grove, 1979: Global maps of lake-level fluctuations since 30,000 BP. Quaternary Research, 12, 83-118.

Street-Perrott, F.A. and N. Roberts, 1983: Fluctuations in closed-basin lakes as an indicator of past atmospheric circulation patterns. Variations in the Global Water Budget. A. Street-Perrott, M. Beran and R. Ratcliffe, Eds., Reidel, Dordrecht, 331-345.

Street-Perrott, F.A. and S.P. Harrison, 1985: Lake levels and climate reconstruction. Palaeoclimate and Modelling, A.D. Hecht, Ed., John Wiley and Sons, New York, 291-340.

Taljaard, J.J., 1981a: The anomalous climate and weather systems of January to March, 1974. South Africa Weather Bureau Technical Paper, No. 90, 92 pp.

Taljaard, J.J., 1981b: Upper-air circulation, temperature and humidity over southern Africa. South Africa Weather Bureau Technical Paper, No. 10, 94 pp.

Tanaka, M., B.C. Weare, A.R. Navato and R.E. Newell, 1975: Recent African rainfall patterns. Nature, 255, 201-203.

Tucker, C.J., C. Vanpraet, E. Boerwinkel and A. Gaston, 1983: Satellite remote sensing of total dry matter production in the Senegalese Sahel. Remote Sensing of Environment, 13, 461-474.

Tucker, C.J., C. Vanpraet, M.J. Sharman and G. Van Ittersum, 1985a: Satellite remote sensing of total herbaceous biomass production in the Senegalese Sahel. Remote Sensing of Environment, 17, 233-249.

Tucker, C.J., J.R.G. Townshend and T.E. Goff, 1985b: African land-cover classification using satellite data. Science, 227, 369-375.

Tyson, P.D., 1980: Temporal and spatial variation of rainfall anomalies in Africa south of latitude 22° during the period of meteorological record. Climatic Change, 2, 363-371.

Tyson, P.D., T.G.J. Dyer and M.N. Mametse, 1975: Secular changes in South African rainfall: 1880 to 1972. Quarterly Journal of the Royal Meteorological Society, 101, 817-833.

Tyson, P.D., 1984: The atmospheric modulation of extended wet and dry spells over South Africa. Journal of Climatology, 4, 621-636.

United Nations, 1977: Desertification Its Causes and Consequences. Pergamon, Oxford, 448 pp.

Walker, J. and P.R. Rowntree, 1977: The effect of soil moisture on circulation and rainfall in a tropical model. Quarterly Journal of the Royal Meteorological Society, 103, 29-46.

Walsh, J.E., W.J. Jasperson and B. Ross, 1985: Influences of snow cover and soil moisture on monthly air temperature. Monthly Weather Review, 113, 756-768.

Weickmann, K.M., G.R. Lussky and J.E. Kutzbach, 1985: Intraseasonal (30-60 day) fluctuations of outgoing longwave radiation and 250 mb stream function during northern winter. Monthly Weather Review, 113.

Wendler, G. and F. Eaton, 1983: On the desertification in the Sahel zone (Part 1: ground observations). Climatic Change, 5, 365-380.

Westphal, D.L., O.B. Toon and T.N. Carlson, 1987: A two-dimensional numerical investigation of the synamics and microphysics of Saharan dust storms. Journal of Geophysical Research, 92, 3027-3049.

Williams, M.A.J. and H. Faure, eds., 1980: The Sahara and the Nile. A.A. Balkema, Rotterdam, 607 pp.

# SENSITIVITY OF CLIMATE MODEL TO HYDROLOGY

Duzheng Ye

Institute of Atmospheric Physics, Academia Sinica, Beijing, China

## Introduction

The importance of land–surface processes in affecting climate change has been analyzed and discussed by Namias [1962,1963]. The physics of the land–surface processes affect the climate because the ground hydrology, together with the vegetation and soil, determine the surface moisture availability which, in turn, controls the partition between the sensible and latent heat fluxes [Rowntree, 1984] and also the transfer of momentum. Further the vegetation cover and the soil moisture content can determine the ground surface albedo in snowless conditions. Therefore the heat balance and water balance in the planetary boundary layer are highly influenced by the hydrological processes. Through the planetary boundary layer, the influence of the ground hydrological processes can be felt through the whole troposphere [Yeh, et al., 1984; Rowntree and Bolton, 1983]. Here a crucial factor is the soil moisture content. In the region of dry anomalies, the following sequence of events tends to occur: a decrease of evaporation, a ground surface warming with an increase of sensible heat flux, a warming of lower layers of the atmosphere with a decrease of relative humidity (due to decrease of evaporation and warming of lower atmophere), a decrease of precipitation with a cooling of the atmosphere higher up (due to decrease of latent heat), and then a change of upper air circulation [Rowntree and Bolton, 1983]. It is the decrease of precipitation which will cause the initial dry anomaly to persist. In the region of moist anomalies, the opposite sequence of events tends to occur.

Besides soil moisture, the vegetation is another factor of importance in the hydrological processes. The role of vegetation is shown schematically in Fig.1 from Dickinson [1984]. Simply speaking this figure shows that the vegetation foliage can prevent part of the precipitation from reaching the ground; part of the intercepted water is re–evaported and part of it drops to the ground. Part of the water reaching the ground is partitioned into the soil and the rest appears as runoff. The water in the soil passes downward and may be tapped by the roots from the ground water reservoirs which will eventually feed the streamflow as another kind of runoff. At the same time, plants extract water from the soil through their roots and move it into the atmosphere as transpiration. The canopy can drastically alter the energy balance processes at the surface and so alter the ground evaporation and snow melting. The plants can also significantly reduce the surface albedo and thus increase the solar radiation received at the ground (if other factors are kept the same). Further the plants will increase the surface roughness which can influence the ground hydrological processes. These processes have been discussed in detail by Dickinson [1984], Eagleson [1982] and others. With no vegetation, only the water in a shallow layer of soil is available for evaporation and this can dry out after a few days of high evaporation typical in the tropics and over the central continents in summer. With deep rooted vegetation some evaporation is likely to occur even during a long dry period [Rowntree and Bolton, 1983]. Only through vegetation can water in the deep soil layers be connected with the atmosphere and take part in the hydrological processes.

Because of the importance of the hydrological processes in climate, a series of simulation experiments on the sensitivities of climate to hydrological processes were done recently by many researchers as reviewed by Mintz [1984] and Rowntree [1984]. This paper will also review briefly the results of some of these experiments. But due to limitation of space, only the essential points drawn from three–dimensional GCM's will be discussed.

## Global Experiments with Prescribed Soil Moisture

The simplest but still illustrative experiment is a global experiment with prescribed soil moisture. Using the general circulation model of NASA Goddard Space Flight Center Laboratory for Atmospheric Sciences, Shukla and Mintz [1981] made experiments with the global soil kept either dry or saturated all the time. Fig.2 gives the distribution of pricipitation (mm / day) in wet–soil (top) and dry–soil case (bottom) for July. The most striking difference between the two cases is the sharp reduction of precipitation over the continents of the Northern Hemisphere. Large parts of these continents are almost void of rainfall except Southeast Asia, tropical Africa and west coast of north America. This lack of rainfall indicates that over the extratropical continents in summer much of the moisture for precipitation is derived from evapotranspiration and the rainfall is mainly of the convective type which derives its moisture from planetary boundary layer. The intercomparison of the two cases also shows that the rainfall in the ITCZ is not much different, indicating that the moisture converging into the ITCZ comes mainly from the oceans.

In the Southern Hemiphere July is, of course, a winter month. It is interesting to note that little reduction of precipitation from the wet to the dry case is found in the winter Southern Hemisphere. This suggests that in winter the continental precipitation mainly comes from large–scale lifting condensation and that the source of moisture is miainly from oceans by advection. Thus the evapotranspiration does not have too much influence on the precipitation in winter.

Mintz [1984] further showed that for the extratropical continents in summer the moisture taken away by evapotranspiration is larger than that added by precipitation. The deficit of soil moisture in summer

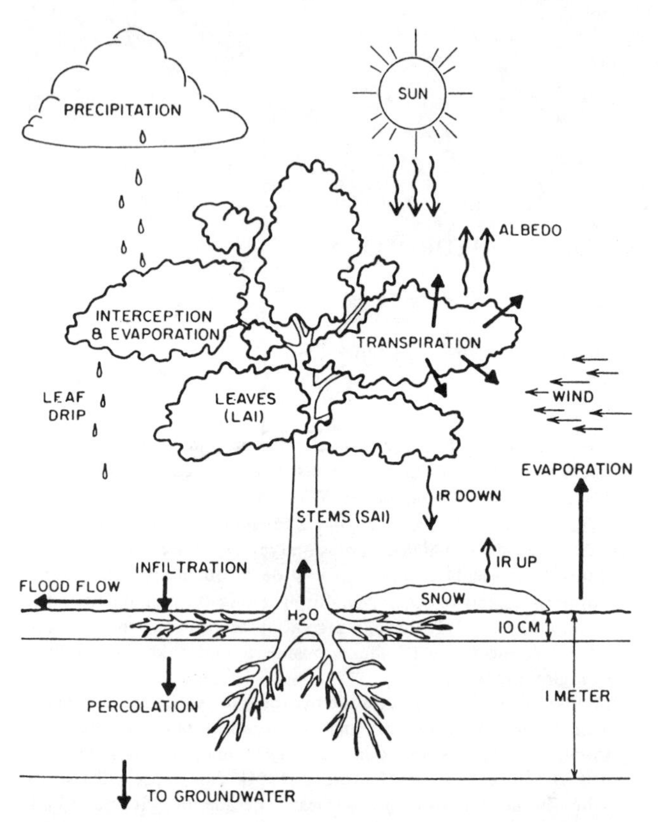


Fig.1 Schematic diagram of the processes that need to be considered in a canopy model of surface evapotranspiration and energy balance [Dickinson, 1984].

must be replenished by winter precipitation. Thus we can see that the annual cycle of soil moisture content must be connected with the oceanic processes. Since there is no evaporational cooling of the ground and since more solar radiation (less precipitation, less cloudness) is absorbed by the ground in the dry case, the ground temperature will be higher than in the wet case. The big difference in the ground temperature between the two cases can easily be seen in Fig. 3. The high ground temperature will obviously lead to a turbulent sensible heat flux and a large loss of long-wave radiation from the ground in the dry case. Thus the heat balance and the moisture balance (more water vapor in the wet case) in the boundary layer will be much different between the wet and dry case.

Shulka and Mintz's experiment revealed the importance of the hydrological processes in the climate. Similar experiments have been done by other authors, for instance, Suarez and Arakawa [see Mintz's review, 1984]. The essential results were similar to those of Shukla and Mintz, but there were large differences between them in the details. One reason for the difference can be attributed to the difference in parameterization schemes in the two models [Mintze, 1984]. The influence of parameterization schemes has been discussed by many authors [laval, et al., 1984; Dickinson,1984; Cunnington and Rowntree, 1986; Warrilow, 1985]. They all found that the parameterization is important for the details of the simulations.

## Experiments with Interactive Soil Moisture – with A Discussion of the Dependence of Hydrological Processes on Climate Regime

Althogh the global experiments with prescribed soil moisture do show the importance of hydrology in climate, the constraints of these experiments are too severe. One step to relax the constraints is to allow the soil moiture to vary with time but on a global scale. Using a M–21 version of the GFDL spectral model [Manbe, et al. 1979] Kurbatkin, et al., [1979] make an experiment with variable soil moisture in which the global continents were dry initially but change with time as predicted by the model. The distributions of the rate of evaporation and the rate of precipitation difference (arid experiment–control experiment ("natural")) are shown in Fig.4. It is seen from the figure that both evaporation and precipitation rates of the global initially dry case were less almost everywhere over the continents. These results were also obtained in other GCM expeiments [e.g. Yamazaki, 1986]. It is worth to note the two difference maps over most of inner parts of the summer continents (NH) are nearly equal. This indicates that over the extratropical continents in summer the moisture for precipitation is chiefly derived form evapotranspiration. But at midlatitude over the east coast of Asia there is an increase of precipitation. It is also interesting to note that decrease in precipitation is not limited to the continents; over the oceans, especially over the western tropical Pacific, the decrease of precipitation is also very pronounced, up to – 5 mm / day in large areas. Kurbatkin et al. [1979] attributed this to suppress of vertical motion over the oceans. The vertical motion over the continents in the initially dry case is increased although the precipitation rate is decreased due to lack of humidity. The increase of the continental vertical motion is probably associated with the decrease in vertical motion over the ocean.

To further relax the constraints of the simulation we discard the global condition and go to regional experiment with time–dependent soil moisture. Many authors have done it, e.g., Rowntree and Bolton [1983], Sud and Fennessy [1982, 1984], Warrilow [1985], Yeh, et al., [1984], and many others. In the global experiments above we have already seen that the climatic response to uniform global condition is

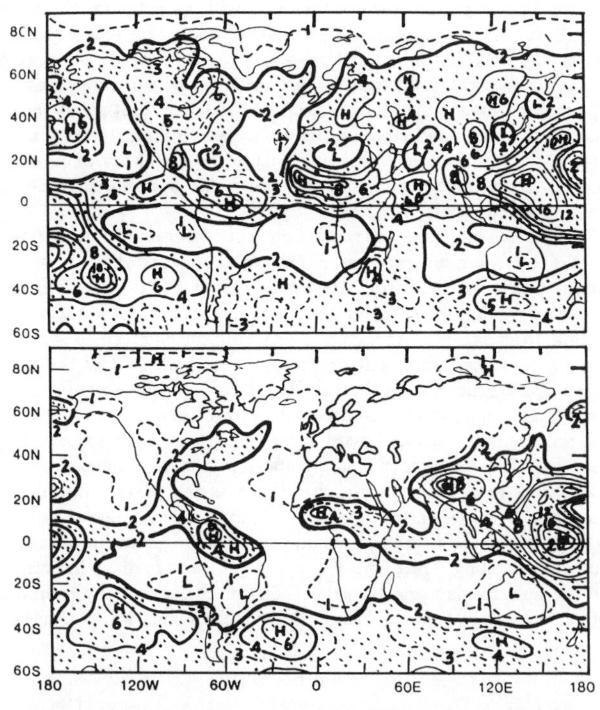


Fig.2. Contours of precipitation from Shukla and Mintz [1982]. For the top case, they assumed soil evaporated as a saturated surface. For the bottom case, they assumed zero evaporation from the soil. Precipitation greater than 2 mm / day is shaded.

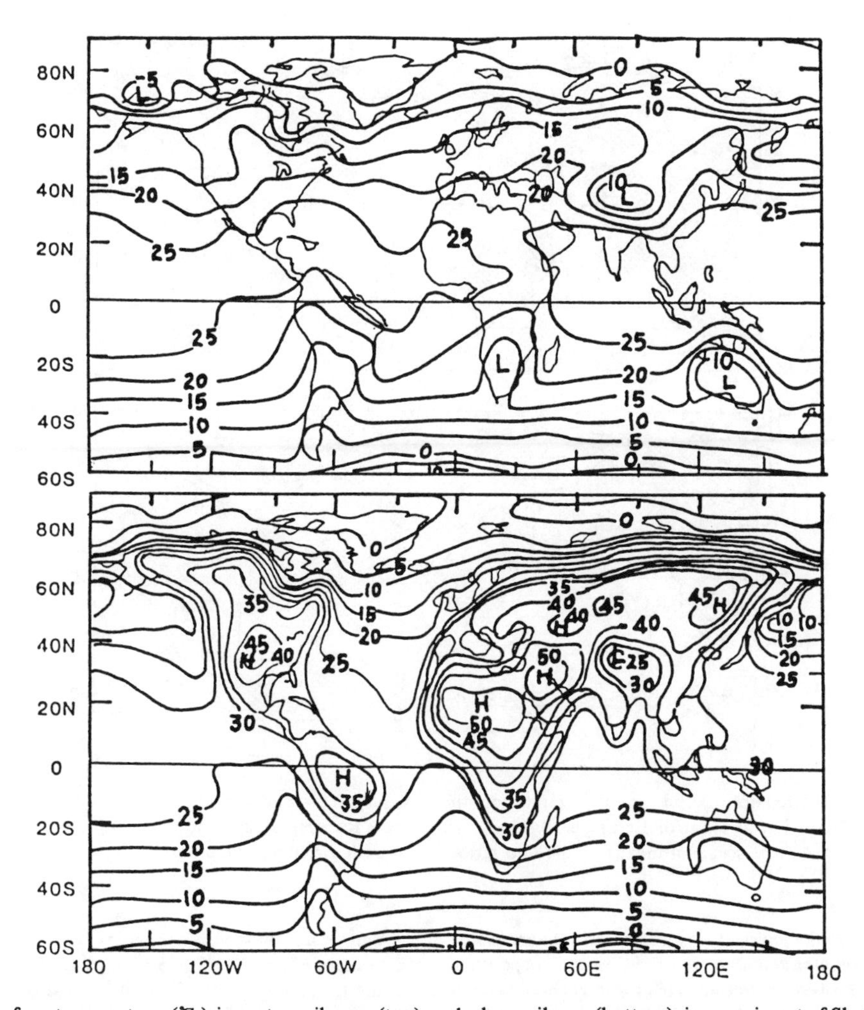


Fig.3. Ground–surface temperature (℃) in wet– soil case (top) and dry–soil case (bottom), in experiment of Shukla Mintz [1981].

not everywere the same. This means that the influence of hydrological processes depends on the climate regimes. We shall further discuss this point by using Yeh, et al.'s [1984] results. They irrigated the continents of 30–60N zone to its saturation (15cm) on July 1 and two other zones, 0–30N and 15S–15N, to saturation on Jan. 1, and then integrated each case for a period of five months. The influence of the large–scale irrigation on climate was obtained by analyzing the difference between the irrigated and the standard run.

Fig.5a–c show the latitude–time distribution of the mean zonal difference in precipitation (cm / day) between the irrigated and control experiments. One common feature in all the three figures is the increase of precipitation in the regions where the soil is instantaneously saturated with water. However, it is clear that the period of enhanced precipitation last longest in the 30–60N case (about 5 months) and shortest in the 15S–15N case (about 3 months). These difference in precipitation among the three cases are qualitatively similar to the difference in the persistence of soil moisture anomaly or enhanced evaporation (not shown here). It can also be seen that in the 30–60N case (Fig.5a) the increase is maximum near the center of the irrigated region and that it decreases both northward and southward. The corresponding distribution for 15S–15N case (Fig.5c) is less symmetric. Its maximum is displaced into the Southern Hemisphere in January and shifts equatorward during February and March. For the 0– 30N case (Fig.5b) there is a notable difference from the other two cases, i.e., the intensity of the increase is much smaller than in the other two cases. Further, there is no well–defined maximum of precipitation increase within the irrgated region; instead, a relatively samall increase takes place throughout the entire zone. But the intensity of the evaporation increase for 0–30N is of comparable value with that for 30–60N or 15S–15N cases. This indicates that the major part of water vapor gained from evaporation in the zone 0–30N is transported out and only a small part is used to increase the precipitation. Yeh, et al. [1984] invoked an increase in the descending branch of the Hadley circulation to explain this phenomenon.

Figs.5a and 5b share another characteristic. Both show two belts of maximum increase, one to the north and the other to the south of the irrigated region. In the 30–60N case the northern belt of maximum increases is found at 60–65N, lasting about two months. The southern belt is in the tropics and persists for a much longer period. It is interesting to note that the northern belts of maximum increase in precipitation in Figs.5a (summer) and 5b (winter) are near the midlatitude rainbelts in summer and winter, respectively, and that the southern belts of maximum increase in precipitation in Figs. 5a (summer) and 5b (winter) are situated over the ITCZ rainbelt which is north of the equator and south of the equator, respectively. This indicates the large-scale increase of soil moisture can induce certain physical processes which tend to enhance the activity of the ITCZ and midlatitude rainbelt even outside the region of soil moisture increase. Yeh, et al.

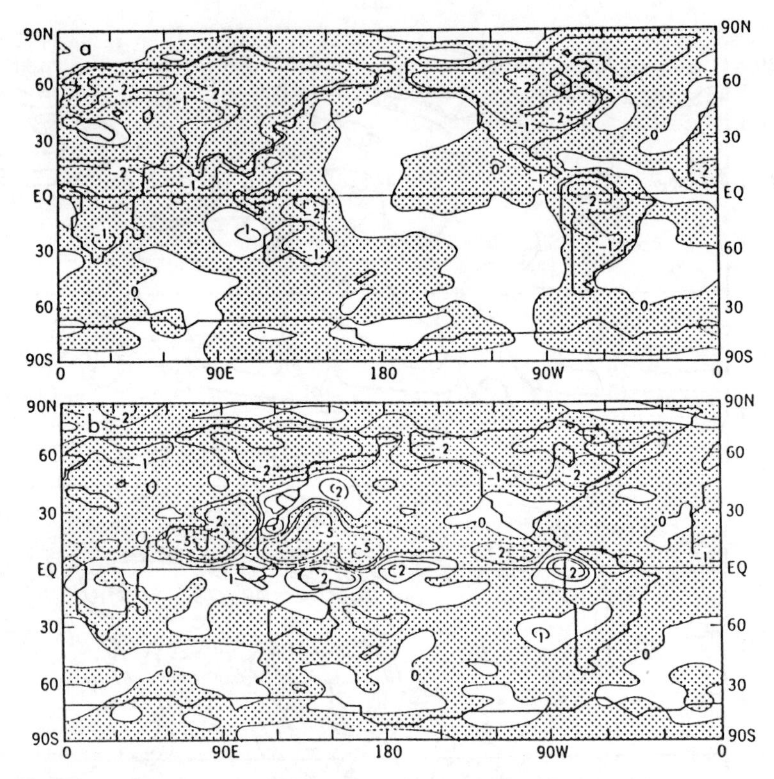


Fig.4. Rate of evaporation difference (a) and rate of precipitation difference (b). Difference (Arid Exp. −Control Exp.) distributions are constructed from the July−August seasonal means of the two model experiments and are horizontally smoothed to emphasize the large-scale features. Isolines are drawn for −5, −2, −1, 0, 1, 2, 5, mm / day in both distributions. [after Kurbatkin, et al., 1979].

[1984] postulated the following physical reasons for the enhancements: The strengthening of evaporation due to irrigation will produce a sheet of moist and cool air near the groud. This sheet will spread from the region and supply moisture to the neighboring rainbelts and enhances the already existing convective activity and precipitation there.

Yeh, et al. [1984] also pointed out that for all the three cases there is a decrease of precipitation just south of the area of maximum increase in the tropical rainblts. They speculated that this reduction might be due to the change of upward motion in these regions. Yeh, et al. [1984] also calculated the total increase of precipation (ΔP) and the corre-

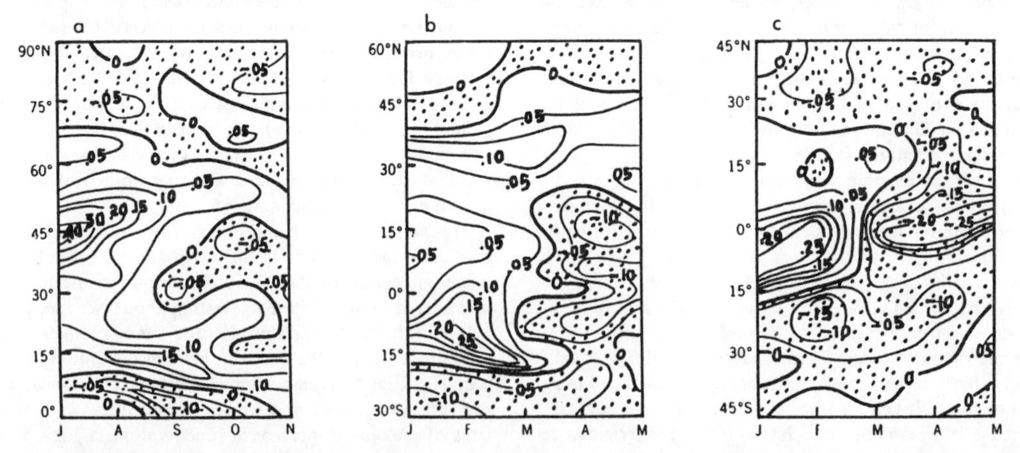


Fig.5. The latitude−time distributin of the zonal mean difference of precipitation rate (cm / day) between the perturbed and the normal experiment. (a) the 30 ° N−60 ° N case; (b) the 0−30 ° N case; (c) the 15 ° S−15 ° N case. [after Yeh, Wetherald and Manabe, 1985].

TABLE 1. The integrated increase in the water balance components ΔE, ΔP, and ΔR due to an increase of soil moisture ΔWs for three latitude zones averaged over the period where the values of ΔP were Positive (unit: cm) and the ratios of water balance components to ΔWs. (After Yeh, Wetherald and Manabe, 1984).

| Latitute Zone | ΔWs | ΔE | ΔP | ΔR | ΔE / ΔWs | ΔP / ΔWs | ΔR / ΔWs |
|---|---|---|---|---|---|---|---|
| 30N–60N | 11.2 | 15.3 | 9.6 | 3.1 | 1.37 | 0.86 | 0.28 |
| 0N–30N | 13.9 | 15.5 | 3.7 | 1.2 | 1.11 | 0.27 | 0.09 |
| 15S–15N | 13.0 | 11.8 | 6.6 | 6.7 | 0.91 | 0.51 | 0.52 |

sponding changes of evaporation (ΔE) and runoff (R) for the whole irrigated zone during the period when the average regional change of precipitation is positive. The results are shown in Table 1. The total water (ΔWs) used to irrigate each zone to saturation is also given in the table. The values of ΔWs of the three zones are comparable as seen in the table. The ratios ΔE / ΔWs, ΔP / ΔWs and ΔR / ΔWs in Table 1 can be used as a measure of efficiency of irrigation in inducing changes in the hydrological processes. The larger the ratio ΔP / ΔWs, the more benefit is gained from irrigation. When $\Delta E / \Delta Ws \approx \Delta P / \Delta Ws$ the benefit attains its optimum. In the region where $\Delta P / \Delta Ws \ll \Delta E / \Delta Ws$ the efficiency of irrigation is least. From Table 1 it is seen that the highest efficiency of irrigation is in the zone 30–60N and the least is in zone 0–30N. The physical reason for the efficiency being least in the 0–30N is that this zone is located under the descending branch of the Hadley cell. Another interesting point seen in the table is that in the 0–30N zone both ΔP / ΔWs and ΔR / ΔWs are the lowest among the three zones. This indicates that the transport of moisture obtained from the irrigation out off the irrigation area should be highest in 0–30N zone. Table 1 also tells that $(\Delta P / \Delta Ws)(30\text{–}60N) > (\Delta P / \Delta Ws)(15S\text{–}15N)$. The reason for this is that $(\Delta R / \Delta Ws)(30\text{–}60N) < (\Delta R / \Delta Ws)(15S\text{–}15N)$. Thus the moisture obtained from irrigation going to runoff is greater in the 15S–15N zone than in the 30–60N zone, making less moisture available for precipitation in the 15S–15N zone. The above discussion clearly shows the high dependence of the influence of the hydrological processes on the climate regimes. Many other authors have obtained similar results but they cannot be disussed here due to limited space.

Although the physics of the hydrological processes mainly lies at the ground surface, the upper air will also be influenced through the atmospheric boundary layer. To illustrate this in an example, we give Fig.6 which shows the latitude–height distribution of zonal mean temperature difference between the irrigated and the standard experiment for the zone 30–60N in the second month of irrigation (August) of Yeh, et al.'s [1984] experiment. The picture shows negative difference in the lower layers of the atmosphere above the irrigted zone and positive difference higher up. This indicates that in the lower atmosphere the increase of latent heat flux by irrigation from the earth decreases the temperatures while at higher levels the increase of latent heat release in precipitation increase the temperature. Even two months after the large– scale irrigation, the maximum decrease of temperature is still nearly 4℃ and the maximum increase in temperature at high levels 1.8℃ The corresponding chang in zonal–mean westly wind is given in Fig.7.

## Sensitivity of Climate Models to Surface Albedo

The presistent drought in Sahel in recent years aroused great concern among the people of the world including scientists and politicians. Meteorologists studied the cause of the drought from various points of view and made simulations by GCM under different land surface forcings. In discussing the simulation of the Sahelian drought we should keep in mind that this region is situated roughly in the descending branch of Hadley cell and that its spatial scale is on the order of a few thousand kilometers.

Charney [1975] first proposed a model for maintaining the drought in the Sahel. Charney's dynamics depend upon a feedback mechanism involving radiation, subsidence and albedo. A lack of vegetation gives a higher surface albedo which leads to a net radiative heat loss at the top of the atmosphere. This will induce a subsidence of dry air aloft with a subsequent reduction of precipitation, thus maintaining the desert. This hypothesis was subsequently tested by a numerical simulation with a general circulation model [Charney, et al. 1977]. Following Charney et al. Chervin [1979], Sud and Fennessy [1982], Planton [1986], Cunnington and Rowntree [1986], Laval [1986], and Carson and Sangster [1981] using different GCM's obtained similar

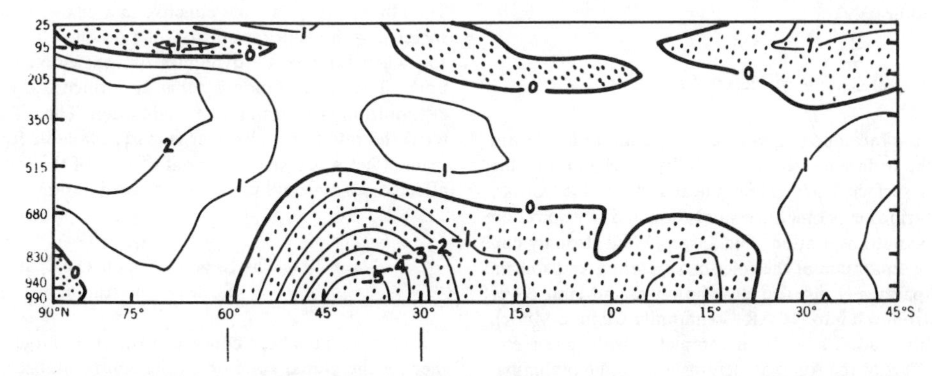


Fig.6. The latitude–height distribution of the zonal mean difference of air temperature (℃) over land between the pertured and normal experiment, August of the 30 ° N–60 ° N case [after Yeh, Wetherald and Manabe, 1985].

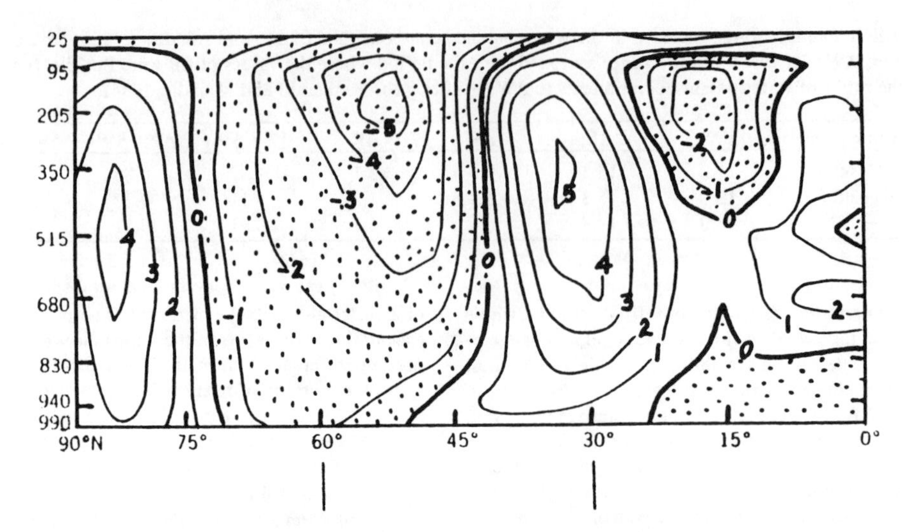


Fig.7. The latitude–height distribution of the difference of the August zonal mean wind (m / s) between the perturbed 30 ° N–60 ° N and the standard experiment. [after Yeh, Wetherald and Manabe, 1985].

results of decreasing precipitation when increasing the surface albedo. The results of these experiments supported Charney's hypothesis.

All above experiments undoubtdly indicate that a high albedo is important in maintaining deserts. However to prove this hypothesis as the main cause of the Sahelian drought, we need observational evidence of the actual albedo change. Charney, et al. [1977] also found that changes in the surface hydrology can have comparable effects as a change in surface albedo from about 0.1 to 0.3, which is the value used by Charney for explaining the Sahelian drought. Rowntree and Sangster [1985] recently also simulated the decrease of sahelian precipitation by only decreasing the soil mosture there.

In 1986 Cunnington and Rowntree [1986] went further to test the influence of the initial atmospheric moisture state on the precipitation and hydrological factors of the Sahara.

To close this section we should say that although the soil moisture content and the albedo are indeed important factors determining the Saharan climate we sould be very cautions to use the above experimental results to state that we actually explained the recent drought in the Sahel. There have been many long–term natural variations of the averaged precipitation over Africa [Nicholson, 1985]. We must be able to detect the albedo– or soil moisture content– induced effects against this large background noise.

## Sensitivity of Climate Models to Vegetation

As pointed out in the Introduction, vegetation is another important factor determining the surface moisture availability which controls the partition between the surface fluxes of latent heat and sensible hent. In all the foregoing described experiments evaporation and transpiration were combined as evapotranspiration. Dickinson [1985] strongly emphasized the need for separation of the roles of vegetation and soils in the GCM's. A comparatively detailed description of the land cover and soil has been introduced into NCAR Community Climate Model, and several sensitivity studies have been completed with major emphasis given to the effect of the Amazon deforestation. The preliminary conclu– sions drawn from comparisons of the control and the deforested runs for the Amazons region were given by Dickinson [1985]. They are: "the control and deforested runs both agree fairly well with the observed seasonal cycle of rainfall. Changes in rainfall with deforestation are not obvious because of high natural variability of model convective rainfall. The deforested case has more runoff, especially during the dry season and less interception of rainfall. The period during the season of driest soil is widened from one month to several months in southern Amazonia. The soil and air become warmer by several degrees, especially during the dry season; sensible heat fluxes are increased and evapotranspiration is reduced by deforestation."

With emphasis on tropical deforestation, Henderson–Sellers et al. [1985] studied the possible climate impacts of land cover transformations. Hansen et al. [1983] have also, but in a much simpler way, incorporated the parameters associated with different vegetation types into their climate model, model II. The vegetation distribution over the earth were compiled by Mathews [1983]. Using this model Rind [1982] investigated the influence of vegetation on the hydrological cycle. In this model the function of vegetation is much simplified and the soil is considered as composed by two layers, an upper layer of 10 cm depth and a lower layer down to 4 m depth.

Using this model Rind [1984] studied further the albedo, water holding capacity and diffusion for the vegetation types in the climate. The albedo effect is most effective in areas with plentiful moisture associated with evaporation from the ocean surface and during the season when large–scale dynamics or convection control the precipitations. The water holding capacity influence dominates where local evaporation is essential for precipitation. The diffusion experiment altered the rate at which water is made available from deeper levels. Its main effect was a stretching out of time of the soil moisture loss, wich affected the seasonal phase of the precipitation.

## The Dependence on The Spatial Scale of The Anomaly

So far we have been concerned with very large– scale anomalies, either on the global scale or on the continental scale. Certainly the importance of the surface hydrology on climate will depend on the spatial scale of the surface anomaly. It may be estimated relative to the role of advection. Rowntree [1984] estimated the importance of advected moisture on continental rainfall under the assumptions of nondivergent flow of air across a land mass with a humidity acquired while crossing the land mass, an evaporation rate of 3 mm / day (appropriate for moist tropics), and a mean flow of 5 m / s. Under these assumptions he showed that for an island of 100km horizontal scale (Lo), the prescribed abvected moisture flux convergence plays a dominant role, and local evaporation a very minor role. However for Lo ≈ 1000 km (e.g., Penisular India), local evaporation is a little more important while for Lo > 3000 km (e.g., Africa), the contribution from the horizontal moisture flux convergence is relatively small. Although Rowntree's calculation was based on very idealized assumptions, they give a qualitative eatimate of the scale effect in the influence of evaporation anomalies.

## Concluding Remarks

The above discussions show that the hydrological porcesses are very important in climate. The inclusion of hydrology in GCM's is not only necessary for climate studies or predictions, it can also improve medium–range weather forecast [Blondin, 1985], a few days forecast [Tada, 1985] and even very short–range weather forecast [Gadd and Keers, 1970].

Another potentially possible practical application of the surface hydrological processes would be the modification of local weather or climate. Anthes [1984] made a study showing the possibility of enhancement of convective precipitation in semi–arid regions by planting alternating bands of dense vegetation with width of 25–50km. Segal et al., [1983] suggested to modify the local climate by filling in the Qattara Deppression (about 18,000 $km^2$ in size) with Mediterranean water which is about 90km from this deppression. There are, of course, many other possible applications.

Since the two crucial factors in hydrology are the soil moisture content and the vegetation, we strongly recommend that soil moisture observations be made by satellite and that the astellite data be validated by ground–truth observations and global mapping of vegetation.

## References


Anthes, R.A. (1984), Enhancement of convective precipitation by mesoscale variations in vegetative covering in semiarid regions, *J. Climate & Appl. Met.*, **23**, 541–554.

Blondin, C.A. (1985),Treatement of land–surface properties in ECMWE model, *Proc. ISLSCP Conference, Rome, Italy, 2–6 Dec. 1985.*

Carson, D.J. and A.B. Sangster (1981), The influence of land–surface albedo and soil moisture on general circulation model simulations, *GAPP / WCRP: Research Activities in Atmospherical and Oceanic Modelling*, (Ed. by J.D. Rutherford) *Numerical Experomentation Programme. Rport No.2*, PP5. 14–5.21.

Charney,J.G., (1975),Dynamics of deserts drought in the Sahel, *Quart. J.R. Met. Soc.*, **101**, 193– 202.

Charney, J.G.,W.J. Quirk, S.H. Chow and J. Kornfield (1977), A comparative study of the effects of albedo change on drought in semi– arid region, *J. Atemos. Sci.*, **34**, 1366–85.

Chervin, R.M. (1979),Respones of the NCAR general circulation model to changed land surface albedo, *Report ofthe JSC study Conference on Climate Models: Performance, Intercomparison and Sensitivity Studies. Washington, DC, 3–7, April, 1978*, GAPP Publ. Series, No. 22, Vol.1 563–81.

Cunnington, W.M. and P.R. Rowntree (1986), Simulations of Shaharan atmosphere–dependence on moisture and albedo, *Quart. J.R. Met. Soc.*, **112**, 971–999.

Dickinson, R.E. (1984),Modelling evapotranspiration for three–dimensional global climate models. 58–72, In *Climate Processes and Climate Sensitivity, Geophys. Mon.*, **29**, J.E. Hansen, and T. Takanhashi editors.

Dickinson, R.E. (1985),GCM sensitivity studies– implications for parameterizations of land surface processes, *Proc. ISLSCP. Conference, Rome, Italy, 2–6, Dec. 1985*, EAS. SP–248, 127– 129.

Eagleson, P.S. (1982), Dynamical hydro–thermal balance at macroscale, In *land Surface Processes in Atmospheric General Circulation*, Ed. by P.S. Eagleason, Cambridge University Press.

Gadd, A.J and J.F. Keers (1970), Surface exchanges of sensitive and lantent heat in a 10–level model atmosphere, *Quart. J.R. Met. Soc.*, **96**, 297–308.

Hansen, J.E., G. Russell, D. Rind, P. Stone, A. Lacis, S. Lebedeff, R. Ruedy and L. Travis: (1983), Efficient three dimensional global models for climate studies: Models I and II. *Mon. Wea. Rev.*, **111**, 609–662.

Henderson–Sellers, A. and H. Gornitz (1985), Possible climatic impacts of land–cover transformations with particular emphasis on tropical deforestation, *Climate Change*, **6**, 231– 257.

Kurbatkin, G.P., S. Manabe and D.G. Hahn (1979), The moisture content of the continents and the intensity of summer monsoon circulation, *Meteorlogiya i Gidrologya*, **11**, 5–11.

Laval, K. (1984), Modeling the impact of soil properties on European climate. In *Current Issues in Climate Research*, A. Ghazi and R. Fantechi, editors. D. Reidel Publishing Company.

Laval. K. (1986), General circulation model experiment with suface albedo changes, *Climate Change*, **9**, Special issue: Climate and Desertification, 91–102.

Laval, K., A. Perrie and Y. Serratini (1984), Effect of parameterization of evapotranspira– tion on climate simulated by a GCM, In *New Perspectives in Climate Modeling*, Ed by A.C. Berger and C. Nicolis, 223–247.

Mathews, E. (1983), Global vegetation and land use: New high resolution data basis for climate studies, *J. Clim. and Appl Meteor.*, **22**, 474– 487.

Mintz, Y. (1984), The sensitivity of numerically simulated climates to land–surface boundary conditions, 79–105, In *The Global Climate*, Ed. by J. T. Houghton.

Namias, J. (1962), Influences of abnormal surface heat sources and sinks on atmospheric behavior, *Proc. Internat. Symp. Num. Wea Predict. Tokyo, Nov. 7–13. 1960*, Met. Soc. Japan, 615–627.

Namias, J. (1963), Surface–atmosphere interactions as fundamental causes of drought and other climatic fluctuations, *Arid Zone Research Changes of Climate Proc, Rome Symp. UNESCO and WMO*, 345–359.

Nicholson, S.E. (1985), African drought: an example of the influence of land–surface processes on climate. *Proc. ISLSCP Conference, Rome, Italy, Dec. 1985*, ESA SP–248, 405–410.

Planton, S. (1986): Sensitivity of the annual cycle simulated by a GCM to a change in land surface albedo, *Proc. ISLSCP Conference, Rome, December, 1985*, ESA SP–248, 135–142.

Rind, D. (1984), The influence of vegetation on the hydrological cycle in a global climate model, 73–91, In *Climate Processes and Climate Sensitivity, Geophys. Mon.*, 29, Ed. by J.E. Hansen and T. Takahashi.

Rowntree, P.R. and J.A. Bolton (1983), Simulation of the atmospheric response to soil moisture anomalies over Europe, *Quart. J. R. Met. Soc.*, **109**, 501–526.

Rowntree, P.R. (1984), Review of general circulation models as a basis for predicting the effect of vegegation change on climate, *Met. 0, 20, Technical Note.* II / 225.

Rowntree, P.R., W.F. Wilson and A.B. Sangster (1985), Impact of land surface variation on African rainfall in general circulation, *DCTN 30 Met. 0.*

Segal, M., R.A. Pielke, and Y. Mahrer (1983), On climatic change due to deliberate flooding Qatara Deppression (Egypt), *Climatic Change*, **5**, 73–83.

Shukla, J. and Y. Mintz (1982),Influence of land– surface evapotranspiration on the earth's climate, *Sciece*, **215**, 1498–1501.

Sud, Y.C. and M. Fennessy (1982), A study of the influence of surface albedo on July circulation in semi–arid regions using GLAS GCM, *J. Climatology*, **2**, 105–125.

Sud, Y.C. and M. Fennessy (1984), Influence of evaporation in semi–arid regions on the July circulation – a numerical study, *J. Climate*, **4**, 383–398.

Tada, K. (1985), Land–suface parameterization in JMA operational spectral model, *Proc. ISLSCP Conference, Rome, Italy, 2–6, Dec. 1985*, ESA. SP–248.

Warrilow, D.A (1985), The sensitivity of the UK Meteorological Office atmospheric general circulation model to recent changes in the parameterization of hydrology, *Proc. ISLSCP Conference, Rome, Italy, Dec. 1985*, ESA. SP– 248, 143–149.

Yamazaki, K. (1986), The sensitivity experiment to land–suface boundary conditions with the M.R.I. G.C.M. *Proc. ISLSCP Conference, Rome, Italy December, 1985*, ESA. SP–248, 151–157.

Yeh, T.–C., R.T. Wetherald and S. Manabe (1984), The effect of soil moisture on the short–term climate and hydrology change–a numerical experiment, *Mon. Wea. Rev.*, **112**, 474–490.

# STABILITY OF TREE/GRASS VEGETATION SYSTEMS

Peter S. Eagleson

Department of Civil Engineering
Massachusetts Institute of Technology, Cambridge, Massachusetts

Abstract. The average annual water balance of tree/grass vegetation systems is modeled as an interactive competition for water and energy. Ecological optimality hypotheses are introduced which allow specification of the fractional area covered by woodland canopy and by grass canopy under conditions of natural equilibrium. Three equilibrium states are found. Two are monocultures, i.e., grassland and closed forest which are shown to be unstable with respect to perturbations of canopy cover. The third, a tree/grass mixture, is shown to be stable with respect to perturbations of canopy cover but metastable with respect to climate change.

## Introduction

We consider a mixed formation of grasses and woody plants which is often referred to as "savanna" [e.g., Dansereau, 1957]. In the absence of photographs, an accurate visual image is evoked by Walter's [1973] description " ... homogeneous grasslands upon which woody plants are more or less evenly distributed". This "even distribution" of the woody plants is indicative of their control by soil moisture availability and is a key to the conceptualization used here. The climate of savanna vegetation systems is characterized by marked seasonality in moisture availability [Monasterio and Sarmiento, 1975]. In a survey of the literature describing savannas in South Africa, West Africa, Sudan and South America, Segarra [1983] found the mean rainy season length (i.e., fraction of year) $m_\tau$ to vary from 0.46 to 0.67.

## Model of Tree/Grass Equilibrium

We first seek the roots of the time-averaged conservation equation for soil water as described in detail by Eagleson and Segarra [1985]. The average annual water balance is modeled one dimensionally as shown schematically for a unit landsurface area in Figure 1. The area of bare soil is indicated by $M_s$. The area shaded by grass (with the sun directly overhead) is $M_g$ and that by the trees is $M_w$. This is often called the "projective foliage cover" or the "canopy density". Following the observations of Sarmiento [1984] we assume no runoff.

We further assume that the taller canopy suppresses the atmospheric water vapor transport capacity $e_p$ that is effective for the shorter canopy. Accordingly and using a simple linear relation, we have for the grass

$$e_{p_g} = (1 - M_w)\, e_{p_w} \tag{1}$$

and for the bare soil

$$e_{p_s} = (1 - M_g)\, e_{p_g} \tag{2}$$

in which the subscripts w, g and s refer to trees, grass and bare soil, respectively, and the $e_p$'s are average annual rates (cm sec$^{-1}$). Actual average evapotranspiration rates $e_T$ (cm sec$^{-1}$) are related to the potential rates $e_p$. For the vegetation components,

$$e_T = k_v e_p \tag{3}$$

where $k_v$ (dimensionless) is called a "plant coefficient", and for the bare soil [see Eagleson, 1978a]

$$e_{T_s} = J\, e_{p_s} = J_o s_o^{5/2} \tag{4}$$

where $s_o$ = time and space average soil moisture concentration in the upper layer (the hydroclimatologic state variable), where J (dimensionless) is the evaporation efficiency, and where $J_o$ (cm sec$^{-1}$) is defined in terms of the properties of the soil, water and climate.

We assume the climate to be seasonal and the soil to be divided into two layers. The top layer contains soil moisture only during the rainy fraction $m_\tau$ of the year and is thus assumed to be the sole source of moisture for the

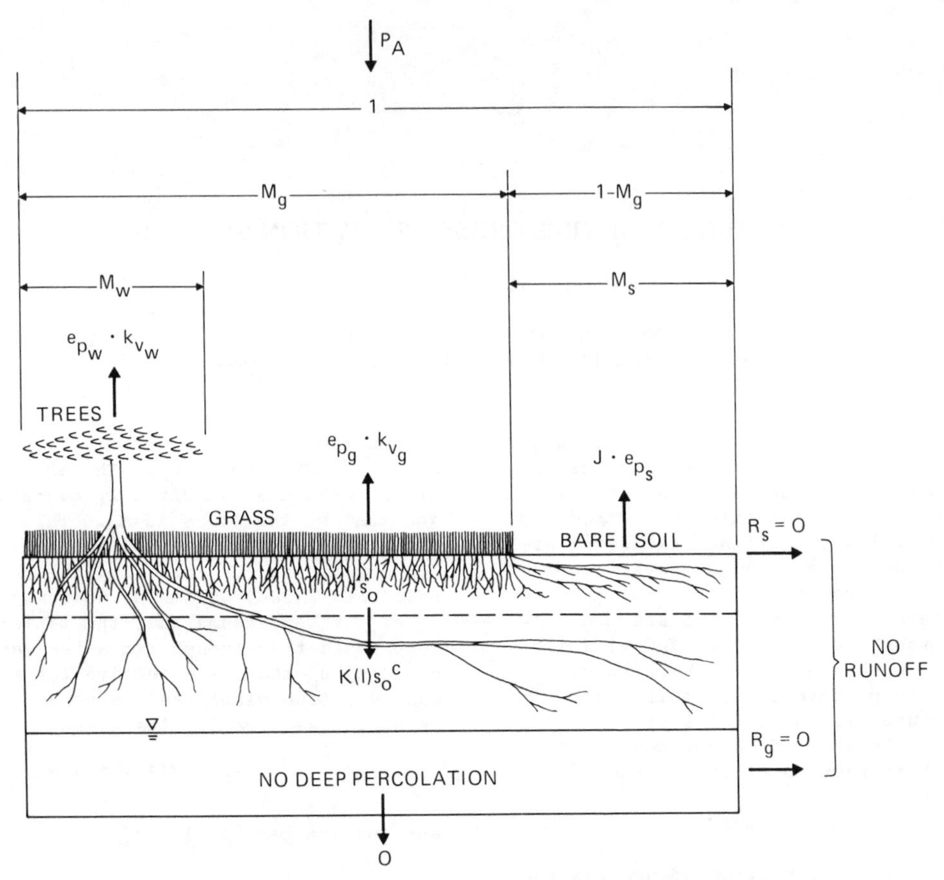


Fig. 1. Annual water balance of tree/grass savanna.

grass. Soil moisture unused by the grass percolates from the upper to the lower soil layer during the rainy season where it provides fraction $\alpha$ of the moisture transpired annually by the trees. The remaining fraction $1 - \alpha$ of the trees' annual moisture use comes from the upper soil layer via shallow roots. We assume the trees are spaced so that they use all the percolated water. On an annual basis we express this as

$$m_\tau K(1) s_o^5 = \alpha M_w m_\delta e_{p_w} k_{v_w} \tag{5}$$

in which

$m_\delta$ = fraction of year during which soil moisture in lower layer can support transpiration by trees;

$K(1)$ = saturated hydraulic conductivity of upper layer soil, cm sec$^{-1}$.

Dimensionlessly, (5) becomes

$$M_w = (KS/\alpha)\ s_o^5 \tag{6}$$

where

$$K \equiv m_\tau\ k_{v_g}/(m_\delta k_{v_w}) \tag{7}$$

and

$$S \equiv K(1)/(e_{p_w} k_{v_g}) \tag{8}$$

With reference to Figure 1 the average annual water balance can now be written in the dimensionless form

$$G = \frac{M_w}{K} + M_g(1 - M_w) + (1 - M_g) R\ s_o^{5/2} \tag{9}$$

where

$$G \equiv P_A/(m_\tau\ e_{p_w} k_{v_w}) \tag{10}$$

and

$$R \equiv J_o/(e_{p_w} k_{v_g}) \tag{11}$$

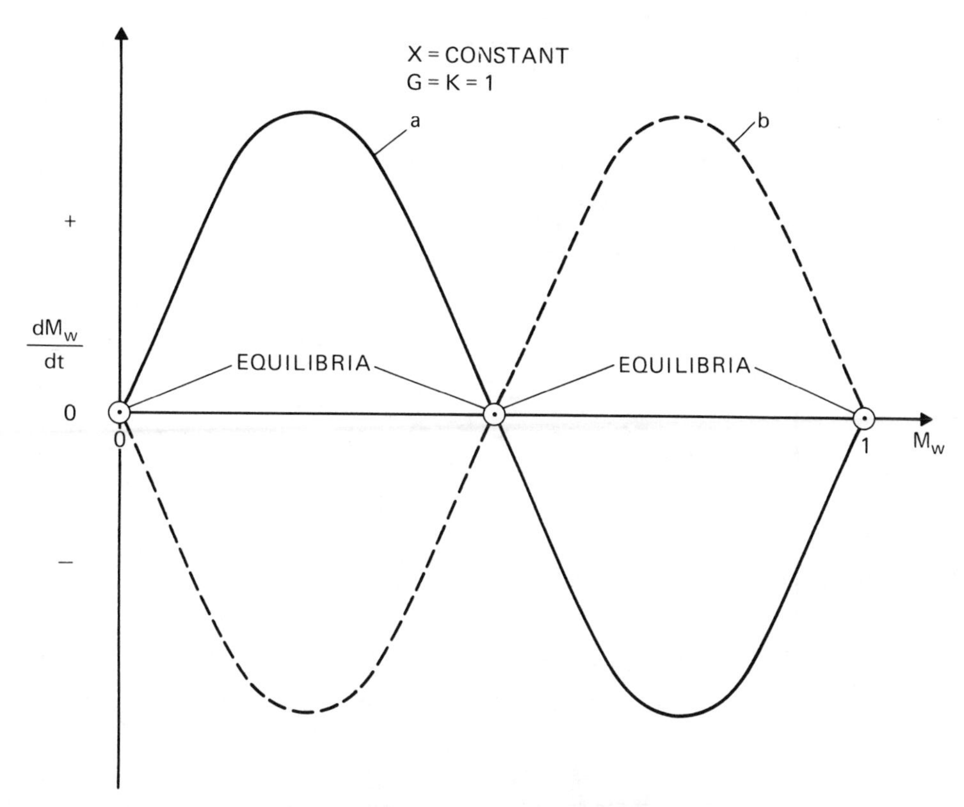


Fig. 2. Stability of savanna woodland equilibria.

in which $P_A$(cm) is the average annual precipitation rate.

Considering $k_{v_g}$ and the product $m_\delta k_{v_w}$ to be known parameters of the system there are three unknowns, $s_o$, $M_w$ and $M_g$ but only two equations, (6) and (9). Invoking the hypothesis [Eagleson, 1978b, 1982] that natural vegetation systems will attempt to minimize water demand stress, we add the third equation

$$\partial s_o/\partial M_g = 0. \tag{12}$$

Eqs. (6), (9) and (12) have a physically feasible solution only for

$$G = K = 1 \tag{13}$$

for which

$$M_w = [-X + (X^2 + 1)^{1/2}]^2 \tag{14}$$

where

$$X = R/[2(S/\alpha)]^{1/2} \tag{15}$$

with $M_g$ being arbitrary.

There are thus three equilibrium states of the savanna system. The condition K = 1 implies that a closed forest ($M_w$ = 1, $M_g$ = 0) will use the the same amount of water annually as will grassland ($M_w$= 0, $M_g$ = 1), and G = 1 specifies that $P_A$ is exactly sufficient for both of these extreme states. An intermediate tree/grass equilibrium is given by (14).

## Importance of Seasonality

Note that as seasonality vanishes (i.e., $m_\tau$ and hence $m_\delta$ approach unity) equations (7), (10) and (13) give

$$k_{v_g}/k_{v_w} \rightarrow 1$$

and

$$P_A/(e_{p_w} k_{v_g}) \rightarrow 1$$

which can be satisfied either for $M_w$ = 0 or for $k_{v_g} = k_{v_w}$. Since under water-limited conditions $k_{v_g} > k_{v_w}$ [Eagleson and Segarra, 1985], the latter condition will occur only with unlimited moisture when $k_{v_g} = k_{v_w} = 1$. Under such conditions the vegetation will probably seek to optimize biomass rather than soil moisture [Eagleson, 1982] and trees will replace the grass. This situation is outside the range of the current analysis.

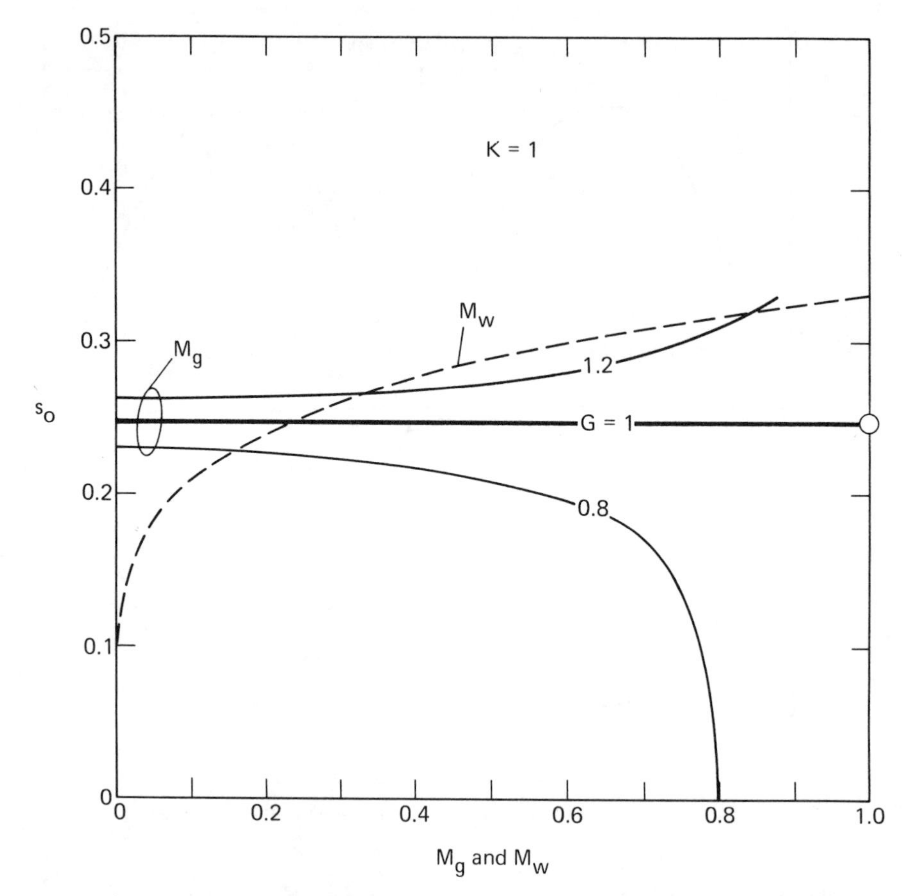


Fig. 3. Sub-optimal equilibria; R = 24, $S/\alpha$ = 250, K = 1.

We thus see that seasonality of the moisture supply is a necessary condition to have an equilibrium tree/grass system.

### Stability to Vegetation Perturbations

We begin by considering perturbations to the equilibrium vegetation under the assumption that these cause no change in the parameters of the climate and/or soil.

In the phase plane diagram of Figure 2 the three equilibria in M are located along the $dM_w/dt = 0$ axis. The solid and dashed lines are qualitative representations of the only two possible forms of $dM_w/dt = f(M_w)$. The slope of $f(M_w)$ at $dM_w/dt = 0$ determines the stability of the associated equilibria. Writing the conservation equation for water in the lower soil layer where $Z_w$ is the water table elevation and differentiating with respect to time gives

$$\alpha m_\delta e_{p_w} k_{v_w} \frac{dM_w}{dt} = 5 m_\tau K(1) s_o^4 \frac{ds_o}{dt} - n \frac{d^2 Z_w}{dt^2} \quad (16)$$

in which n is the soil porosity. From an order-of-magnitude analysis we assume the absolute value of the first term on the right hand side of (16) exceeds that of the second. The sign of $dM_w/dt$ will then be that of $ds_o/dt$.

Should we be at the $M_w = 0$, $M_g = 1$ equilibrium and somehow decrease $M_g$ we replace grass, transpiring at the potential rate, by bare soil of lower evaporation efficiency. The soil moisture $s_o$ will thus rise making $ds_o/dt$, and hence $dM_w/dt$ positive. Should we be at the $M_w = 1$, $M_g = 0$ equilibrium and somehow decrease $M_w$ we will have moisture loss by bare soil evaporation and from growth of grass thus decreasing $t_o$ and making $dM_w/dt$ negative. This reasoning establishes $f(M_w)$ as the solid curve (a) in Fig. 2 and the intermediate tree/grass equilibrium is seen to be unconditionally stable to perturbations in $M_w$. If fire should decrease $M_w$ from this equilibrium value, $M_w$ would be in a region of positive $dM_w/dt$ and the system would return to equiilibrium. Such restoration would not follow perturbations of the $M_w = 1$ and $M_w = 0$ equilibria and they are therefore unstable equilibria under this analysis. Slash-and-burn agriculture would thus be expected to create savanna from forest and grassland under the appropriate (G = K = 1) climatic conditions.

### Stability to Climate Change

Consider Fig. 3 in which for K = 1 and for representative constant values of R and $S/\alpha$ various solutions of the equilibrium equations (6) and (9) are presented. The dashed curve represents $M_w(s_o)$ as given by (6) and is independent of the climate parameter G. The solid curves represent $M_g(s_o;G)$ as given by (6) and (9). The horizontal line G = 1 is the only solid line that also satisfies (12) and thus its intersection with the dashed line represents the intermediate tree/grass equilibrium discussed above.

For G > 1 we see that $M_g$ increases with $s_o$ until $s_o$ reaches the value ($s_o$ = 0.33 for G = 1.2) at which $M_w$ becomes unity. At this point $M_g$ is discontinuous since there, due to (1), $M_g$ must drop to zero. We note therefore that as $s_o$ increases under nonoptimal G > 1, the equilibrium tree density is driven to unity and the equilibrium grass density to zero. The observation of savanna for G > 1, K = 1 must therefore be due to interference by another agent such as fire.

For G < 1 we see that $M_g$ decreases with increasing $s_o$ until the grass disappears. At $M_g$ = 0, $s_o$ reaches its maximum value which is less, as is the associated $M_w$, than that for optimum G = 1. We should therefore not expect to find closed forest ($M_w$ = 1) where G < 1 and K ≈ 1.

We thus conclude that if we accept the ecological optimality criterion of (11) as the determinant of tree/grass equilibrium, climates with G > or < 1 cannot have stable tree-grass savannas. For G > 1 pressure to develop toward increasing $s_o$ will lead eventually to closed forest while for G < 1 this pressure will lead to low density trees and an absence of grass. The flatness of the driving gradient $\partial s_o/\partial M_g$ in the small $M_w$ range of most savannas indicates, however, that there is ample opportunity for other stress-producing factors (such as nutrition, pests, fire, etc. that have been neglected in this water-based analysis) to play a stabilizing role.

This suggests the monitoring of savanna tree density changes as a visible indicator of climatic change.

### Conclusions

When the tree/grass (i.e., savanna) vegetation system is modeled as a competition for water and solar energy there are three equilibrium states: closed forest, grassland, and a tree-grass mixture. Only the last of these appears stable with respect to perturbation in the vegetation components but it is metastable with respect to climate change.

*Acknowledgments*. This work was supported by the National Science Foundation under Grant No. ATM-8114723. Prior publication of these results has been made first, in a more complete form by the American Geophysical Union [Eagleson and Segarra, 1985], and second, essentially as presented here by NASA [Eagleson, 1986].

### References

Dansereau, P., *Biogeography, An Ecological Perspective*, Ronald, New York, 1957.

Eagleson, P. S., Climate, soil, and vegetation; 4. The expected value of annual evapotranspiration, *Water Resources Research*, *14*(5), 731-739, 1978a.

Eagleson, P. S., Climate, soil, and vegetation; 6. Dynamics of the annual water balance, *Water Resources Research*, *14*(5), 749-764, 1978b.

Eagleson, P. S., Ecological optimality in water-limited natural soil-vegetation systems; 1. Theory and hypothesis, *Water Resources Research*, *18*(2), 325-340, 1982.

Eagleson, P. S., Stability of tree/grass vegetation systems, in *Climate-Vegetation Interactions*, *NASA Conference Publication No. 2440*, [C. Rosenzweig and R. Dickinson, Eds.], Greenbelt, Maryland, pp. 149-155, 1986.

Eagleson, P. S., and R. I. Segarra, Water-limited equilibrium of savanna vegetation systems, *Water Resources Research*, *21*(10), 1483-1493, 1985.

Monasterio, M. and G. Sarmiento, A critical consideration of the environmental conditions associated with the occurrence of savanna ecosystems, in *Tropical America*, edited by F. B. Golley and E. Medina, Springer-Verlag, New York, 1975.

Segarra, R. I., Stability of natural savanna ecosystems, Civil Engineer thesis, Dept. of Civil Eng., Massachusetts Institute of Technology, Cambridge, MA, 1983.

Sarmiento, G., *The Ecology of Neotropical Savannas*, translated by O. Solbrig, Harvard Univ. Press, Cambridge, MA, 1984.

Walter, H., *Vegetation of the Earth*, vol. 15, Springer-Verlag, New York, 1973.

# Section V

# TROPICAL OCEAN AND GLOBAL ATMOSPHERE

## TOGA AND ATMOSPHERIC PROCESSES

Kevin Trenberth

National Center for Atmospheric Research, Boulder, Colorado

*Abstract.* A brief outline is given of the TOGA (Tropical Oceans and Global Atmosphere) Program along with a more detailed discussion of the relationship of the atmospheric circulation to sea surface temperatures (SSTs) in the tropics. The best known phenomenon that is part of TOGA is El Niño-Southern Oscillation (ENSO) and this paper focuses on the atmospheric component of ENSO. The similarities and differences among different ENSO events are reexamined as seen through a Southern Oscillation index and indices of SST in the tropical Pacific. Variations from event to event are marked and phase locking of ENSO events to the annual cycle is only weak. The 1986-87 ENSO has been especially anomalous with regard to timing. However, it provides an excellent illustration of the link between atmospheric convection and SSTs. The reasons why convection occurs where it does and the importance of warm water greater than 28°C are discussed. It is shown that there is a need to better understand the atmosphere-ocean links in the tropics and a need to measure more accurately and understand changes in SSTs.

### Introduction

The TOGA Program is an international program under the World Climate Research Program (WCRP) focused on the interannual variability of the Tropical Oceans and Global Atmosphere (TOGA) as a coupled climate system. The goals of TOGA are:

- To gain a description of the tropical oceans and the global atmosphere as a time dependent system, in order to determine the extent to which this system is predictable on time scales of months to years, and to understand the mechanisms and processes underlying its predictability.
- To study the feasibility of modeling the coupled ocean-atmosphere system for the purpose of predicting its variations on time scales of months to years.
- To provide the scientific background for designing an observing and data transmission system for operational prediction if this capability is demonstrated by coupled atmosphere-ocean models.

Formally, TOGA began in 1985 and consists of a decade-long program with several components.

1) Long-term observations: To obtain better global atmospheric data sets and tropical ocean data sets that document the phenomenon of interest and can be used to force models and to validate models.
2) Empirical studies
3) Modeling: Atmospheric, with specified sea surface temperatures (SSTs); oceanographic, with a specified surface atmosphere (or fluxes); and coupled ocean-atmosphere models.
4) Prediction: Statistical-dynamical, and using models.
5) Data Management

These aspects are dealt with in depth by WCRP (1985a, b) and National Academy of Sciences (1986).

The best known phenomenon relevant to TOGA is El Niño-Southern Oscillation (ENSO). In this paper the focus will be on the atmospheric component of ENSO and its link to the tropical SSTs and, in particular, why convection occurs where it does. Many recent studies have used the tool of compositing to bring out the common features of past ENSO events [e.g. Rasmusson and Carpenter, 1982]; [van Loon, 1984]; [van Loon and Shea, 1985, 1987]. Here we will emphasize the differences among the events and focus especially on recent developments, the 1986-87 ENSO event, which turns out to have some unique features, but which nicely illustrates the relationships between SSTs and atmospheric convection.

### El Niño-Southern Oscillation Events

ENSO events are made up of components from both the atmosphere and the ocean. The SO is the atmospheric component of ENSO. It has a time scale of 2-7 years [Trenberth, 1976] and consists of a global-scale, predominantly standing wave with centers of action in surface pressure over Indonesia and the tropical South Pacific, see Fig. 1 [Trenberth and Shea, 1987]. Within this large pattern, there are seasonal teleconnections to higher latitudes that tend to be strongest in winter. These teleconnections are thought to arise as Rossby waves propagating out of the tropics combined, especially in the Northern Hemisphere, with interactions with the stationary planetary waves and "basic state" flow [e.g. Branstator, 1985]. However, there is evidence indicating statistically significant although weak relations in the summer season too, as shown for instance by the correlation maps of [Trenberth and Paolino, 1981]. This is important in the current context because of the very strong anomalies during the northern summer of 1987 associated with the current ENSO.

The El Niño is thought of as the oceanic component of ENSO. It is manifest as major increases in SSTs in the tropical Pacific. The terminology is confusing, since El Niño was originally a warm current off the coast of Peru which penetrated south about Christmastime (hence Niño, the Christ child). The term has since become associated with the occasional unusually large warmings "El Niño events", which, in turn, are associated with basin scale anomalously high SSTs. Strictly speaking, the term is used only to refer to the original South American coastal phenomenon, but in the ENSO context it has come to mean a basin-scale warming of the tropical Pacific.

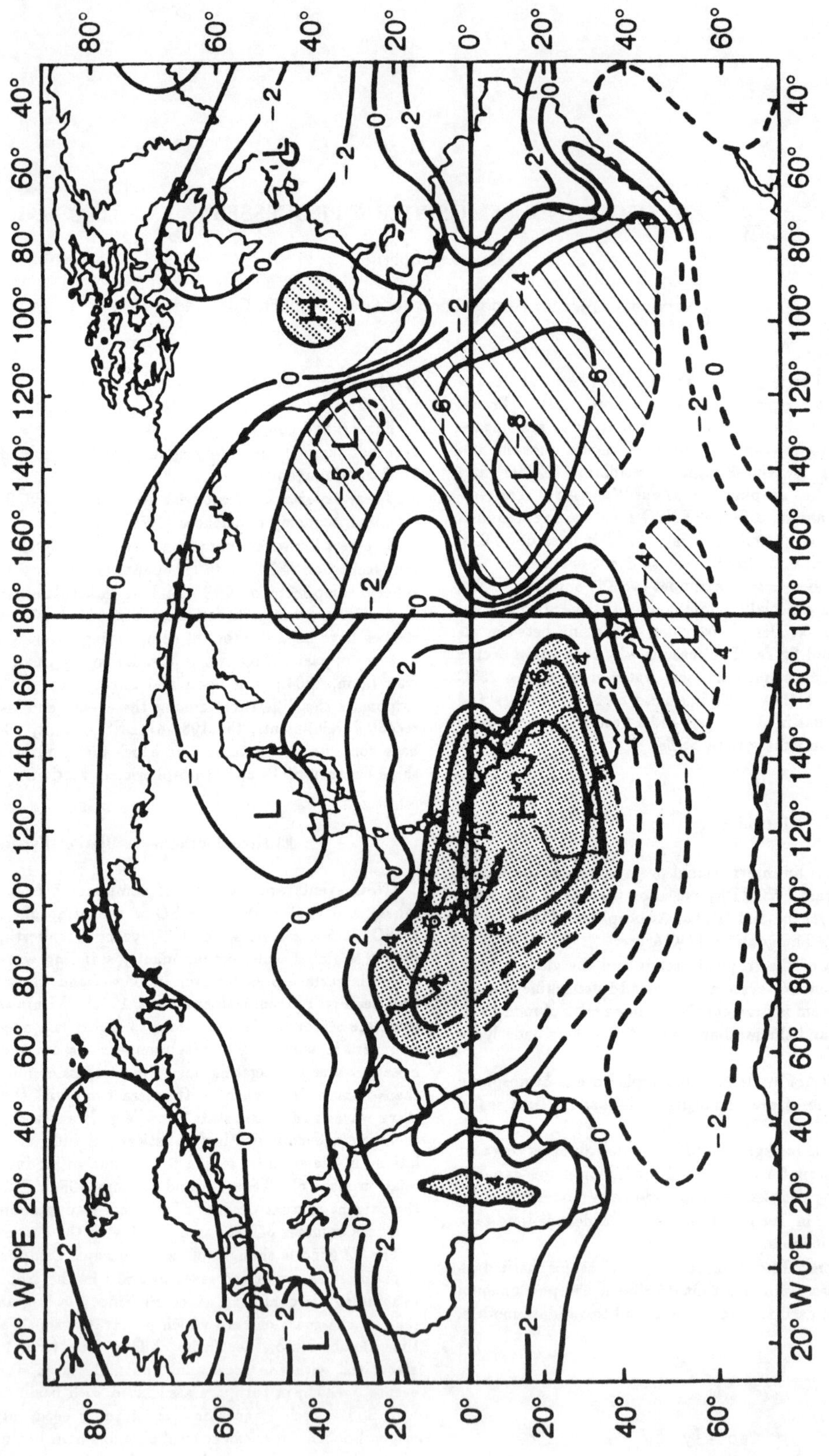


Fig. 1. Correlations of the annual mean sea level pressures with Darwin, from Trenberth and Shea (1987).

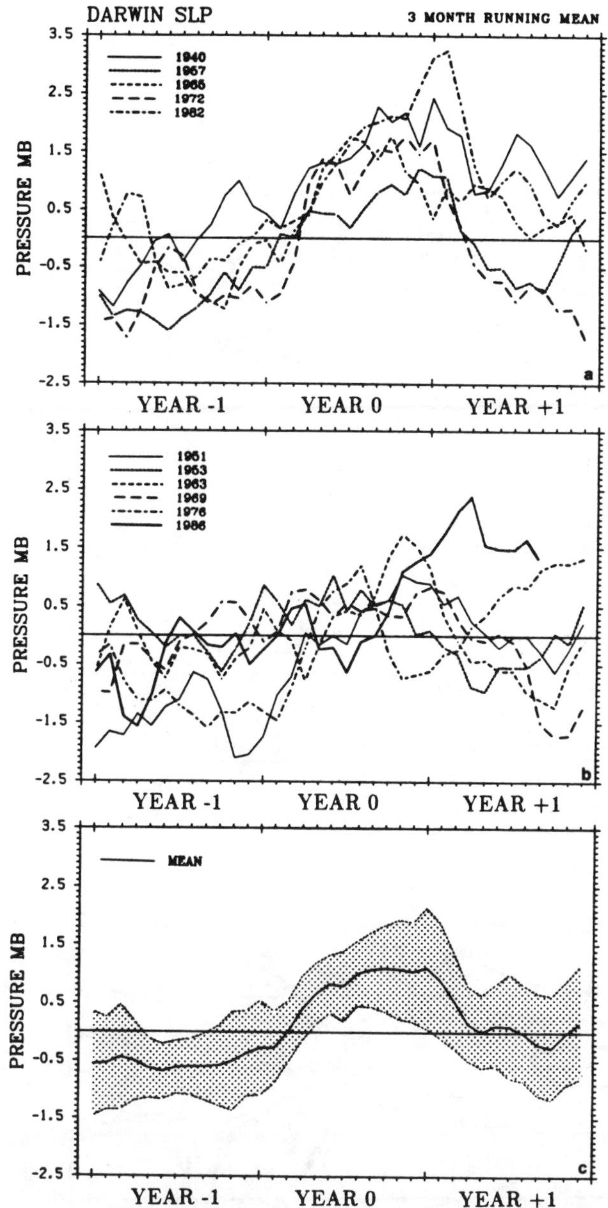


Fig. 2. Sequences of anomalies in sea level pressure at Darwin over all El Niño events since 1939. Each event is labelled by the central year (year 0). The larger events are in a, smaller events in b, and the composite mean (excluding 1986) plus and minus one standard deviation are in c.

Thus ENSO events are those in which both an SO extreme (Darwin pressure high, Tahiti low) and an EN occur together. A point made by Trenberth and Shea [1987] is that the SO and EN are not necessarily linked on a one-to-one basis. Above normal SSTs can occur without an SO swing, such as during 1979. In particular, South American coastal events are not very closely linked to the SO except during the major El Niños where SSTs in the central Pacific are affected. This is an aspect illustrated later here. ENSO events are also referred to as "Warm Events" by van Loon and Shea [1985, 1987].

We use Darwin as an index of the SO to show two points 1) the extent to which the phenomenon is phase locked to the annual cycle, as implied in many studies following the landmark paper by [Rasmusson and Carpenter, 1982]; and 2) the extent to which the events are similar or differ from one another through their life cycle. Subsequently, these points are reexamined using SST data.

Figure 2 shows the three year sequences of anomalies of Darwin sea level pressure, as 3-month running means, during all ENSO events since 1939. Each event is labelled by the central year (year 0) and extends from the year before (year-1) to the year following (year+1), using the terminology in Rasmusson and Carpenter. The large events are shown in the top panel, the smaller events in the central panel, and the composite plus and minus one standard deviation ($\sigma$) (excluding 1986) is given below. Of note is the lack of significance of the composite which is nowhere more than $2\sigma$ from the mean. This is due mostly to the inclusion of the smaller events in the central panel, which are of the order of one $\sigma$ events, both in the SO and SST fields. However, another factor is that the phase locking with the annual cycle is not very strong. Phase locking seems reasonable in Fig. 2a, but the weaker events do not follow that pattern and the 1986-87 sequence is the most unusual on record. It did not really begin until late in 1986, six months later than usual, and is the most extreme case on record in May-July of year +1. It has been associated with a very poor summer monsoon in India, one of the worst in several decades.

## Relationships with SSTs

Although many studies have used atmospheric and oceanic parameter anomalies in their statistical analyses, the link between the two is complex. Instead, the simplest link appears to be between the atmospheric anomalies and the total SST field, especially in models [Shukla and Wallace, 1983]. This is illustrated in a general way in Fig. 3 which shows the annual mean SST and outgoing longwave radiation (OLR) fields from [Shea, 1986] and [Janowiak et al., 1985]. Low OLR in the tropics is associated with high top clouds, and OLR is therefore an index of convection in the tropics. The stippled regions correspond to the high SSTs (water warmer than 28°C) and low OLR (the convergence zones). There is a close relationship over the oceans which is also present, although with migration north and south, in the individual seasons.

Low OLR values also occur over the warm tropical continents. In fact, the absence of a tropical continent in the Indonesian region is one thing that makes the Pacific unique. The warmest water on the globe, averaging over 29°C, occurs in the tropical western Pacific and is actually located slightly south of the equator, corresponding to the northernmost part of the South Pacific Convergence Zone (SPCZ).

The 1986-87 ENSO has been relatively modest by some standards. The SST anomalies have not been huge, or even close to the magnitudes experienced during 1982-83. However, they have been sufficient to significantly change the location of the warmest water.

This is illustrated in Fig. 4 which shows the total SST field and the anomalies for April 1987. The pattern is typical of that for January-July 1987. Positive anomalies along the equatorial zone of the Pacific are only ~1°C and are largest in the east. However, in the Pacific the warmest water of ~30°C is shifted from west of the dateline to ~170°W. The corresponding OLR fields in Fig. 5 show that the SPCZ is shifted north and east, as is characteristic of ENSO events [Trenberth, 1976], and merges with the ITCZ (InterTropical Convergence Zone) near 170°W. The maximum tropical convection in the western Pacific has shifted from 150°E to 170°W, as is shown also by the OLR anomaly field. The OLR anomalies exceed 45 W/m$^2$ and are highly significant. Note the close link with the total SST field but the relationship with the SST anomaly field is less clear.

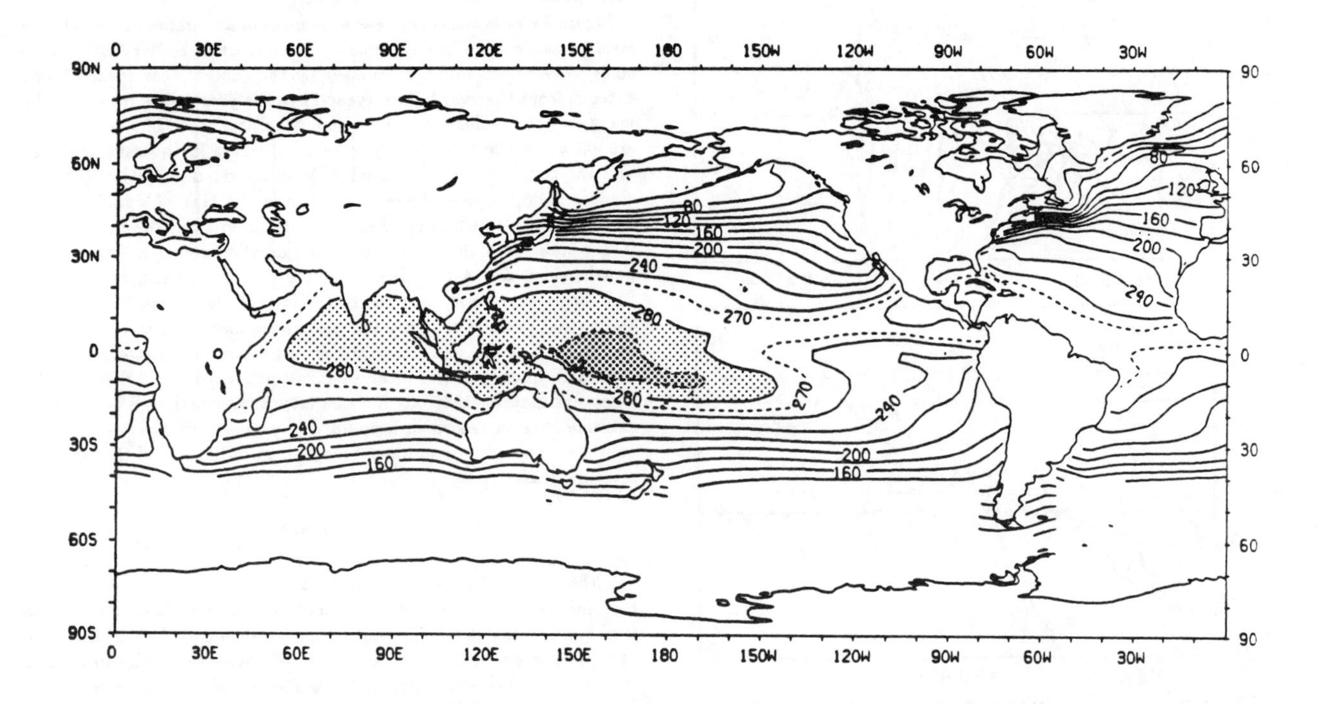


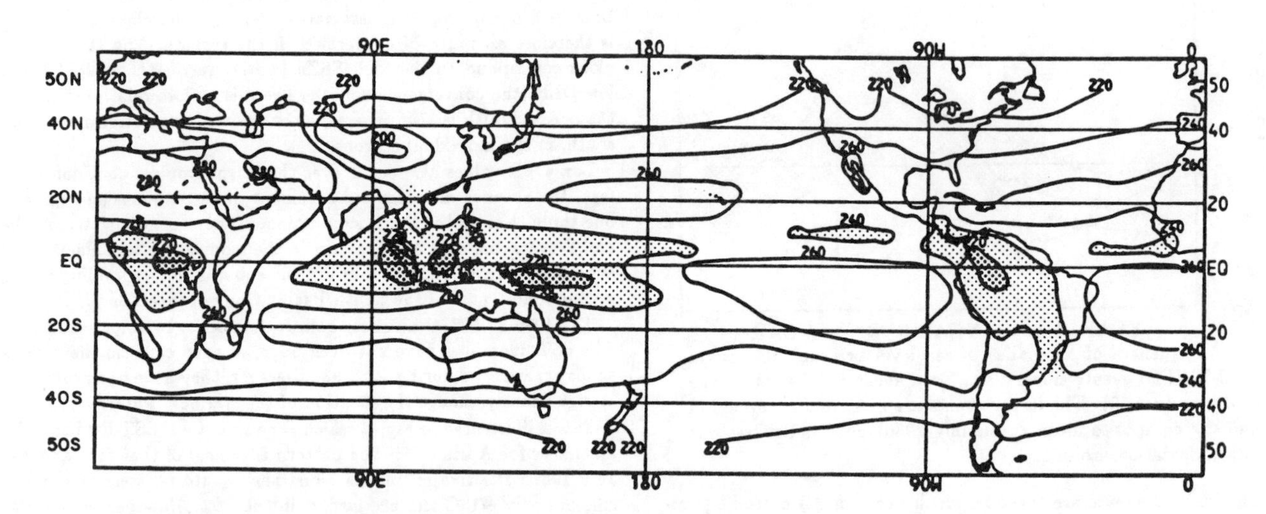


Fig. 3. (i) Annual mean SST for 1950-79 from Shea (1986). SSTs greater than 28°C are stippled. (ii) Annual mean OLR for 1974-83 from Janowiak et al. (1985). In the tropics, values less than 240 W/m$^2$ are stippled and correspond to regions of persistent high cloud.

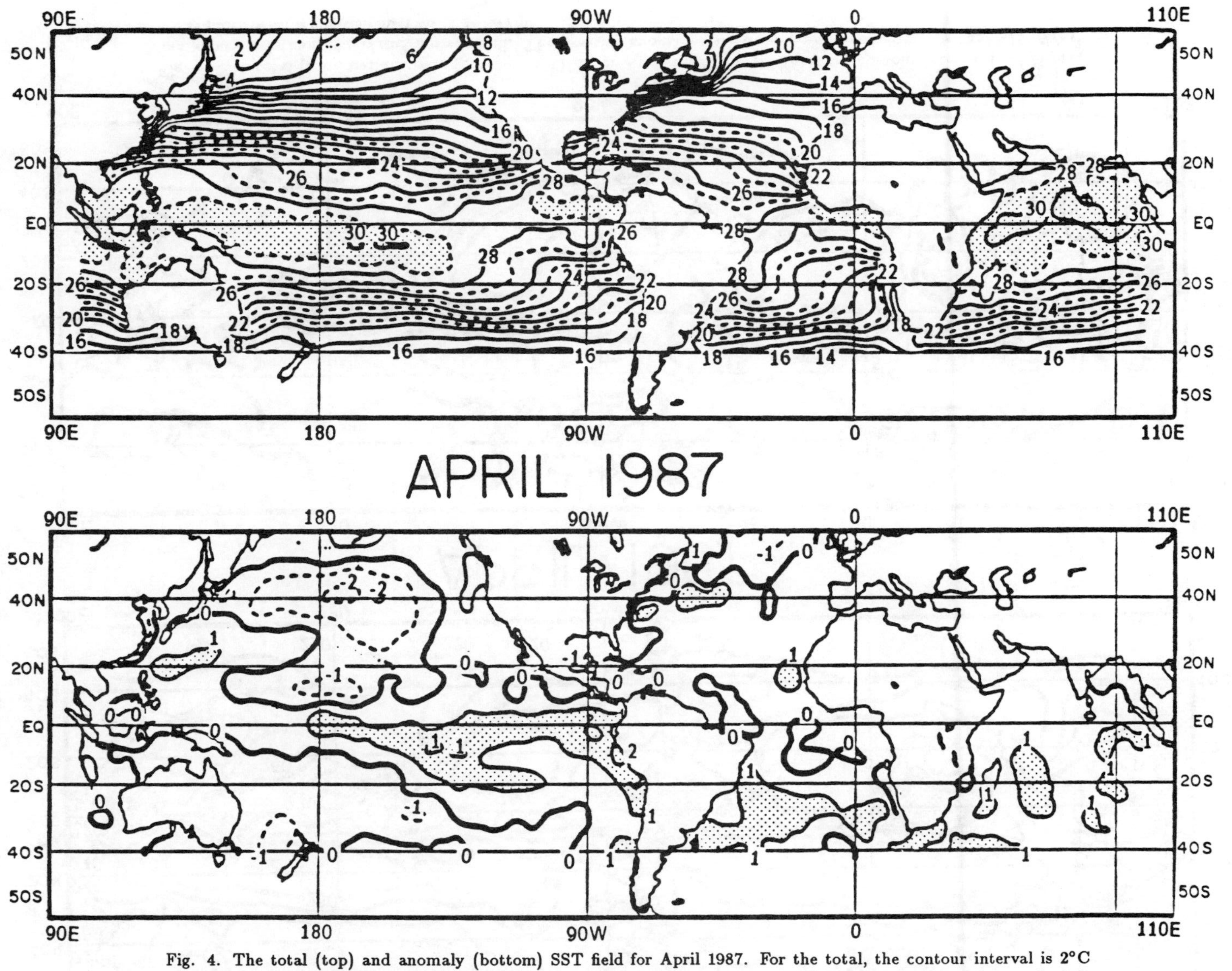


Fig. 4. The total (top) and anomaly (bottom) SST field for April 1987. For the total, the contour interval is 2°C and values exceeding 29°C are stippled. For the anomaly, the contour interval is 1°C and values > 1°C are stippled. Adapted from the *Climate Diagnostics Bulletin*, April 1987.

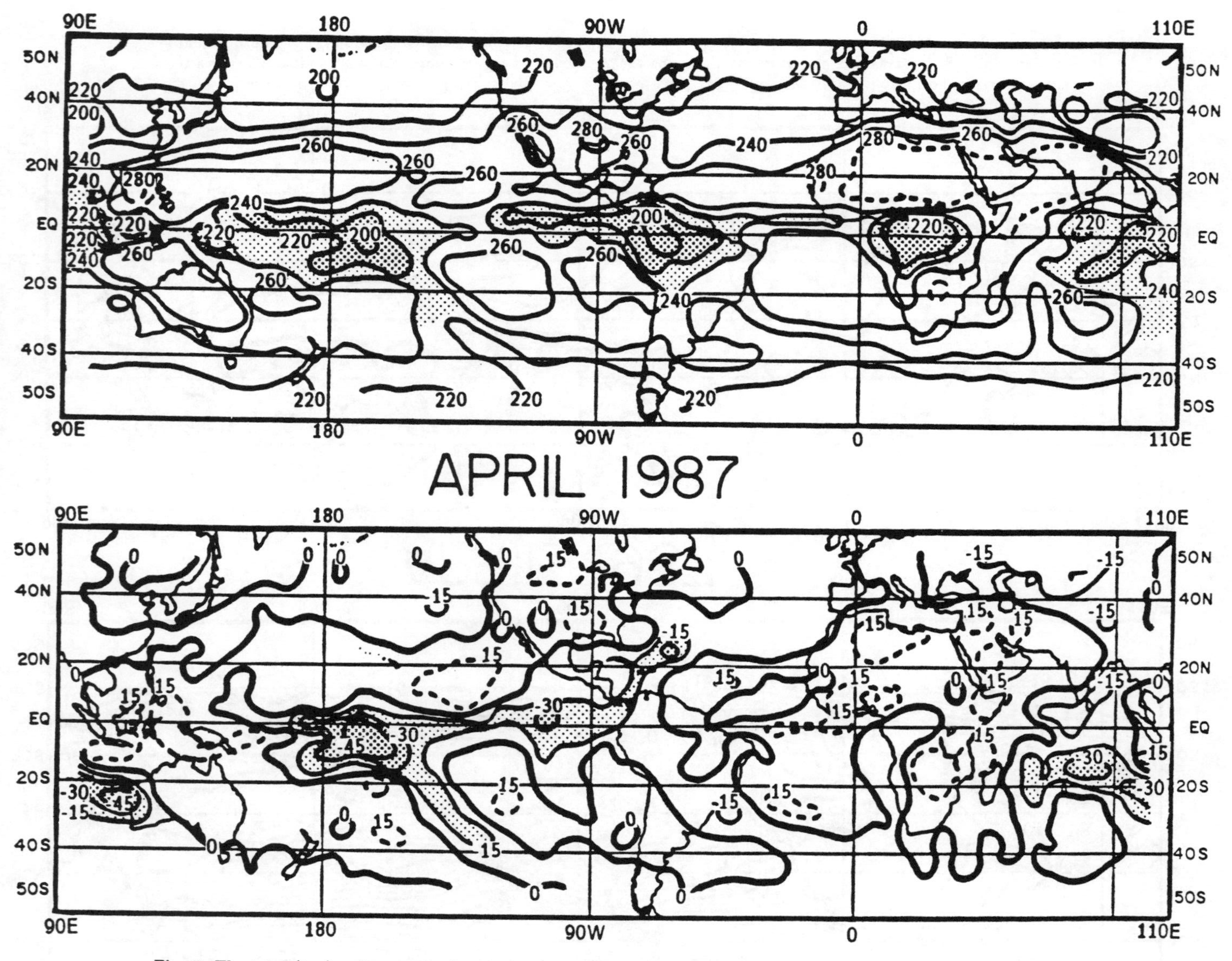


Fig. 5. The total (top) and anomaly (bottom) of OLR for April 1987. For the total the contour interval is 20 W/m$^2$ and values less than 240 W/m$^2$ are stippled. The positive anomalies are dashed and anomalies less than -15 W/m$^2$ are stippled with a contour interval of 15 W/m$^2$.

The evolution of the SST field in several critical regions is monitored by the U.S. National Meteorological Center's Climate Analysis Center (CAC) through averages over regions labeled Nino 1+2 (0-10°S, 90-80°W), Nino 3 (5°N-5°S, 150-90°W) and Nino 4 (5°N-5°S, 160°S-150°W), see Fig. 6.

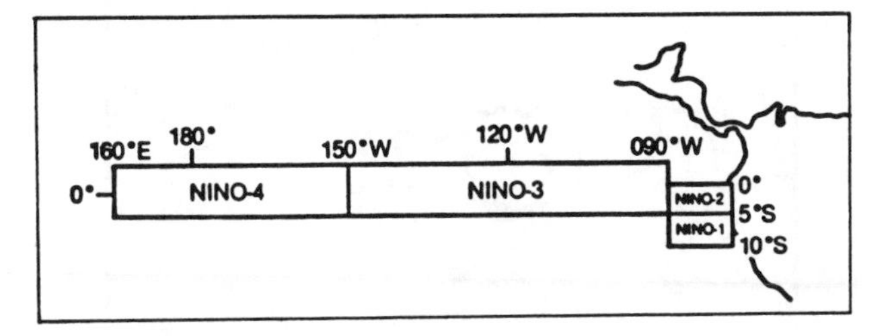


Fig. 6. Areas averaged to produce SST indices for the Niño 1+2 region and for the Niño 3 and Niño 4 regions.

In both Figs. 7 and 8 the SST anomalies in these three regions are shown for all the El Niño events since 1970 (which is the current limit to reliable values), plus estimates for the 1940 event for Nino 1+2 [from Rasmusson and Wallace, 1983]. The anomalies have been computed relative to the overall means 1970-1986 (which differ somewhat from those used in the CAC *Climate Diagnostics Bulletin*).

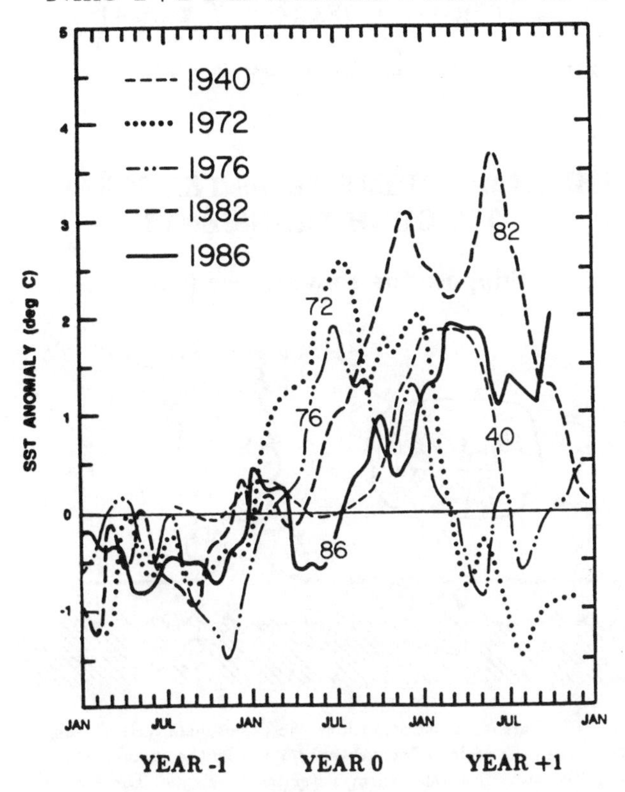


Fig. 7. Sequences of SST anomalies for the Niño 1+2 region, for all El Niño events since 1970 plus 1939-41.

As in Fig. 2, there is some evidence for similar evolution of SST anomalies in each event but the differences among the events are also marked. Once again, phase locking with the annual cycle is not very strong. The 1986-87 event is quite modest along the coast of South America (Fig. 7). Until the middle of 1987, it also featured mediocre anomalies in the Niño 3 region, but by July anomalies were exceeding those of the 1982-83 event. Although the anomalies in Niño 4 are small (Fig. 8), for 1987 they are the largest by far of all the ENSO events shown and, as seen from Fig. 4, they are sufficient to move the location of the warmest water into this region. Consequently, it is the modest anomalies in Niño 4 which turn out to be critical to the current event.

This makes a very strong case for knowing the details of the SST patterns well. It is especially important to know SSTs accurately, to better than 0.5°C, in regions of water warmer than 28°C since subtle changes can alter the region where the warmest water occurs by thousands of km. Gadgil et al. [1984] noted that deep convection mainly occurs when SSTs exceed 28°C.

## The Importance of the Warmest Water

Why is the warmest water, water at temperatures higher than 28°C, so important? An alternative question is why does the atmospheric convection occur where it does? We have implied from the above discussion that the atmospheric convection preferentially tends to occur over the region of warmest water. In other words, the atmospheric convergence zone tends to be so located. In order to understand this, at least qualitatively, there are a number of points that must be taken into account.

1) Mass continuity in the atmosphere implies that convection cannot occur everywhere. There are preferred scales.
2) There is a competition for organized convection. Since the warmest water tends to be ~29°C (Fig. 2), SSTs greater than about 28°C are needed before a region can be a competitor. Moreover, the region of warmest water is likely to "win" owing to effects arising from the Clausius-Clapeyron equation. The latter expresses the saturation mixing ratio $r_s$ as a nonlinear function of temperature. For SSTs of 20 and 30°C, at 1000 mb, $r_s$ increases from 15 to 28 g/kg. Moreover, the increase in $r_s$ per degree increase in SST at 30°C is 1.7 times that at 20°C. The consequences are illustrated in Fig. 9 and discussed by Neelin and Held [1987]. Convective instability is greatly and nonlinearly enhanced over warmer water owing to Clausius-Clapeyron effects on the gross moist static stability which leads to deeper convection and excites a positive feedback loop that is critical in determining the ultimate outcome (Fig. 9).
3) There is also a question concerning how the atmosphere 'feels' the ocean and thus knows where the warmest water is. This is best understood through a perturbation approach. Suppose that the convergence zone is moved for some reason, then the low level atmospheric flow that provides the convergence will immediately result in huge sensible and latent heat fluxes into the atmosphere in the vicinity of the warmest water (Fig. 10). The result is that convection tends to break out in that region and the convergence zone moves back over the warm water. Then the wind is reduced over the warm water and the surface fluxes actually become minimized, which explains why it is not possible to understand the answer to this question from a budget study standpoint. In the real atmosphere, there are many transients, ranging over time scales from individual convection, to easterly waves, to 40-50 day oscillations. We suspect, however, that the nature of the transients in atmo-

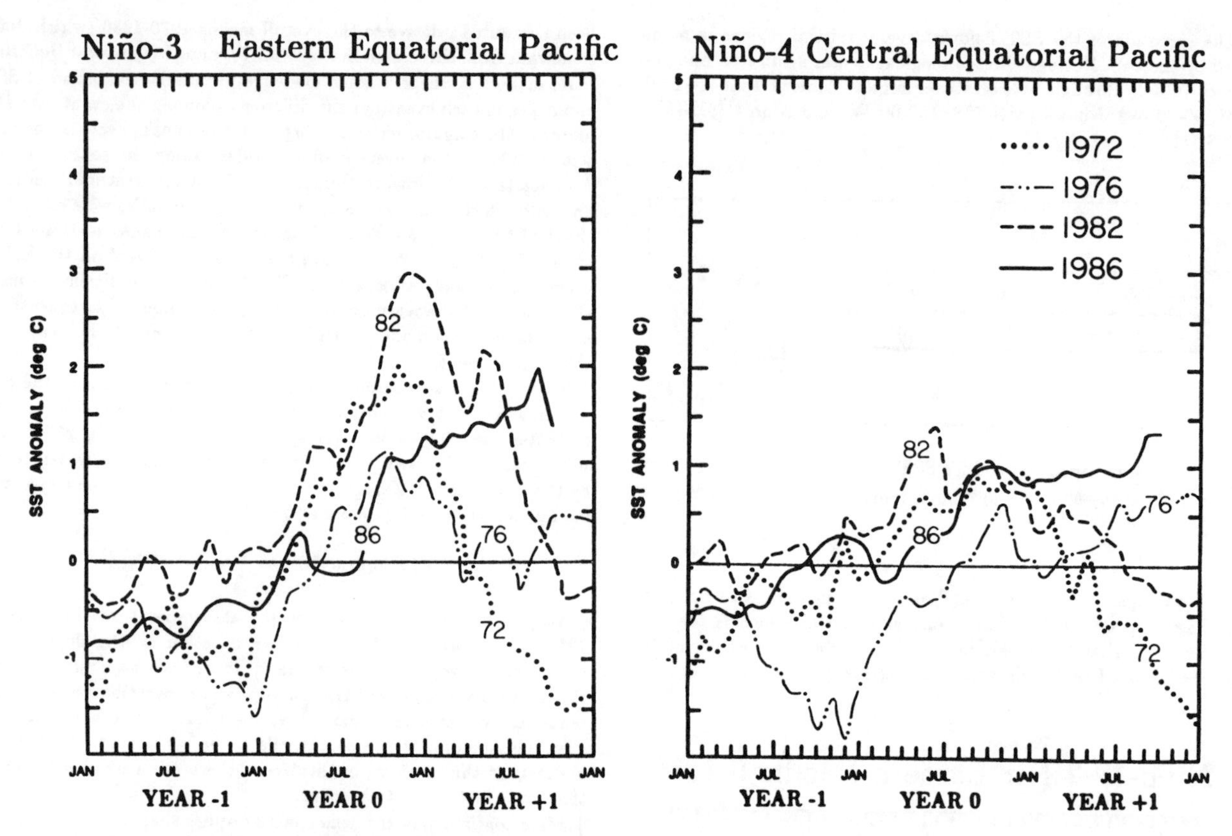


Fig. 8. Sequences of SST anomalies for Niño 3 and Niño 4 regions (see Fig. 6) for all El Niño events since 1970.

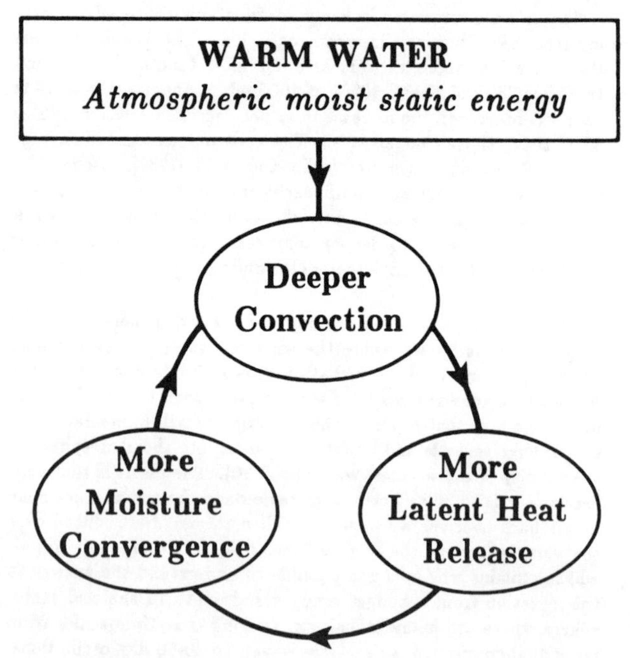


Fig. 9. Feedback cycle showing why warm water is important.

Fig. 10. Cartoon illustrating how the atmosphere feels the high SSTs when the region of convergence, given by the cumulus cloud, is shifted away from the warm water by transients. The vertical arrows indicate enhanced sensible and latent heat fluxes into the atmosphere and the thin arrow shows the atmospheric flow.

spheric general circulation models (GCMs) is rather different and GCM transients tend to be dominated by the "convective adjustment" process. The impression is that in GCMs, the convection is tied to the warmest water even more strongly than in the real atmosphere.

The main point to be made here then, is that the response of the atmosphere to SSTs is a highly nonlinear process. Since the area of fairly warm (>28°C) water is extensive a small change in SST can alter the configuration and is capable of producing a major shift in the location of the convergence zones with consequences for where the release of latent heat occurs. In turn, this changes the atmospheric forcing of Rossby waves and the teleconnections into mid-latitudes. The need for accurate SST fields, and the need to understand changes in SSTs can be seen to be paramount, and thus these are central to the TOGA program. However, it should be noted that other, in particular radiative, effects are also important.

Finally, since atmospheric winds drive the ocean and, as discussed here, SSTs are critical in determining the atmospheric response, it is essential to consider the coupled system in order to fully understand atmosphere-ocean interactions.

*Acknowledgments.* I am grateful to Vern Kousky and CAC for supplying me with the SST index values. The National Center for Atmospheric Research is sponsored by the National Science Foundation.

References

Branstator, G., Analysis of general circulation model sea surface temperature anomaly simulations using a linear model. Pt. 1. Forced solutions. *J. Atmos. Sci.*, *42*, 2225-2241, 1985.

*Climate Diagnostics Bulletin*, Climate Analysis Center, NOAA, Washington, D.C.

Gadgil, S., P. V. Joseph and N. V. Joshi, Ocean-atmosphere coupling over monsoon regions. *Nature, 312*, 141-143, 1984.

Janowiak, J. G., A. F. Kruger, P. A. Arkin and A. Gruber, Atlas of outgoing long-wave radiation derived from NOAA satellite data. *NOAA Atlas No. 6*, U.S. Dept. of Commerce, 44 pp, 1985.

National Academy of Sciences, *U.S. participation in the TOGA Program: A research strategy.* National Academy Press, 24 pp, 1986.

Neelin, J. D. and I. M. Held, Modeling tropical convergence based on the moist static energy budget, *Mon. Wea. Rev., 115*, 3-12, 1987.

Rasmusson, E. M. and T. H. Carpenter, Variations in tropical sea surface temperature and surface wind fields associated with the Southern Oscillation/El Niño, *Mon. Wea. Rev.*, *110*, 354-384, 1982.

Rasmusson, E. M. and J. M. Wallace, Meteorological aspects of the El Niño/Southern Oscillation. *Science*, *222*, 1195-1202, 1983.

Shea, D. J., Climatological Atlas 1950-79: Surface air temperature precipitation, sea-level pressure, and sea-surface temperature (45°S-90°N), *NCAR Tech. Note NCAR/TN-269+STR*, 1986.

Shukla, J. and J. M. Wallace, Numerical simulation of the atmospheric response to equatorial sea surface temperature anomalies. *J. Atmos. Sci.*, *40*, 1613-1630, 1983.

Trenberth, K. E., Spatial and temporal variations of the Southern Oscillation. *Quart. J. Roy. Meteor. Soc.*, *102*, 639-653, 1976.

Trenberth, K. E. and D. A. Paolino, Jr., Characteristic patterns of variability of sea level pressure in the Northern Hemisphere. *Mon. Wea. Rev.*, *109*, 1169-1189, 1981.

Trenberth, K. E. and D. J. Shea, On the evolution of the Southern Oscillation, *Mon., Wea. Rev.*, *115*,3078-3096, 1987.

van Loon, H., 1984: The Southern Oscillation. Part III. Associations with the trades and with the trough in the westerlies of the South Pacific Ocean. *Mon. Wea. Rev.*, *112*, 947-954, 1984.

van Loon, H. and D. J. Shea, The Southern Oscillation. Part IV. The precursors south of 15°S to the extremes of the oscillation. *Mon. Wea. Rev.*, *113*, 2063-2074, 1985.

van Loon, H. and D. J. Shea, The Southern Oscillation. Part VI. Anomalies of sea level pressure on the Southern Hemisphere and of Pacific sea surface temperature during the development of a Warm Event. *Mon. Wea. Rev.*, *115*, 370-379, 1987.

WCRP, Scientific plan for the Tropical Ocean and Global Atmosphere Programme. *WCRP No. 3*, WMO/TD No. 64, 147 pp, 1985a.

WCRP, International conference on the TOGA Scientific Program. *WCRP No. 4*, WMO/TD No. 65, 1985b.

# TOGA REAL TIME OCEANOGRAPHY IN THE PACIFIC

David Halpern

Earth and Space Sciences Division, Jet Propulsion Laboratory
California Institute of Technology, Pasadena, CA 91109

Abstract. Reliable estimates of the evolution of large scale sea surface temperature (SST) variations up to several months in advance is a primary goal of the 1985-94 Tropical Oceans & Global Atmosphere (TOGA) Program. Since the beginning of the TOGA, significant innovative accomplishments include (1) an increase in the quantity of in situ data and efficiencies of data management, (2) rapid distribution of real time ocean products, (3) effective utilization of global observations from satellites, and (4) assimilation of data into an ocean general circulation model to simulate monthly mean features of upper ocean thermal and flow fields. TOGA accomplishments are demonstrated with a discussion of oceanographic conditions during June 1987.

## Introduction

Episodes of anomalous warm surface water, which typically exist for about 1-year, occur in the equatorial Pacific Ocean at irregular intervals every 4-7 years. During the extremely warm episode of 1982-83, the monthly mean equatorial sea surface temperature (SST) rose above its climatological-mean monthly value by as much as 4°C. This high SST covered a wide area (perhaps 40° longitude by 10° latitude) centered approximately along the equator, intensified the Hadley circulation, and displaced the Walker circulation (Rasmusson and Wallace, 1983). The upward, eastward slope of the thermocline along the equator decreased (Halpern, 1987). Sea level decreased in the western equatorial Pacific, and increased by more than 20 cm in the eastern region (Wyrtki, 1984). The strengths of the eastward flowing Equatorial Undercurrent and westward flowing South Equatorial Current were substantially reduced (Halpern et al., 1983; Halpern, 1987). The zone of intense atmospheric convection, which normally resided west of about 160°E, moved eastward and traversed the entire width of the Pacific (Liu, 1988).



Episodes of high SST in the equatorial Pacific are strongly correlated with regional and, oftentimes, global atmospheric circulation changes (Rasmusson and Wallace, 1983). The decade-long (1985-94) Tropical Ocean & Global Atmosphere (TOGA) program, which is one of six subprograms of the World Climate Research Program, was formed to focus upon the coupled ocean-atmosphere system. The primary aim of the TOGA program is the development of an operational capability for dynamical prediction up to several months in advance of the time averaged (month-to-season) anomalies of the coupled tropical ocean-atmosphere system.

Variations of SST in low latitudes are related to the responses of upper ocean thermal and flow fields to large scale changes in surface winds, in addition to local influences due to air-sea momentum and surface heat fluxes. Prediction of the onset of large scale, long period ocean-atmosphere interactions up to several months in advance requires knowledge of the evolution of equatorial SST variations on time scales of days to a month. This time scale is short compared to the one applicable for middle latitudes. Philander (1979) reported that basin wide density gradients in the upper portion of a resting ocean would be established in about a decade in middle latitudes and only weeks near the equator. Remote wind effects in the tropical ocean are important on time scales of weeks because of the rapid propagation of dynamic signals, such as Kelvin wave motion along a narrow equatorial zone (Knox and Halpern, 1982).

At the time of onset of El Nino of 1982-83, surface wind measurements from the World Meteorological Organization volunteer observing ship (v.o.s.) network and SST observations from v.o.s. and satellite radiance retrievals were the only oceanographic parameters recorded throughout the equatorial Pacific and distributed as data or data products in (oceanographic) real time. Within the ocean sciences, real time means approximately 30-days (sometimes up to 60-days) of measurement. Expendable bathythermograph (XBT) measurements of the vertical profile of temperature between the surface and about 450 m were made from a small subset of about 25 ships within the vast v.o.s.

network, but five years ago these data were not distributed in real time. To achieve the TOGA objective of dynamical prediction up to several months in advance, two activities required development: (1) an operational capability of recording, transmitting, and distributing data on sea level, subsurface thermal and flow fields, and net air-sea heat and momentum fluxes; and (2) assimilation of oceanographic data into an operational ocean general circulation model (until a reliable interactive ocean-atmosphere general circulation model becomes available) for simulation and prediction of oceanographic conditions.

It was recognized immediately by TOGA scientists that the oceanographic data base required improvements. One of the perpetual contributions of TOGA will be the substantial increases in quantity and quality of the subsurface oceanographic fields. TOGA fostered an awareness of real time oceanography and of prediction of large scale, long period SST variations and related features such as thermocline depth, zonal slope of thermocline along equator, and anomalous behavior of surface current. This situation was virtually nonexistent in oceanography before TOGA. This too is an everlasting contribution of TOGA. As a result of the emphasis upon rapid transmission of TOGA data, a number of monthly publications or bulletins regularly display oceanographic data. Examples are listed in Table 1.

As an indication of the progress due to TOGA, many oceanographic aspects of the evolution of the warm episode of 1986-87 were described every month from its beginning via data products displayed in several publications (Table 1). This innovative development of real time viewing of oceanic conditions will be discussed in this paper. The Pacific Ocean is emphasized because distribution of oceanographic data products in real time has developed more rapidly there than in the Atlantic or Indian Oceans. This paper is not an exhaustive treatment of all oceanographic research within the TOGA program.

## TOGA Real Time Oceanography

The fullness of TOGA real time oceanography was demonstrated at the XIX General Assembly of the International Union of Geodesy and Geophysics (IUGG) in August 1987 when a preliminary version of this paper was presented. Attention was focused upon June 1987 oceanographic conditions because real time data products become available with a delay of about 1-month, making June the period closest in time to the IUGG Assembly when data would be available. Also, the warm episode in the Pacific, which began in 1986 (Bergman, 1987) was continuing, which provided an opportunity for discussion of ocean dynamics associated with this moderate El Nino.

The 16 June - 1 July 1987 averaged SST anomaly pattern (Figure 1) resembled the August - October anomaly distribution of the composite El Nino described by Rasmusson and Carpenter (1982), but the 16 June - 1 July 1987 maximum amplitude of 2°C was more than 0.5°C larger. Levitus (1987) showed that the magnitude of the SST anomaly was dependent upon the climatology and data analysis tech-

TABLE 1. Regularly occurring bulletins documenting upper tropical ocean and global/regional surface meteorological conditions, and addresses to secure addtional information.

| Bulletin | Address |
|---|---|
| 1. Climate Diagnostic Bulletin | Climate Analysis Center/NMC, National Weather Service, National Oceanic and Atmospheric Administration, Washington, DC 20233, UNITED STATES OF AMERICA |
| 2. Climate Monitoring Bulletin - Southern Hemisphere | Bureau of Meteorology, National Climate Centre, P.O. Box 1289K, Melbourne 3001, AUSTRALIA |
| 3. Climate System Monitoring Monthly Bulletin | World Climate Program, WMO Secretariat, Case Postale No. 5, CH-1211 Geneva 20, SWITZERLAND |
| 4. Darwin Tropical Diagnostic Statement | Bureau of Meteorology, P.O. Box 735, Darwin, Northern Territory 5794, AUSTRALIA |
| 5. ERFEN Boletin de Analisis Climatico | Comision Permanente del Pacifico Sur, Calle 76 No. 9-88, Apartado 92292, Bogota, COLUMBIA |
| 6. MEDS Realtime Data Monthly Monitor Report/Drifting Buoys | Marine Environmental Data Service, 200 Kent Street, Ottawa, Ontario K1A 0E6, CANADA |
| 7. Monthly Report on Climate System | Long-Range Forecast Division, Japan Meteorological Agency, 1-3-4 Ote-machi, Chiyoda-ku, Tokyo, JAPAN |
| 8. Oceanographic Monthly Summary | National Ocean Service, National Oceanic and Atmospheric Administration, 5200 Auth Road, Camp Srings, MD 20746, UNITED STATES OF AMERICA |
| 9. Veille Climatique Satellitaire | Centre de Meteorologie Spatiale, B.P. 147, Lannion 22303, FRANCE |

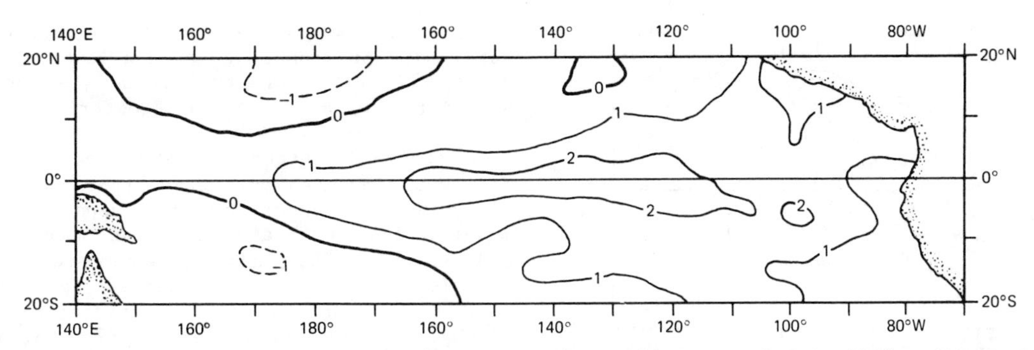


Fig. 1. Sea surface temperature (SST) anomaly during 16 June - 1 July 1987. Climatological-mean monthly SST was defined by Reynolds (1982). Contour interval is 1°C and a dashed line represents a negative value, which means that SST was lower than normal. The 15-day SST product represents a blend of in situ, but not drifting buoy SST observations, and satellite data recorded during the previous 15 days. Redrafted from a chart distributed by the NOAA Ocean Product Center.

nique used to compute the climatology. Comparing the SST anomaly distributions for the 15-day intervals before and after the 16 June - 1 July 1987 period indicated that the location of the SST anomaly was not constant.

Near surface currents are recorded by the movements of freely drifting buoys drogued at about 15 m depth and tracked several times each day by the Argos data collection and platform location system on-board NOAA polar orbiting satellites. This technique was used successfully during El Nino of 1982-83 (Halpern et al., 1983). During June 1987 there were 29 drifters afloat in the 20°N - 20°S Pacific (Figure 2). The comparison between the May 1987 surface current field and the June 1987 distribution indicated that the near equatorial eastward current observed in May between 140°W and 110°W was now confined to a smaller region near 130°W. The eastward flow in May 1987 may have contributed to the development of the SST anomaly in June by advection of warm water from the west, a situation similar to that found during the 1982-83 El Nino (Halpern, 1987).

Since about 1978, numerous temperature profiles within the upper 500 m have been routinely measured with XBTs launched from the v.o.s. network. As a result of the TOGA program, XBT lines are now more uniformly distributed throughout the tropical Pacific. Nearly 10,000 XBTs are launched annually in the tropical Pacific. Although there seems to be enough XBT observations to produce an adequate realization of the monthly mean thermal field, bimonthly data products are generated by Pazan et al. (1987). The depth of the 14°C isotherm, which usually occurs near the bottom of the thermocline, is an indicator of thermocline depth. During May - June 1987 the thermocline along the equator was shallower (deeper) than a 4-year mean depth, which was computed from data recorded during 1979, 1980, 1981, and 1984, in the western (eastern) region (Figure 3). Throughout the region 10°N - 10°S westward (eastward) of 130°W, the thermocline depth was less (greater) than the mean depth.

In the tropical oceans a close relationship exists between sea level and the depth of the thermocline. Along the equator, changes in zonal sea level slope mirror, to a large degree, variations in zonal slope of the thermocline (Wyrtki, 1984). Two sources of sea level data existed: in situ and satellite data. Tide gauge measurements

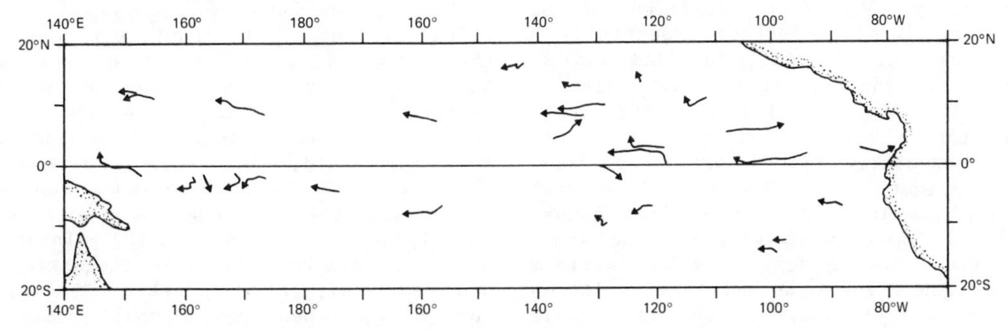


Fig. 2. Trajectories of satellite-tracked drifter buoys drogued at about 15 m depth during June 1987. Redrafted from a chart which appeared in the Climate Diagnostic Bulletin.

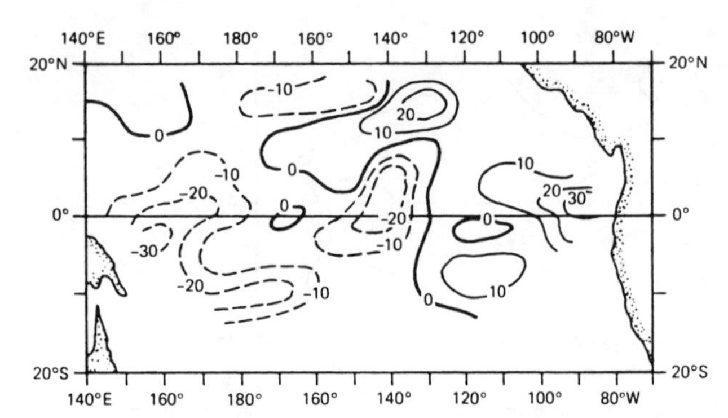


Fig. 3. Normalized depth of the 14°C isotherm during May - June 1987. Anomaly was defined as departure from average monthly mean distribution computed during 1979, 1980, 1981, and 1984. Dashed contours represent negative values, which mean that the 14°C isotherm was shallower than normal. Redrafted from a chart which appeared in the Climate System Monitoring Monthly Bulletin.

showed that the June 1987 sea level deviation from the 7-year (1975-81) mean sea level was below average in the western Pacific with a -27 cm minimum in the Solomon Islands and was above average in the eastern Pacific (Figure 4A). The previous month's distribution was very similar. Along the equator, the June 1987 sea level deviation represented a decrease in the zonal pressure gradient, a decrease of near surface westward flow, and a reduction in the Equatorial Undercurrent. A second sea level data set is produced by the U.S. Navy's Geosat mission, which measures sea surface topography with a root-mean-square error in the western equatorial Pacific approaching 4 cm for 2° latitude by 8° longitude regions and time scales longer than about 1-2 weeks (Cheney et al., 1986; Wyrtki, 1987). The 1 June 1987 sea level departures from a 1-year (April 1985-March 1986) annual mean distribution based upon Geosat data (Figure 4B) has features both different and similar to the sea level pattern estimated from in situ data (Figure 4A). Along the equator, the reduced zonal slope of sea level was produced in both data sets. In the central Pacific the positions of the zero isoline were very different: essentially a north-south trend for the in situ data and an east-west trend for the satellite data. This was due in part to the nearly total lack of tide guage data from the central and eastern Pacific in contrast to the regularly spaced altimeter data. Within the approximate 5°N - 10°N North Equatorial Countercurrent, sea level departures from annual mean sea surface topographies were not the same. Dissimilar definitions of long term sea surface topography used in Figures 4A and 4B contributed to the contrasting patterns, and the tide guage data should be analyzed relative to an April 1985-March 1986 mean value. In the eastern and central regions, the much reduced spatial structure of the in situ data was caused by the absence of station data. The prominent -20 cm sea level departure at about 5°N, 135°W was typical of the annual cycle associated with the North Equatorial Countercurrent.

Much of the structure portrayed by the Geosat data was simulated by Leetmaa (personal communication) with a wind-driven ocean general circulation model. This primitive equation model (Philander and Seigel, 1985) has been compared on several occasions with tropical surface and subsurface ocean observations with satisfactory results (Richardson and Philander, 1987; Philander and Seigel, 1985; Garzoli and Philander, 1985). The modeled surface dynamic height anomaly for June 1987 (Figure 4C) was defined as the difference between the June 1987 simulation forced with June 1987 surface winds determined from v.o.s. wind data and the climatological-mean monthly simulation forced with Hellerman and Rosenstein's (1983) winds for June. The model result and Geosat data showed a similar pattern of sea surface height variation in the North Equatorial Countercurrent where heights were less than the mean (although different mean values were used in Figures 4B and 4C). The model simulation of sea level and in situ sea level measurements (Figures 4A and 4C) had excellent correspondence in the southwest Pacific with mimimum sea level near 10°S, 160°E. The in situ tide guage network did not contain any spatial structure of sea level eastward of 140°W (Figure 4A) because of the absence of data, in contrast to the modeled surface dynamic height (Figure 4C), which contained fewer occurrences of alternating positive and negative departures from mean sea level than the Geosat data (Figure 4B). The magnitude of the differences between Figures 4B and 4C were about as large as the monthly mean values. Because the pattern of monthly mean differences varied from month-to-month, it is not clear which of these two sea level products is more representative of the natural state. The reasonable agreement between model simulated surface dynamic height, Geosat data, and in situ sea level measurements represented an advance towards the TOGA objective of developing an operational capability for dynamical ocean prediction.

One reason for the success of ocean general circulation models in tropical regions is that the thermal and flow variations are not very intense and essentially linear. Current and temperature variations in low latitudes are more strongly related to wind forcing than to instabilities of the flow field, which commonly occur in middle latitudes. In tropical regions, most of the mesoscale variability is associated with long (approximately 1000-km wave length) linear or weakly nonlinear wavelike features (Legeckis, 1977; Hansen and Paul, 1984; Halpern et al., 1988), in contrast to the intense small scale eddies and rings prevalent throughout the mid-latitude ocean. Essential ocean dynamical features in low latitudes, except for microscale mixing processes,

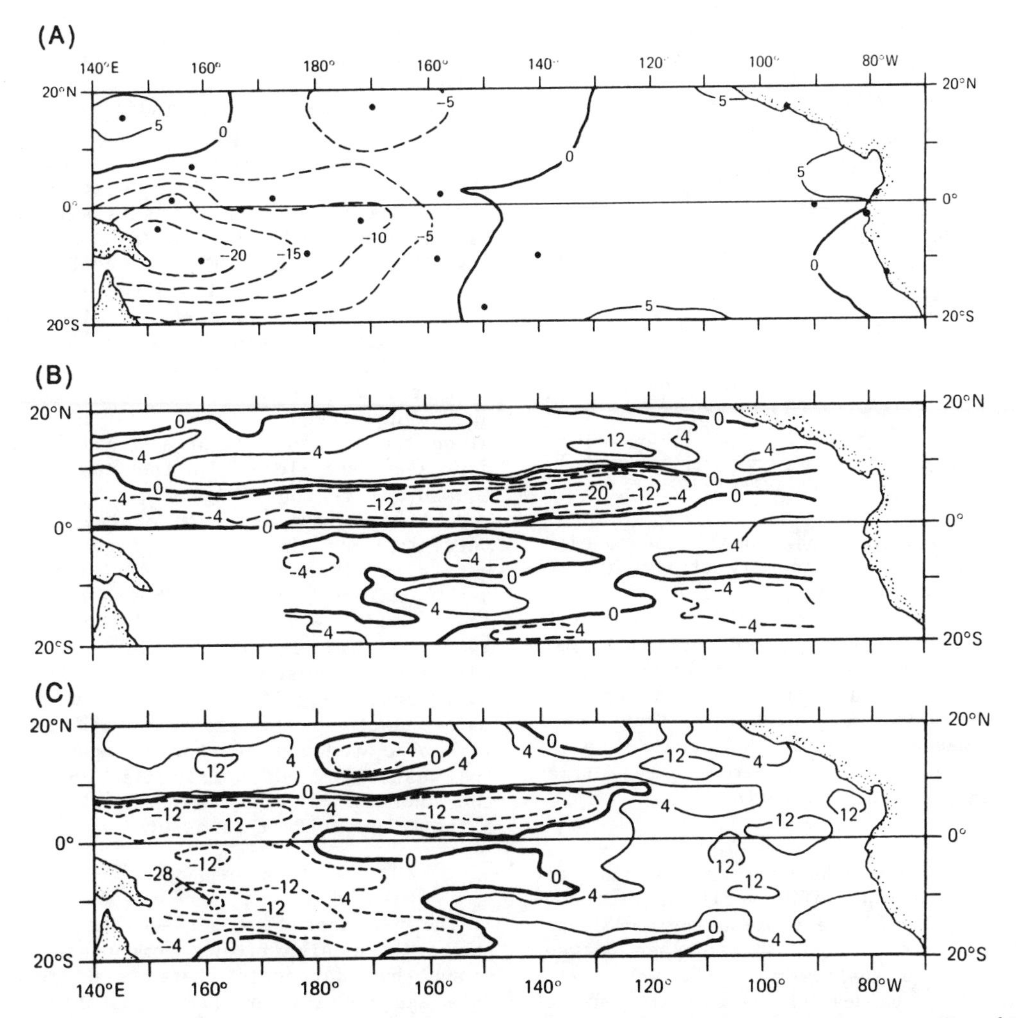


Fig. 4. (A) Sea level during June 1987 estimated from tide guage measurements recorded at island and coastal stations. This map represents departure of sea level from 7-year (1975 - 81) mean. Solid dots represent in situ sea level stations. Dashed lines represent negative values, which mean that sea level was lower than the mean. Redrafted from a chart appearing in the Climate Diagnostic Bulletin. (B) Sea level anomaly during 1 June 1987 estimated from Geosat sea surface topography data using daily data, a 15-day decorrelation time, and a grid of 8° longitude by 2° latitude. Anomaly is defined as difference from 1-year (April 1985 - March 1986) mean. Dashed contours represent negative values. Redrafted from a chart which appeared in the Climate Diagnostic Bulletin. (C) Surface dynamic height anomaly during June 1987 estimated from an ocean general circulation model. Anomaly defined as difference between monthly mean distribution and climatological-mean distribution determined from Hellerman and Rosenstein's (1983) wind field. Dashed contours represent negative values. Redrafted from a diagram kindly provided by Dr. Ants Leetmaa (Climate Analysis Center).

are resolvable by primitive equation models run on currently available supercomputers. Equatorial wave-like structures were simulated by Philander et al. (1986) with a 33-km latitudinal by 100-km longitudinal model geometry near the equator.

The depth of the 20°C isotherm along the equator, which usually occurs in the middle of the thermocline, is a suitable indicator of thermocline depth and of the zonal pressure gradient force in the upper ocean. Simulations of the monthly difference of the depth of the 20°C isotherm relative to the climatological-mean monthly distribution computed with the Hellerman and Rosenstein (1983) wind field were computed by Leetmaa (personal communication). Figure 5 indicated that maximum departures from the mean may have been

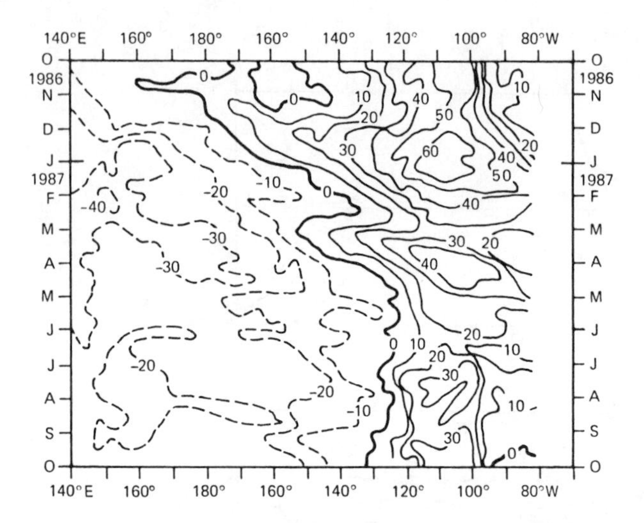


Fig. 5. Time - longitude section of the model simulation of monthly mean anomaly of the 20°C isotherm depth. Anomaly defined as difference between monthly mean depth of 20°C isotherm and climatological-mean monthly depth determined from a wind-driven ocean model using Hellerman and Rosenstein's (1983) wind field. Dashed contours represent negative values, which mean that the 20°C isotherm was shallower than normal. Redrafted from a chart which appeared in the Climate System Monitoring Monthly Bulletin.

reached in June 1987; subsequent model runs for several months after June 1987 verified this conclusion. In June 1987, the thermocline was deeper east of 130°W (i. e., about 30 m deeper between 105°W and 95°W) and shallower to the west, with anomalous negative values (i. e., shallower) of more than 20 m west of 175°W. This model result is sensitive to the surface winds: a 20% change in the surface wind stress will produce a thermocline depth difference of about 25 m (Harrison et al., 1988), which is about 50% of the absolute range displayed in Figure 3.

In the equatorial zone, the east-west surface wind stress component is very important in the generation of thermal and flow variations (Gill, 1982). Many surface wind products exist at present, but intercomparison tests revealed large differences (Halpern and Harrison, 1982), primarily because of the scarcity of in situ surface wind data. As a result of the TOGA program, the number of island and moored buoy stations reporting wind data in real time was increased dramatically. At Christmas Island (2°N, 157°W), the April 1987 zonal surface wind component was essentially the same as the 25-year average April wind computed by Wyrtki and Meyers (1976). However, in May the zonal component became a westerly anomaly (i. e., a reduction in speed of the normally occurring easterly or westward wind) of 4-5 m $s^{-1}$ and the wind even blew towards the east (Figure 6A). In April 1987 the westerly wind anomaly was confined to the 170°E to 170°W region, but in May the westerly wind anomaly increased in strength and longitudinal coverage from 160°E to 140°W (Figure 6B). In June the westerly wind anomaly was limited to the region from the date line to 160°W and the easterlies in the far western Pacific were stronger than normal. If this situation had continued and expanded towards the east, it would signal the demise of El Nino conditions: this did not occur and SST continued to be warm.

It is tempting to associate April and May 1987 surface wind patterns with the June 1987 oceanographic conditions. Conventional wisdom dictated that reduced easterlies in May would produce in June a decreased east-west tilt of the thermocline along the equator (Figures 3 and 5), elevated and depressed sea level in the eastern and western equatorial regions (Figure 4), respectively, and a reduction in the strength of the South Equatorial Current along the equator (Figure 2). The eastward direction of the buoy drifting near 0°, 130°W was anomalous because the South Equatorial Current is usually well developed in June (Halpern, 1987). This eastward motion was indicative of advection of warmer water from the west, which contributes significantly to the generation and maintenance of the anomalous SST pattern (Figure 1). The reduced equatorial upwelling caused by the westerly wind anomaly also contributed toward the anomaly of high SST, though no measurements exist to quantify this assertion.

## Discussion

The net heat flux between the ocean and atmosphere is a significant component of SST variations, but currently there is no readily accessible and reliable product. Monthly mean net surface heat flux variations estimated from v.o.s. data (e. g., wind, air and sea surface temperatures, humidity, cloud amount, etc.) are uncertain to ±30 to ±50 W $m^{-2}$. A monthly mean net surface heat flux uncertainty of 40 W $m^{-2}$ corresponds to a monthly mean SST variation of 1.2°C for an isothermal mixed layer of 20 m thickness.

Much progress is being made towards development of satellite analyses of the constituents of the net surface heat flux. For instance, the surface latent heat flux, which is a large contributor of air-sea heat flux, can be estimated from satellite measurements of surface wind speed and total water content in the atmosphere (Liu, 1988). Short wave radiation flux, which is also a large component of the net surface heat flux variation between clear and cloudy regions, seems to be measurable from space (Gautier, 1988), and there is expectation that long wave radiative flux might also be determined from satellite data (Frouin et al., 1988).

Variability in the tropical ocean is primarily atmospherically forced, yet little is known about the structure of the surface wind field. The sensitivity of monthly mean SST simulations to the

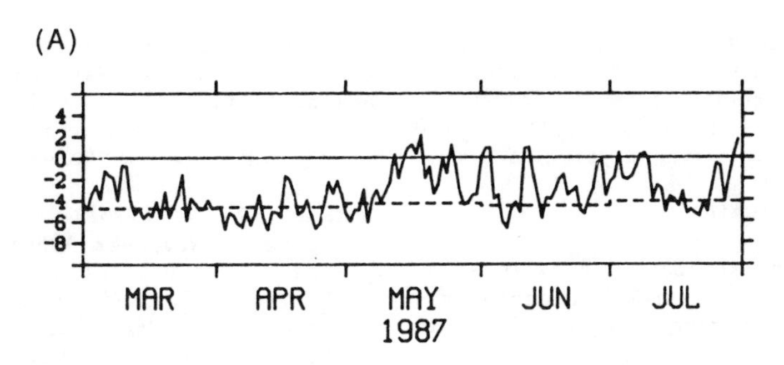


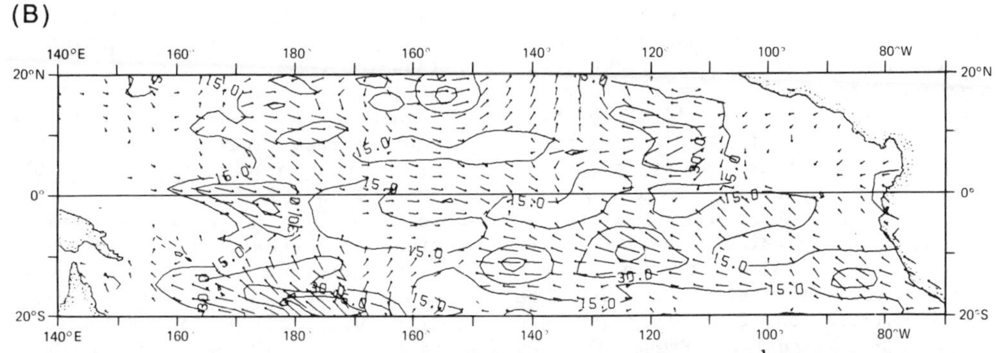


Fig. 6. (A) Time variations of surface zonal wind component (m $s^{-1}$) measured at Christmas Island (2°N, 157°W), which is one of several operational, real time TOGA equatorial Pacific island and moored buoy stations. Dashed lines represent 25-year (1947-72) climatological-mean monthly zonal component wind determined by Wyrtki and Meyers (1976). Diagram kindly produced by Paul Freitag (Pacific Marine Environmental Laboratory). (B) Surface wind stress anomaly during May 1987. A wind stress vector is defined as the wind components multiplied by the wind magnitude. A stress vector of 60 $m^2$ $s^{-2}$ corresponds to 1 dyne $cm^{-2}$ assuming a drag coefficient of $1.4 \times 10^{-3}$ and air density of 1.2 kg $m^{-3}$. Monthly mean anomaly defined as departure from 23-year (1961 - 83) monthly mean distribution. A westerly wind anomaly means a reduction in the speed of the normally occurring easterly or westward wind. Adapted from a chart which appeared in the Climate Diagnostic Bulletin.

monthly averaged surface wind was demonstrated by Harrison et al. (1988). A 20% error in wind stress, which is about the minimum error expected with present technology, was equivalent to a 2°C uncertainty in SST. Large wind speed variations occur on short time scales of 1- to 5-days (Figure 6A). The accuracy of monthly mean v.o.s. data to 0.5 m $s^{-1}$, which is necessary for climate studies, is doubtful, except along a few well-traveled shipping routes because of inadequate sampling (Halpern, 1988). Rosati and Miyakoda (1988) improved the SST simulation by using winds with a 1-day Nyquist period, which substantially reduced the underestimation of the net surface heat flux.

In the early 1990s, every day there will be more than 50,000 surface wind vectors with a resolution of nearly 50-km available from the National Aeronautic and Space Administration's (NASA) scatterometer (called NSCAT), which may be launched in 1993 on Japan's ADEOS satellite, and from the European Space Agency's (ESA) scatterometer scheduled to be launched in 1991 on the ERS-1 satellite. The number of satellite surface wind vectors to be recorded in one month will be greater than the total number of v.o.s. surface wind observations measured throughout history. Unfortunately, these satellite scatterometers do not yield accurate estimates of the wind speed below about 3 m $s^{-1}$. A new, innovative satellite scatterometer designed to measure the tropical surface wind field with a root-mean-square accuracy of ±20% for wind speeds less than 3 m $s^{-1}$ is currently being considered for development by NASA.

A major obstacle remaining in the development of ocean models for simulation of SST is the parameterization of horizontal and vertical turbulent mixings in the upper ocean. Philander and Seigel (1985) used a vertical mixing parameterization based upon the Richardson number. Rosati and Miyakoda (1988) demonstrated that a turbulence closure scheme was more appropriate than constant eddy viscosity. Evaluating different mixing parameterizations is exceedingly difficult because of the scarcity of suitable in situ measurements. Peters et al. (1988) reported substantial differences between Philander and Seigel's (1985) Richardson number formulation and in situ turbulent mixing data recorded at 0°, 140°W. However, the in situ mixing observations should not be applied beyond the time and place of their measurement until more is learned about the causes of their variability because dissipation rates were 100 times larger at night than during daytime.

## Conclusion

TOGA ocean studies are the latest beneficiaries of numerous multidisciplinary tropical oceanography projects conducted during the previous 15 years. Since the beginning of the TOGA decade in January 1985, significant innovative accomplishments include (1) increase in the quantity of in situ data and efficiencies of data management, (2) real time ocean products and their rapid distribution, (3) effective utilization of global observations from satellites to estimate sea surface topography and latent heat flux, and (4) numerical model simulation of oceanographic features.

The possibility of the TOGA program achieving measurable success in the prediction of time averaged anomalies up to several months in advance rests with effective utilization of satellite observations and assimilation of observations into coupled ocean-atmosphere models. Presently, the differences between model simulation and measurements were equivalent to uncertainties caused by inadequate wind field and net surface heat flux. A 20% error in wind stress produces uncertainties of 2°C in SST and 25 m in the depth of the thermocline along the equator. Assimilation of XBT measurements partially compensates for the lack of adequate wind data. A 40 W $m^{-2}$ error in the net surface heat flux yields a 1.2°C SST error for a 20 m thick mixed layer.

The future challenge is to use satellite data, in situ observations, and general circulation models synergistically. Perhaps our complete understanding of the physics of tropical SST variations awaits analyses of model results and a few specialized observations. The TOGA decade, which has already altered perceptions of tropical oceanography and ocean-atmosphere interactions, will undoubtedly have an everlasting impact upon global geosciences.

Acknowledgements. This paper is dedicated to the memory of Adrian Gill in recognition of his essential contributions to the development of the TOGA program. I am grateful to many colleagues who regularly send their data product in advance of publication in the bulletins: R. Cheney and L. Miller (Figure 4B); P. Freitag (Figure 6A); D. Hansen (Figure 2); A. Leetmaa (Figures 4C and 5); J. O'Brien (Figure 6B); D. Rao (Figure 1); K. Wyrtki (Figure 4A). This manuscript has benefited from comments from many TOGA colleagues, in particular M. Cane, R. Cheney, J.-R. Donguy, J. Fletcher, C. Gautier, R. Lambert, A. Leetmaa, G. Meyers, K. Takeuchi, Y. Tourre, K. Trenberth, and K. Wyrtki. The research described in this paper was performed, in part, by the Jet Propulsion Laboratory, California Institute of Technology, under contract with the National Aeronautics and Space Administration.

## References

Bergman, K. H. (1987) The global climate of September - November 1986: a moderate ENSO develops in the equatorial Pacific. Monthly Weather Review, 115, 2524-2542.

Cheney, R., B. Douglas, R. Agreen, L. Miller, D. Milbert and D. Porter (1986) The Geosat altimeter: a milestone in satellite oceanography. Transactions American Geophysical Union, 67, 1354-1355.

Frouin, R., C. Gautier and J.J. Morcerette (1988) Downward long wave irradiance at the ocean surface from satellite data: methodology and in situ validation. Journal of Geophysical Research, 93, 597-620.

Garzoli, S. and S. G. H. Philander (1985) Validation of an equatorial Atlantic simulation model using inverted echo sounder data. Journal of Geophysical Research, 90, 9199-9201.

Gautier, C. (1988) Surface solar irradiance in the central Pacific at 10°S, 140°W during Tropic Heat: in situ and satellite measurements. Journal of Climate and Applied Meteorology, in press.

Gill, A.E. (1982) Atmosphere-Ocean Dynamics. Academic Press, New York, 662 pp.

Halpern, D. (1988) On the accuracy of monthly mean wind speeds over the equatorial Pacific. Journal of Atmospheric and Oceanic Technology, 5, 362-367.

Halpern, D. (1987) Observations of annual and El Nino thermal and flow variations at 0°, 110°W and 0°, 95°W during 1980-1985. Journal of Geophysical Research, 92, 8197-8212.

Halpern, D. and D. E. Harrison (1982) Intercomparison of tropical Pacific mean November 1979 surface wind fields. Report 82-1, Department of Meteorology and Physical Oceanography, Massachusetts Institute of Technology, Cambridge, 40 pp.

Halpern, D., S. P. Hayes, A. Leetmaa, D. V. Hansen and S. G. H. Philander (1983) Oceanographic observations of the 1982 warming of the tropical eastern Pacific. Science, 221, 1173-1175.

Halpern, D., R. A. Knox and D. S. Luther (1988) Observations of 20-day period meridional current oscillations in the upper ocean along the Pacific equator. Journal of Physical Oceanography, in press.

Hansen, D. V. and C. A. Paul (1984) Genesis and effects of long waves in the equatorial Pacific. Journal of Geophysical Research, 89, 10431-10440.

Harrison, D. E., W. S. Kessler and B. S. Giese (1988) Ocean circulation model hindcasts of the 1982-83 El Nino: thermal variability along the ship of opportunity tracks. Journal of Physical Oceanography, in press.

Hellerman, S. and M. Rosenstein (1983) Normal monthly wind stress over the world ocean with error estimates. Journal of Physical Oceanography, 13, 1093-1104.

Knox, R. A. and D. Halpern (1982) Long range Kelvin wave propagation of transport variations in Pacific Ocean equatorial currents. Journal of Marine Research, 40 Supplement, 329-339.

Legeckis, R. (1977) Long waves in the eastern equatorial Pacific Ocean: a view from a geostationary satellite. Science, 197, 1197-1181.

Levitus, S. (1987) A comparison of the annual cycle of two sea surface temperature climatologies of the world ocean. Journal of Physical Oceanography, 17, 197-214.

Liu, W. T. (1988) Moisture and latent heat flux variabilities in the tropical Pacific derived from satellite data. Journal of Geophysical Research, 93, 6749-6760.

Pazan, S. E., W. B. White and Y. He (1987) Annual report on tropical Pacific subsurface thermal structure - 1985. SIO Reference Number 87.1, Scripps Institution of Oceanography, La Jolla, California, 66 pp.

Peters, H., M. C. Gregg and J. M. Toole (1988) On the parameterization of equatorial turbulence. Journal of Geophysical Research, 93, 1199-1218.

Philander, S. G. H. (1979) Variability of the tropical oceans. Dynamics of Atmospheres and Oceans. 3, 191-208.

Philander, S. G. H. and A. D. Seigel (1985) Simulation of El Nino of 1982-83. In: Proceedings of the 16th International Liege Colloquium on Ocean Hydrodynamics, edited by J. Nihoul, Elsevier, New York, 517-541.

Philander, S. G. H., W. J. Hurlin and R. C. Pacanowski (1986) Properties of long equatorial waves in models of the seasonal cycle in the tropical Atlantic and Pacific Oceans. Journal of Geophysical Research, 91, 14207-14211.

Rasmusson, E. M. and T. H. Carpenter (1982) Variations in tropical sea surface temperature and surface wind fields associated with the Southern Oscillation/El Nino. Monthly Weather Review, 110, 354-384.

Rasmusson, E. M. and J. M. Wallace (1983) Meteorological aspects of the El Nino/Southern Oscillation. Science, 222, 1195-1202.

Reynolds, R. W. (1982) A monthly averaged climatology of sea surface temperature. NOAA Technical Report NWS 31, National Oceanic and Atmospheric Administration, Rockville, Maryland, 35 pp.

Richardson, P. L. and S. G. H. Philander (1987) The seasonal variations of surface currents in the tropical Atlantic Ocean: a comparison of ship drift data with results from a general circulation model. Journal of Geophysical Research, 92, 715-724.

Rosati, A. and K. Miyakoda (1988) A GCM for upper ocean simulation. Journal of Physical Oceanography, in press.

Wyrtki, K. (1984) The slope of sea level along the equator during the 1982/83 El Nino. Journal of Geophysical Research, 89, 10419-10424.

Wyrtki, K. (1987) Comparing Geosat altimetry and sea level. Transactions American Geophysical Union, 68, 731.

Wyrtki, K. and G. Meyers (1976) The trade wind field over the Pacific Ocean. Journal of Applied Meteorology, 15, 698-704.

# Section VI

# MODELLING CLIMATE, PAST, PRESENT AND FUTURE

# AERONOMY AND PALEOCLIMATE

J.-C. Gérard

Institut d'Astrophysique, Université de Liège
B-4200 Ougrée-Liège, Belgium

*Abstract.* Solar-Terrestrial and aeronomic interactions with the Earth's global climate played a key role throughout the evolution of the planet. The evolving composition of the terrestrial atmosphere due to geochemical and biospheric interactions strongly controled the paleoclimate. An example is the apparent absence of glaciation in a period when the solar luminosity was substantially smaller than at the present time (young sun luminosity paradox). The most convincing explanation proposed so far assumes the existence of larger amounts of infrared radiatively actives gases ($CO_2, CH_4$ ) in the ancient atmosphere.

The vertical thermal structure of the atmosphere was notably different from the contemporary one, due to the lower abundance of oxygen and ozone before the development of green-plant photosynthesis. For $O_2$ mixing ratios less $10^{-4}$ time the present atmospheric level, the presence of 11-year and longer periodicities has been observed in precambrian and more recent varve formations. It has been interpreted as an indication of a stronger response of the ancient atmosphere to solar cycle activity. This increased sensitivity was possibly the consequence of a lower $O_2$ atmospheric mixing ratio and a less developed stratosphere.

## Introduction

The evolution of the global climate system during the past 4.5 billion years was determined by both endogenic and exogenic forcings which are still controling today's climate. Amongst the external factors, the dominant one is the solar energy input and its temporal variations. The solar radiation reaching the planet's atmosphere is likely to have varied in time both in total power (solar luminosity) and in spectral composition (solar irradiance). Following the sun formation 4.6 Ga ago and the T-Tauri phase where both the solar luminosity and the ultraviolet radiation increased substantially during approximately 10 Ma, the solar luminosity increased steadily. It rose from a value about 25-30% less than the present one as the sun evolved on the main sequence of the Herzprung-Russel diagram. This nearly linear increase is an unescapable consequence of the progressive conversion of hydrogen into helium in classical theories of the stellar interiors. If the atmospheric composition was unchanged during this period, energy-balance models (North et al., 1981) and radiative-convective models (Wang and Stone, 1980; Gérard and François, 1988) predict that, as a consequence of the ice-albedo feedback, the Earth's climate would adopt a stable condition where the planet is totally ice-covered if the solar constant is decreased by only a few percent. However, geological evidence indicates that no world-wide glaciation occured until 2.3 Ga ago, at the time of the Huronian glaciation (Crowley, 1983). Moreover, models suggest that global glaciations of this type would be irreversible and luminosities higher than the present one would be required to unfreeze the system.

Changes in the continental mass and geographic distribution probably took place and possibly counteracted in part the lower solar luminosity (Endal and Schatten, 1982; Schatten and Endal, 1982). However, variations in the abundance of infrared-active gases has been suggested as the most likely mechanism to maintain the global temperature above the freezing point. It is now generally believed that carbon dioxide probably played the major role in the thermostatic control of the Earth's climate.

In the present atmosphere, ozone plays a dominant effect in the determination of the thermal structure of the stratosphere and a secondary role in the greenhouse heating of the mesosphere and the troposphere. The stratospheric rise in temperature is associated with the absorption of solar radiation shortward of 300 nm. In the primitive atmosphere, the only source of oxygen was the photodissociation of water vapor (Kasting et al., 1980). This mechanism was shown to yield only very small ($10^{-8}$) of the present atmospheric level or PAL) amounts of free oxygen and negligible quantities of ozone. Later, as photosynthetic activity began to develop at the ocean surface about 3.5 Ma ago. (Schopf, 1983), the oxidation of the reducing seawater probably acted as an efficient sink for atmospheric oxygen and prevented the accumulation of $O_2$ in the atmosphere. The simultaneous presence of ferrous and ferric iron observed in the banded iron formations (BIFs) is usually interpreted as an indication of the existence of small amounts ($< 10^{-6}$ PAL) of atmospheric $O_2$ during the precipitation of the oxidized iron (Holland, 1984, François and Gérard, 1986). Photooxidation of Fe $^{2+}$ ions may also have been a significant source of a abiotic formation of the BIFs (Braterman et al., 1983; François, 1987 ). More recently, oxidation of continental crust and formation of red rocks contributed to limit the level of oxygen in the atmosphere. Roughly 400 Ma ago, the oxygen reached approximately its present atmospheric mixing ratio (Cloud, 1983). Therefore, throughout most of its history, the Earth's atmosphere has varied continuously in composition and thermal structure. The abundance of other minor constituents ($N_2O$, $NO_2$, $CH_4$, $NH_3$) has probably also changed substantially during the Earth's history but they appear to have played a less crucial role in the climatic evolution of the planet.

Finally, if evidence of a control of solar cycle activity over the climate has remained elusive in the comtemporary atmosphere, precambrian sedimentary rocks have revealed a clear signature of the ancient solar variations. It is therefore important to examine whether the different atmospheric composition which prevailed in the past may have favored a stronger response of the global climate to solar forcing.

## Carbon Dioxide and Precambrian Climate

The response of the surface temperature to variations in solar luminosity was examined in details with one-dimensional energy balance models (North et al., 1981; Endal and Schatten, 1982). More recently, Gérard and François (1988) used a radiative-convective model to test the sensitivity of the climate response to decreases of the solar luminosity. In particular, they showed that, in agreement with energy-balance models, irreversible global glaciations would be predicted if the solar constant dropped by a few percent below its present value. The exact position of the discontinuity from the unfrozen to the totally frozen solution depends on several factors such as the global albedo-temperature function and the tropospheric lapse rate (figure 1). Sagan and Mullen (1972) initially suggested that large amounts of $NH_3$, an active greenhouse gas, would be able to elevate the surface temperature and compensate for the past lower solar luminosity. However, due to the short lifetime of ammonium (Kasting, 1982) in the primitive atmosphere, this molecule was abandoned and $CO_2$ appeared as the most likely infrared absorber. Hart (1978) calculated a scenario for the $CO_2$ atmospheric level which is compatible with the absence of global glaciation in the Precambrian. His work was based on a simplified global climatic model coupled to a geochemical approximation. This $CO_2$ history was frequently used in conjonction with radiative-convective models to test the validity of the thermostatic effect of this constituent (Owen et al., 1979; Kasting et al., 1984; Kuhn and Kasting, 1983; Kiehl and Dickinson, 1987). The results of Kasting et al. (1984) and Kiehl and Dickinson (1987)(figure 2) show that the combination of increasing solar luminosity and decreasing $CO_2$ mixing ratio is able to maintain the global temperature of the planet within $\pm$ 5 K. However, the treatment by Owen et al. (1979) yields an average surface temperature 3.5 Ga ago approximately 7 K warmer than Kiehl and Dickinson who found that the pressure scaling of the mean band halfwidth of the $CO_2$ bands at 961 and 1064 $cm^{-1}$ accounts for most of the temperature difference with Owen et al.'s calculations.

Adopting the same scenario, we use the one-dimensional radiative- convective global model described by Gérard and François (1988) and François (1988) to calculate the evolution of the mean surface temperature. Briefly, this model solves the thermodynamic equation using a forward time-marching method. The absorption of solar radiation by $CO_2, H_2O, O_3$ and clouds is calculated using the Lacis and Hansen (1974) formulation, including the effects of Rayleigh scattering and surface albedo. The cloud cover is characterized by a single layer with fixed altitude and optical depth. Rossow et al. (1982) demonstrated the important potential role of cloud feedback on the stabilization of the Earth's climate during its evolution. In particular, they demonstrated that the strong negative cloud feedback present in their model is able to partly compensate the lower ancient sun luminosity. The greenhouse effect due to infrared absorption by $O_3$, $N_2O$ and $CH_4$ is treated following the method by Ramanathan (1976) and Donner and Ramanathan (1980). The contribution of water vapor is calculated from the parameterization given by Sasamori (1968). The variation of the $CO_2$ absorptance with the $CO_2$ column given by Kasting et al. 1984) is adopted. This expression was

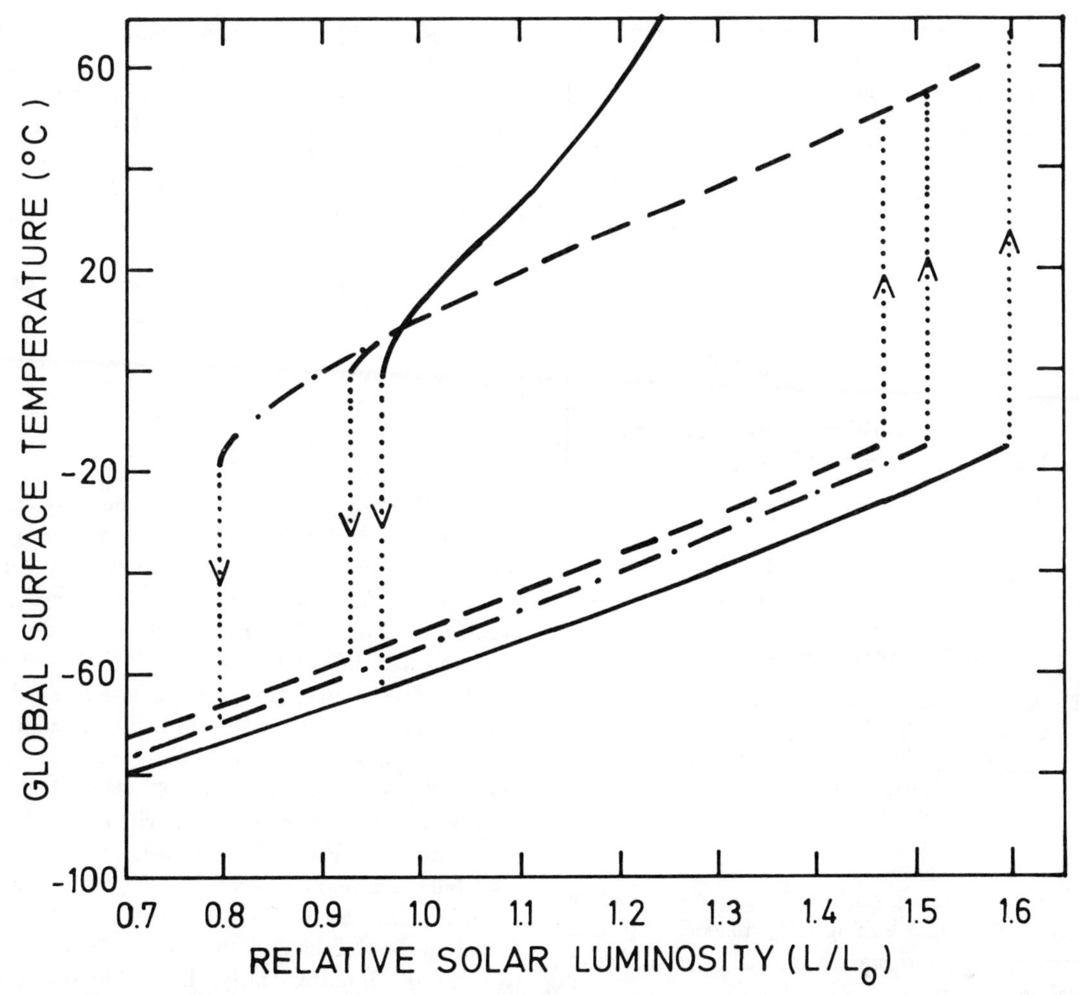


Fig. 1. Global mean surface temperature as a function of the luminosity relative to the present one for different cases. In the first case, the tropospheric temperature gradient is fixed to 6.5 K/km (full line), while in the second curve it is fixed to the moist adiabatic value (dashed line). For both curves, the temperature dependence of the surface albedo is adopted from Gérard and François (1988) and corresponds to an ice albedo of 0.5. In the third case (dash-dotted line), a moist adiabatic gradient is used and the albedo-temperature relation from Wang and Stone (1980) is adopted. The locations of the jumps to and from the frozen solutions are also indicated.

obtained by parameterizing results of laboratory absorption measurements. Whenever the temperature gradient tends to exceed the moist adiabatic lapse rate, a convective adjustement is applied to restore the convective profile. This convective adjustement is made following the method of Manabe and Wetherald (1967). Our calculated surface temperature $T_s$ also remain within 5 K of the value calculated for present conditions, but the variations are slightly larger than those obtained in previous studies It is likely that differences in the radiative code, and in the convective adjustment scheme, amplified by the water vapor feedback, account for the differences in $\Delta T_s$ . Overall, all models lead to the conclusion that it is possible to find a plausible $CO_2$ scenario able to counteract the effect of the reduced past solar luminosity and thus to avoid global glaciation. Yet, it is important to stress that, at this point, the only detailed geochemical model extends back in time only 100 Ma ago (Lasaga et al., 1985). These global carbonate-silicate geochemical cycle calculations predict a significant increase in the past $CO_2$ atmospheric content and a parallel rise in the global surface temperature reaching 8 K, 100 Ma ago. Devel-

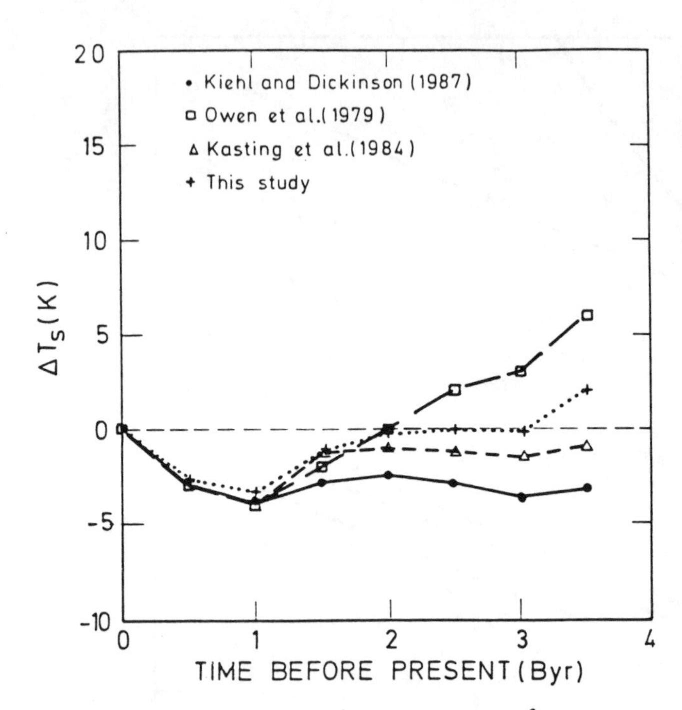


Fig. 2. Change in mean surface temperature from present conditions calculated with various models for the time period 0 to 3.5 Ma before present using the $CO_2$ and solar luminosity scenario from Hart (1978).

opment of such numerical geochemical models extending further back in time and linking $CO_2$, paleoclimate and tectonic activity is unfortunately limited by the lack of geochemical constrains.

## The Climatology of the Oxygen-Poor Atmosphere

In the present atmosphere, ozone plays a dominant role in the determination of the thermal structure of the stratosphere and a secondary role in the mesosphere and troposphere. This peak in temperature is associated with the absorption of solar radiation shortward of 300 nm. Presently, the contribution of ozone to trapping of thermal infrared radiation amounts to 2.3 $W.m^{-2}$ out of a total of 148 $Wm^{-2}$ (Dickinson and Cicerone, 1986). It is believed that, in the ancient atmosphere, containing no or trace amounts of oxygen, the thermal structure was significantly different from the contemporary one. The climatology of these phases of the Earth evolution when the $O_2$ level was substantially less than presently is discussed below.

The model used to investigate this question is a coupled radiative-convective model coupled to a diffusive-photochemical calculation. The first aspect was presented above and the diffusive-chemical model is similar to that described by Gérard and François (1987) and François and Gérard (1988). The continuity and turbulent flux equations are solved for O, $O_2$, $O(^1D)$, $O_3$, H, OH, $HO_2$, $H_2O_2$, $N_2O$, NO, $NO_2$, $NO_3$, $N_2O_5$, $HNO_2$, $HNO_3$, $N_2O_4$ , and CO. Short-lived species ($H_2CO$, $CH_3$, $CH_3O_2$, $H_3CO$, $CH_3OOH$, HCO) are assumed to be in photochemical equilibrium. The vertical mixing ratio distributions of $CO_2$, $CH_4$ and $H_2$ observed in the present atmosphere are adopted. The $H_2O$ profile is consistent with the calculated verticaltemperature distribution. To account for the feedback between the water vapor and ozone distributions and the thermal structure, numerical integration is alternated between radiative and photochemical calculations until convergence is achieved.

The effect of the $O_2$ rise during the Earth's evolution on the ozone column and on the average surface temperature is shown in figure 3, ignoring the variation of the solar luminosity. Our model calculations yield a surface temperature $T_s$ slightly larger than the observed mean global temperature, due to the particular choice of the cloud cover characteristics. However, the absolute value of $T_s$ is not essential here since we are only interested in the effect of $O_2$ variations on the mean surface temperature changes. When $O_2$ is decreased for 1 to $10^{-1}$ PAL, this model predicts a 3-4 K drop of $T_s$. This change is the result of two combined effects. First, as $O_2$ is decreased, the ozone distribution is reajusted vertically and the ozone column is slightly increased. Consequently, the efficiency of the greenhouse is larger and the surface temperature tends to increase. Simultaneously, the visible albedo decreases due to changes in the ozone and water vapor concentration. Another source of temperature changes amounting to about 5 K is due to the total pressure drop accompanying the $O_2$ decrease. This is the result of the reduction of the line broadening of $H_2O$ and $CO_2$ infrared bands. The relative importance of these competing effects is dependent on the assumption concerning the tropospheric lapse rate. In this case, the use of a moist adiabatic lapse rate stabilizes the surface temperature and causes the $T_s$ change to be smaller than in other studies (Levine and Boughner, 1979).

When the $O_2$ level is further decreased, the reduction of the ozone column and its vertical redistribution causes the temperature drop. This reduction is particularly efficient between $10^{-3}$ and $10^{-4}$ PAL. This is due to the localization of the ozone peak between 10 and 20 km for this $O_2$ level. Indeed, sensitivity studies (Wang et al., 1980; François, 1988) have shown that the sensitivity of the climate to ozone perturbations maximizes when the perturbations are located in this altitude range. A $T_s$

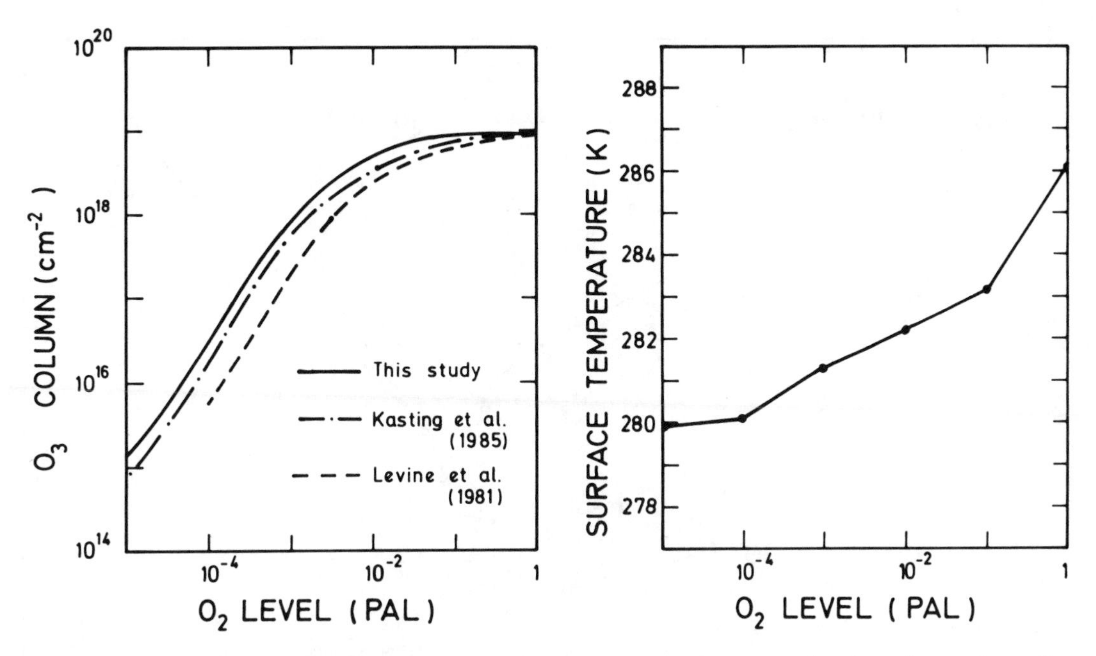


Fig. 3. Left : dependence of the ozone vertical column upon the oxygen level at the surface. Comparison is made with previous studies. Right : calculated variation of the global surface temperature with the $O_2$ atmospheric level.

drop by another 0.9 K is predicted when the $O_2$ level in the atmosphere is set to a vanishingly small value. Consequently, a global temperature decrease of 8.6 K is predicted if the oxygen was removed from the atmosphere, all other conditions being held constant.

The vertical distribution of the global mean temperature calculation with this model levels of $O_2$ from $10^{-4}$ to 1 PAL are shown in figure 4. The progressive development of the stratosphere is observed as the oxygen and the ozone abundances are increased. At 0.1 PAL, the stratospheric temperature peak is lower in magnitude and altitude. This change is due to the decrease of ozone in the upper stratosphere. By contrast, a small increase of the temperature is obtained in the region near 20 km as a result of the slightly enhanced ozone concentrations in the lower stratosphere. At $10^{-2}$ PAL of $O_2$, no stratospheric maximum is predicted by the model due to the low stratospheric ozone density. As the $O_2$ level is further decreased to $10^{-3}$ PAL, an inflexion in the temperature near 20 km is the only effect of vanishingly small $O_3$ concentrations. Finally, at $10^{-4}$ PAL, the role of ozone in the thermal structure of the atmosphere becomes negligible.

It is also noted that, in spite of the important changes in the vertical temperature distribution, the top of the convective region is located in each case between 10 and 14 km.

## Response of the Precambrian Atmosphere to Solar Cycle Activity

Analysis of periodicities observed by Williams (1981) and Williams and Sonett (1985) in the annual deposits (varves) of a precambrian lake in South Australia strongly suggest a solar control of the mean annual temperature by solar activity. These 680 Ma old deposits were found on the Elatina formation in the Flinders Range, North of Adelaide. They were formed during the Marinoan glaciation, one of the series of glacial. episodes which characterize the end of the Precambrian era. The paleographic setting of the site indicates a marked seasonal, arid, periglacial climate. These circumstances imply a strong seasonal control of the glacier meltwater discharge into the periglacial lake. Laminae characteristic of distal clastic varves were obtained, avering a total period of 19, 000 va (varve years). Non-random cyclicities in the thickness of the annual varves were found in the sample showing characteristic periods of 8 to 16 va, with a mean period of 12 years. Recent re-analysis by Sonett and Williams (1987) indicate that the actual average period close to 11.7 va. Longer periods of 22 and a 314 years were also obtained in the Fourier analysis of the time series. Comparison with modern analogues indicates that the Elatina varve thickness may be interpreted as an in-

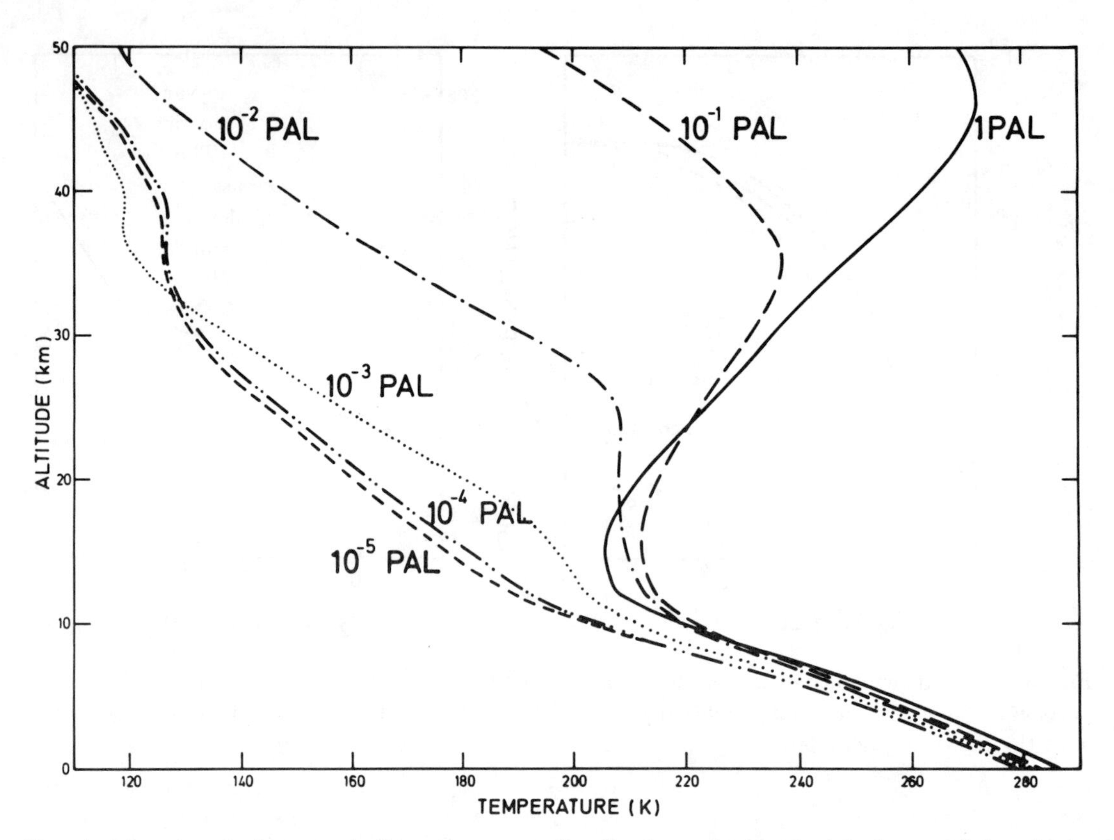


Fig. 4. Mean vertical atmospheric temperature distribution calculated with the coupled radiative-convective photochemical model for different mixing ratios of $O_2$ at the Earth's surface.

dicator of the mean annual temperature. Therefore, the Elatina series can be considered as a proxy of the time-variation of the mean annual temperature.

A weak solar influence on glacial climate was also discovered in the analysis of a 236-year varve sequence from Skilak Lake, Alaska (Sonett and Williams, 1985) . The geographic location of the lake, the climatic environment and seasonality of meltwaters make it a good analogy with what appear to have been the conditions prevailing at the time of the Elatina deposition. However, the Elatina periglacial environment was dryer and probably free of glacial ice whereas Skilak Lake is in a relatively humid periglacial climate with permanent glaciers as a source of meltwater. A correlation study between the varve sequence thickness and the sunspot indices between years 1700 to 1930 reveals a good correlation between the two series with common periods of 11 and 22 years. By contrast, analysis of three time series of varves from the Eocene Green River formation (Crowley et al., 1986) showed only a weak 11-year signal during a restricted portion of the 7496-year sample core deposited approximately 50 million years ago.

Finally, 2 Ga-old varve records from the BIFs found in the Harmersley basin in Australia, show clear evidence of a 23.3 yr cycle which may be associated with the Hale (22-year) solar cycle. However, Walker and Zahnle (1986) interpreted this periodicity as reflecting the climatic influence of the lunar nodal tide, which is weakly present in modern climate records with a period of 18.6 years. In this theory, it would arise from the precession of the Moon's orbital plane occuring with a longer period at the Precambrian as a result of the smaller Moon-Earth distance in the past.

The clarity of the solar-cycle signature in the Elatina Formation and the remarkable length of the record make this set of data unique at the present time. Compared to the general absence of solar signal in published varve series, it is legitimate to wonder whether specific environmental conditions prevailed near the end of the Precambrian era which could explain the specificity of the Elatina response to the solar cycle forcing.

It is striking that the Elatina varves were formed during a period characterized by a series of glacial ages occurring between 0.9 and 0.6 Ma ago: a most unusual climatic event. The distribution of continental masses is difficult to reconstruct for this remote period. However, paleomagnetic studies indicate that the three major phases of extensive glaciations occured at low latitudes. Indeed, most regions of the Earth, with the possible exception of Antarctica, containing Precambrian rocks show evidence for glaciation during this phase of the Earth's history (Crowley, 1983, Christie-Blick, 1982). More specifically, a recent study of paleomagnetic measurements of the remanent magnetization of the Elatina Formation was made by Embleton and Williams (1986). The results clearly demonstrate that the deposition was made at a paleomagnetic latitude of 5 degrees, in agreement with previous determinations from other paleomagnetic Australian late Precambrian rocks (Huimin and Wenzhi, 1985). Thus, the occurence of ice sheets and periglacial climate near sea level at low latitudes, considered together with the apparent absence of glaciation during the late Precambrian is in itself, as a major enigma. The climatic seasonality responsible for the deposition of varves is also difficult to explain if the region was located at low latitudes at this period.

The periodicities associated with the Elatina varves have been interpreted as being of solar origin since the 11-yr and longer periods known to occur in the modern sun were found in the sediments. Could the different atmospheric conditions at the late Precambrian be responsible for the larger climatic response to the solar forcing? A possible mechanism linking the global climate and the solar activity cycle was analyzed by Gérard and François (1987). In this scenario, the lower ozone level in the atmosphere 680 Ma ago and its different vertical distribution could possibly have made the troposphere more responsive to the 11-yr modulation of the UV solar irradiance. Indeed, fluctuations of the temperatures associated with the 27-day and 11-yr cycles are observed today in the stratosphere. (Keating et al. 1986; Chandra, 1984). It may be expected that in an atmosphere poorer in ozone and with an ozone peak closer to the ground, the response of the surface temperature would be amplified compared to the modern atmosphere. This idea was quantitalively examined using the coupled chemical-R-C model described above. Since the Elatina varves were deposited during a periglacial period, the model cloud layer is fixed at low altitude with an optical thickness $\tau = 10$ to produce surface temperatures in the range 268-275 K, about 15 to 20 K lower than for the present atmosphere.

The response of the global surface temperature to an imposed variation of the solar UV irradiance associated with the 11-yr solar cycle is strongly dependent on the spectral distribution of the modulation. Therefore, two models of UV solar cycle modulation were tested. The first one (case A), assumed no variability above 200 nm and a max/min irradiance ratio varying linearly from 2.6 at 120 nm to 1 at 200 nm. The mean solar irradiance distribution reported by Brasseur and Simon (1981) was adopted. In this case, the calculated surface temperature variation remains negligibly small ( 0.03 K). In the second model, the solar cycle modulation extends to 300 nm. In this case (case B) the surface temperature response is much larger and exceeds 0.5 K at 3 x 10 $^{-3}$ PAL of $O_2$. A significant response is only obtained in case A if the variability ratio max/min is substantially increased.

Figure 5 illustrates the dependence of $T_s$ and the ozone vertical column on the $O_2$ level for both cases. At low $O_2$ levels, the peak of ozone is located near 10 km, independently of the oxygen concentration. The magnitude of the temperature increases in parallel to the ozone column since the "greenhouse" fluctuation is progressively enhanced. However, for $O_2 > 10^{-3}$ PAL, the ozone maximum and the altitude of the maximum ozone variation shift toward higher altitudes. This altitude change of the solar cycle response tends to reduce the amplitude of the surface temperature response until, at still higher levels, the sign of the $T_s$ modulation itself is inverted. This dependence on $O_2$ is complex and depends on the thermal structure as well as the altitude of the ultraviolet energy deposition. Numerical tests also indicate that these results are nearly unaffected by an increase of the $CO_2$ partial pressure. By contrast, it is found that during cold climatic conditions, corresponding to a dryer troposphere and to a larger ozone tropospheric content, the solar ultraviolet radiation penetrates less deep in the atmosphere. Both effects contribute to increase the variation of the ozone column during a solar cycle. The model indicates that the temperature variation response increases by a factor of 3.5 when the mean surface temperature drops from 18° C to 0° C. This factor may explain, in part, the specificity of the Elatina periglacial environment to the solar cycle activity.

## Summary

Energy-balance and radiative-convective models show that the past reduced solar luminosity should have generated global irreversible glaciations as a result of the ice-albedo feedback. Geologic and paleontologic evidence indicates the absence of a totally frozen Earth at any stage of the planet's evolution. Increased levels of active greenhouse gases such as $CO_2$ appear as the most plausible compensating factor to maintain the global climate within the limited range by paleoclimates. Numerical models

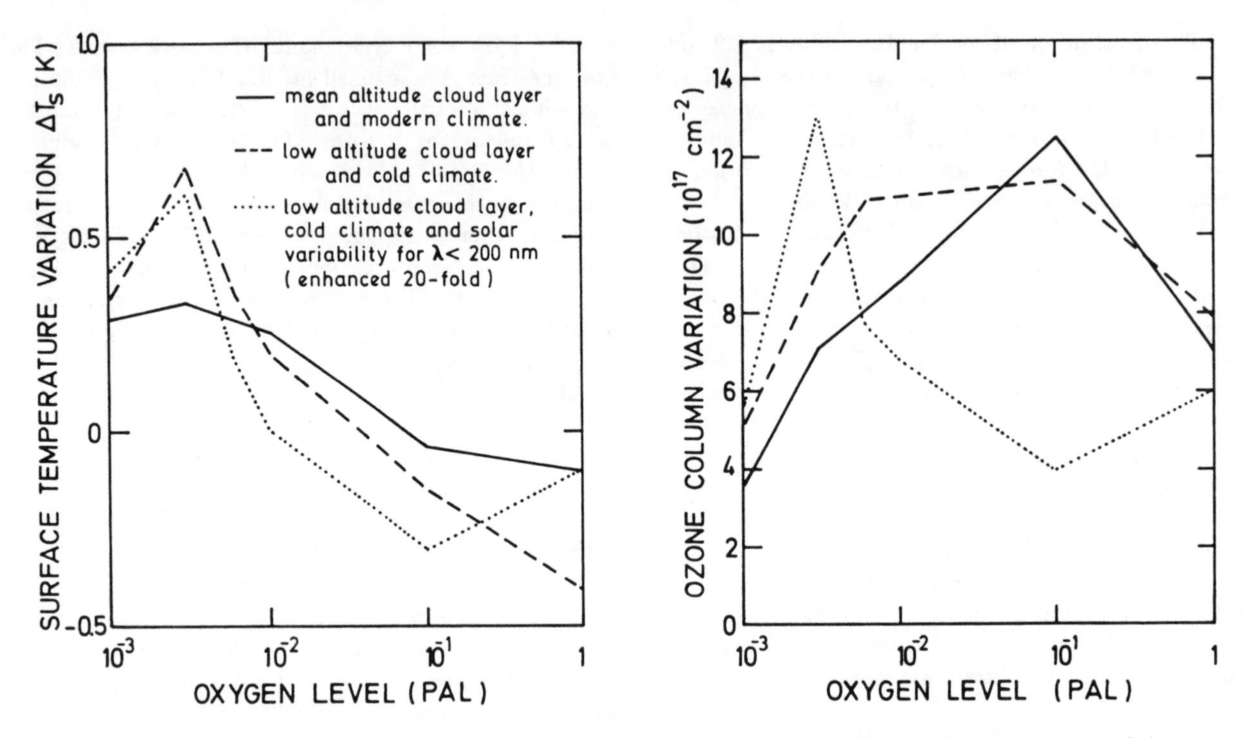


Fig. 5. Variations in surface temperature (left) and vertical ozone column (right) as a function of the oxygen level in the atmosphere from $10^{-3}$ to 1 PAL. The dashed line is obtained for a single cloud layer located at 750 mbar and a mean surface temperature between - 5 and + 2°$C$. The second curve (dotted line) is for a solar UV variability confined to wavelengths below 200 nm (see text).

show that it is possible to find atmospheric carbon dioxide evolution able to keep the global surface temperature within a few degrees of its contemporary value. The accumulation of photosynthetic oxygen in the atmosphere was another important source of climatic forcing. Its presence and the formation of the ozone layer significantly perturbed the thermal structure of the atmosphere and contributed to rise the surface temperature. The lower ozone abundance of the Precambrian atmosphere may have been an important factor to increase the climate response to solar cycle activity observed in the varves of the Elatina Formation. Numerical simulations indicate that the difference in solar irradiance between solar maximum and minimum conditions could have generated surface temperature variations substantially larger than in the present atmosphere.

*Acknowledgments.* I am especially grateful to my colleague Louis François for valuable discussions and critical reading of the manuscript. This research was supported by the Belgian National Foundation of Scientific Research (FNRS).

References

Brasseur G. and P.C. Simon, Stratospheric chemical and thermal response to long-term variability in solar UV irradiance, *J. Geoph. Res., 86,* 7343-7362, 1981.

Braterman P.S., A.G. Cairns-Smith and R.W. Sloper, Photo-oxidation of hydrated $Fe^{2+}$ - significance for banded iron formations, *Nature, 303,* 163-164,1963.

Chandra S., An assessment of possible ozone-solar cycle relationship inferred from NIMBUS 4 BUV Data, *J. Geophys. Res., 89,* 1373-1379 ,1984.

Christie-Blick N., Pre-Pleistocene glaciation on earth: Implications for climatic history of Mars, *Icarus, 50* , 423-443, 1982.

Cloud P., Aspects of Proterozoic biogeology, *Geol. Soc. Am. Mem., 161,* 245-251, 1983.

Crowley K.D., C.E. Duchon and J. Rhi, Climate record in varved sediments of the Eocene Green River Formation, *J. Geophys. Res., 91,* 8637-8647, 1986.

Crowley T.J., The geologic record of climate change, *Rev. Geophys. and Space Phys., 21,* 828-877, 1983.

Dickinson R.E. and R.J. Cicerone, Future global warming atmospheric trace gases, *Nature, 319,* 109-115, 1986.

Donner L. and V. Ramanathan, Methane and nitrous oxide: their effects on the terrestrial climate, *J. Atmos. Sci., 37*,119-124, 1980.

Embleton B.J.J. and Williams G.E., Low palaeolatitude of deposition for late precambrian periglacial varvites in South Australia: implications for palaeoclimatology, *Earth Planet. Sci. Letters,79* , 419-430 , 1986.

Endal A.D. and K.H. Schatten, The faint young sun-climate paradox: continental influences. *J. Geophys. Res., 87,* 7295-7302, 1982.

François L.M., Reducing power of ferrous iron in the Archean Ocean, 1. Role of $FeOH^+$ photooxidation, *Paleoceanography, 2* , 395-408, 1987.

François L.M., The effect of cloud parametrization and temperature profile on the climate sensitivity to ozone perturbations, *J.Atmos. Res., 21,* 305-321, 1988.

François L.M. and J.C. Gérard, Reducing power of ferrous iron in the Archean Ocean, 1. Contribution of photosynthetic oxygen, *Paleoceanography, 1,* 355-368, 1986.

François L.M. and J.C. Gérard, Ozone, climate and biospheric environment in the ancient oxygen-poor atmosphere, *Planet.Space Sci.,* in press ,1988.

Gérard J.C. and François L.M., A model of solar-cycle effects on palaeoclimate and its implictions for the Elatina Formation, *Nature, 326,* 577-580, 1987.

Gérard J.C. and L.M. François, A sensitivity study of the effects of solar luminosity changes on the Earth's global temperature, *Ann. Geophys., 6,* 101-112, 1988.

Hart M.H., The evolution of the atmosphere of the Earth, *Icarus, 33,* 23-39, 1978.

Holland H.D., *The chemical evolution of the atmosphere and oceans* , Princeton University Press, Princeton, 1984.

Huimin Z. and Z. Wenzhi Z., Paleomagnetic data, late precambrian magnetostratigraphy and tectonic evolution of eastern China, *Precambrian Res., 29,* 65-75, 1985.

Kasting J.F., Stability of ammonia in the primitive terrestrial atmosphere, *J. Geophys. Res., 87* , 3091-3098 ,1982.

Kasting J.F., Theoretical constraints on oxygen and carbon dioxide concentrations in the Precambrian atmosphere, *Precambrian Res., 34,* 205-229, 1987.

Kasting J.F. and T.M. Donahue, The evolution of atmospheric ozone, *J. Geophys. Res., 85,* 3255-3263, 1980.

Kasting J.F., J.B. Pollack, T.P. Ackerman, Response of the Earth's atmosphere to increases in solar flux and implications for loss of water from Venus, *Icarus, 57,* 335-355, 1984.

Keating G.M., M.C. Pitts, G. Brasseur and A. De Rudder, Response of middle atmosphere to short-term solar ultraviolet variations, 1, observations, *J. Geophys. Res., 92,* 889-902, 1986.

Kiehl J.T. and R.E. Dickinson, A study of the radiative effects of enhanced atmospheric $CO_2$ and $CH_4$ on early earth surface temperatures, *J. Geophys. Res., 92,* 2991-2998, 1987.

Kuhn W.R. and J.F. Kasting, Effects of increased $CO_2$ concentrations on surface temperature of the early earth., *Nature, 301,* 53-55, 1983.

Lacis A.A., J.E. Hansen, A parameterization for the absorption of solar radiation in the Earth's atmosphere, *J. Atmos. Sci., 31,* 118-133, 1974.

Lasaga A.C., R.A. Berner and R.M. Garrels R.M., An improved geochemical model of atmospheric $CO_2$ fluctuations over the past 100 million years, in *The carbon cycle and atmospheric $CO_2$: natural variations Archean to present*, E.T. Sundquist and W.S. Broecker eds., pp. 397-411, Geophysical Monograph 32, American Geophysical Union, Washington, 1985.

Levine J.S. and R.E. Boughner, The effect of paleoatmospheric ozone on surface temperature, *Icarus, 39,* 310-314, 1979.

Levine J.S., T.R. Augustsson, R.E. Boughner, M. Natarajan and L.J. Sacks, Comets and the photochemistry of the paleoatmosphere, in *Comets and the origin of life.*, edited by C. Ponnamperuma, pp. 161-190, D. Reidel, Dordrecht, 1981.

Manabe S. and R.T. Wetherald R.T., Thermal equilibrium of the atmosphere with a given distribution of relative humidity, *J. Atmos. Sci., 24,* 241-251 ,1967.

North G.R., R.F. Cahalan and J.A. Coakley Jr., Energy balance climate models, *Rev. Geophys. Space Phys., 19,* 91-121, 1981.

Owen T., R.D. Cess and V. Ramanathan, Enhanced $CO_2$ greenhouse to compensate for reduced solar luminosity on early earth, *Nature, 277,* 640-642, 1979.

Ramanathan V., Radiative Transfert within the earth's troposphere and Stratosphere: a simplified radiative-convective model, *J. Atmos. Sci., 33,* 1330-1346, 1976.

Rossow W.B., Henderson-Sellers A. and Weinreich S.K., Cloud feedback : a stabilizing effect for the early Earth?, *Science, 217,* 1245-1247, 1982.

Sagan C. and G. Mullen, Earth and Mars: Evolution of atmospheres and surface temperatures, *Science, 177,* 52-56, 1972.

Sasamori T., The radiative cooling calculation for application to general circulation experiments, *J. Appl. Meteorol., 7,* 721-729, 1968.

Schatten K.H. and A.S. Endal, The faint young sun - climate paradox: volcanic influences, *Geophys. Res. Lett., 9,* 1309-1311, 1982.

Schopf J.W., *Earth's earliest biosphere: its origin and evolution,* Princeton University Press, Princeton, N.J., 1983.

Sonett C.P. and G.E. Williams, Solar affinity of sedimentary cycles in the late precambrian elatina formation, *Aust. J. Phys. 38,* 1027-1043, 1985.

Sonett C.P. and G.E. Williams, Frequency modulation and stochastic variability of the Elatina varve record : a proxy for solar cyclicity?, *Solar Phys., 110,*397-410, 1987.

Walker J.C.G. and K.J. Zahnle, Lunar nodal tide and distance to the Moon during the precambrian,*Nature, 320,* 600-602, 1986.

Wang W.C. and P.H. Stone, Effect of ice-albedo feedback on global sensitivity in a one-dimensional radiative-convective climate model, *J. Atmos. Sci., 37,* 545-552, 1980.

Wang W.C., J.P. Pinto and Y. L. Yung, Climatic effects due to halogenated components in the Earth's atmosphere, *J. Atmos. Sci., 37,* 333-338, 1980.

Williams G.E., Sunspot periods in the late precambrian glacial climate and solar-planetary relations, *Nature,* 291, 624-628, 1981.

Williams G.E. and C.P. Sonett, Solar signature in sedimentary cycles from the late precambrian Elatina Formation, Australia, *Nature, 318,* 523-527, 1985.

# STUDIES OF CRETACEOUS CLIMATE

Eric J. Barron

Earth System Science Center, Penn State University,
University Park, Pennsylvania

Abstract. The warm Cretaceous climate is characterized by (1) warm oceanic deep water, >12°C, (2) tropical surface ocean temperatures similar or slightly warmer than today's values, (3) warm poles as evidenced by abundant high latitude floras and faunas and no direct evidence for permanent ice, (4) continental interiors with a smaller degree of seasonality, (5) a globally averaged surface temperature 6-12°C higher than the present day value, and (6) unusually abundant coals, evaporites, bauxites, black shales and other rock types.

A significant role by external forcing factors is required to explain the Cretaceous warmth. The leading candidates for explaining warm global temperatures are the very different Cretaceous continental geometry, the decrease in continental area due to high eustatic sea level, higher atmospheric $CO_2$ concentrations in the Cretaceous atmosphere, and amplification of these factors through a greater role of the ocean in poleward heat transport.

A variety of climate models, including Energy Balance models, mean annual atmospheric General Circulation Models (GCM's), annually varying atmospheric GCMs with simple coupled oceans and ocean GCMs have been utilized to study Cretaceous climate. Among many different insights from model simulations, several are notable: (1) multiple forcing factors including a higher atmospheric $CO_2$ level are required to achieve Cretaceous warmth if the spectrum of simulations are reasonable estimators of climate sensitivity, (2) a dramatic change in the role of the ocean would create additional problems rather than solutions to the question of Cretaceous warmth, (3) some permanent ice in Antarctica may be easier to justify than the assumption that the globe was entirely ice free and (4) in focusing attention solely on temperature considerable climate variation has been ignored which will provide important constraints on understanding the record of Cretaceous climate.



## Introduction

Paleoclimatology has been revolutionized by the theory of plate tectonics and the development of methods to reconstruct global geography. First, paleogeography provides the framework for interpreting environmental data. Without this framework climatic reconstruction on long time scales is untenable. Second, plate tectonics and a constantly changing geography provide a spectrum of reasonably well-defined causes for climatic change. A series of papers ascribing climate change to changes in geography, in particular explaining the contrasts between major glacial episodes and warm, apparently ice-free climates (Crowell and Frakes, 1970; Luyendyk et al., 1972; Frakes and Kemp, 1972, 1973; Hays and Pitman, 1973; Berggren and Hollister, 1974; Kennett, 1977; Donn and Shaw, 1977; Tarling, 1978; Beaty, 1978) appeared in concert with the first publications of past continental reconstructions and major efforts to determine the evolution of the major ocean basins.

Each of these contributions focussed on one of a few simple, but compelling arguments on how geography might alter the global climate: (1) land provides a surface for the accumulation of high albedo snow, and consequently greater land area at the pole promotes glaciation, (2) the distribution of land can greatly influence, by barriers or gateways, the poleward heat transport by the oceans or (3) because of the different thermal and albedo properties of land and sea, land-sea distribution and total land area will modulate climate.

Paleoclimatology is also experiencing great advances through the development and application of comprehensive models of the climate system. A hierarchy of climate models have been applied to test specifically whether paleogeography explains the major changes in climate found in the geologic record. The majority of these applications have considered the Cretaceous time period (~65 to 140 million years ago) because of the record of warmth and large differences in global geography (e.g. Barron, 1983). The first efforts to explore geography as an explanation of

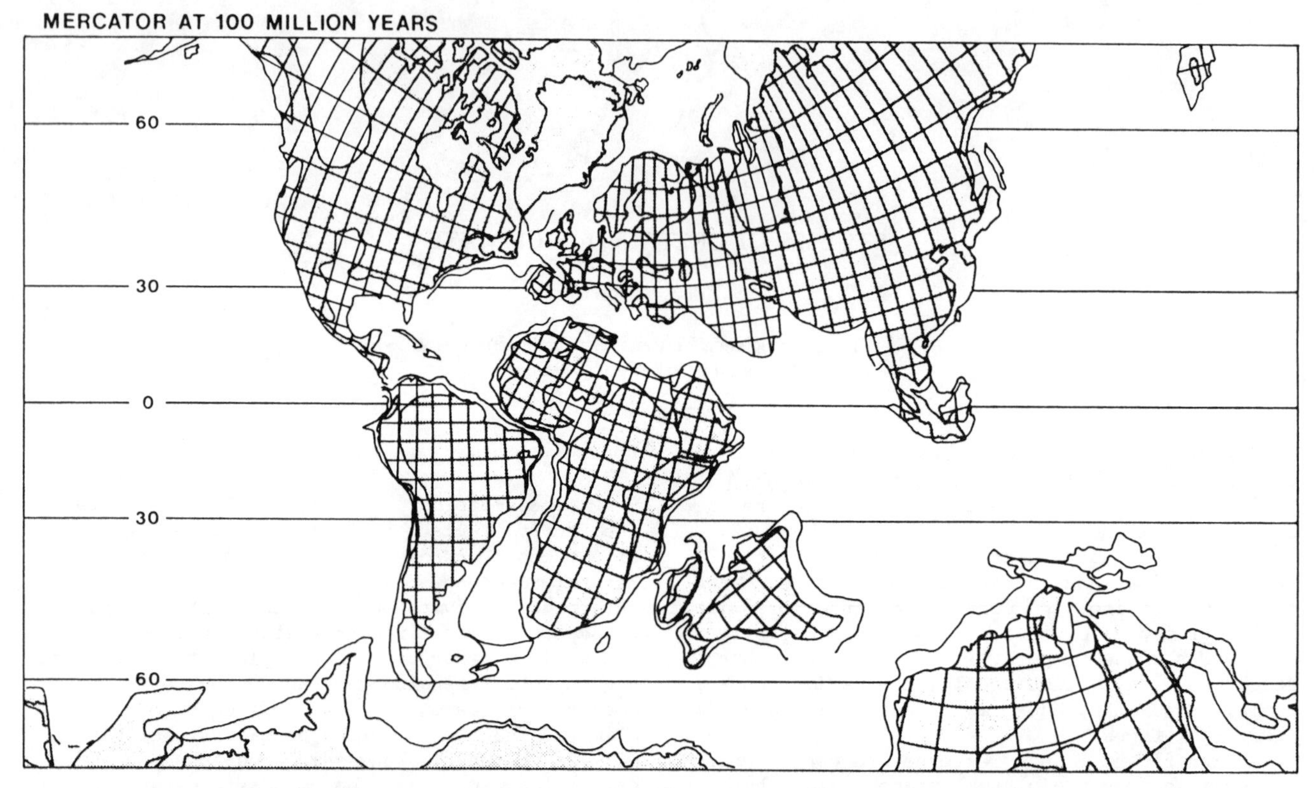


Fig. 1. A reconstruction of mid-Cretaceous geography extending to 70°N and S (after Barron et al., 1981b). Present continental outlines and the 2000m isobath are included with mid-Cretaceous continental area shaded. Thirty degree paleolatitude lines are indicated.

Cretaceous warmth (Barron et al., 1981a; Barron and Washington, 1984) demonstrated that a hierarchy of models fail to achieve Cretaceous warmth given paleogeography as the primary forcing factor.

However, these contributions did not refute hypotheses that geography explained Cretaceous warmth. Rather, they left as an open question whether the discrepancy between the model experiments and climatic reconstructions could be explained because of (1) model limitations, (2) the need for climatic forcing factors in addition to geography and/or (3) the need to reinterpret Cretaceous paleoclimatic data. This contribution summarizes a series of subsequent efforts to demonstrate or eliminate one or more of these points as an explanation for the discrepancy between model results and climatic data. At the end of this effort the same three questions are pertinent, however the evidence is clearer: (1) geography is a substantial climatic forcing factor but (2) geography is unlikely to be the sole explanation for Cretaceous climate.

## Climate Models and Cretaceous Temperatures: The Major Discrepancies

The initial efforts to determine the role of Cretaceous geography (Figure 1) in explaining Cretaceous temperatures (Figure 2) focussed on experiments with an Energy Balance Model (Barron et al., 1981a) and with a mean annual General Circulation Model (Barron and Washington, 1984).

Barron et al. (1981a) utilized a model based on a zonally averaged energy balance of the vertically integrated earth-atmosphere system computed for the annual cycle. The energy transported poleward was approximated as a diffusion process. The zonally averaged land distribution was derived from Figure 1. Two conclusions were apparent. First, without major prescribed changes in cloud cover or some additional net global heating, the Cretaceous planetary temperature could not be achieved (a 6-12°C increase in comparison with the present day). The EBM achieved a 1.6°K warming as a function of geography alone. Second, in each

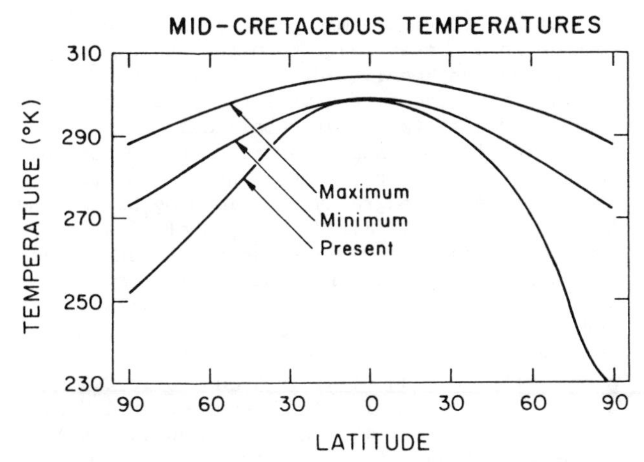


Fig. 2. Estimates of mid-Cretaceous mean annual temperatures (° Kelvin) (maximum and minimum interpretations) based on data described in Barron (1981a), and given in reference to present day values.

sensitivity experiment, the greatest response in temperature occurred in high latitudes, but a similar sign (less magnitude) change occurred as well in the tropics. Evidently to achieve sufficiently warm poles the model would generate tropical temperatures which would exceed observations and some life limits.

Barron and Washington (1984) completed a similar experiment utilizing a version of the Community Climate Model (CCM), a spectral General Circulation Model of the atmosphere at the National Center for Atmospheric Research. The associated global grid has 40 latitudes and 48 longitudes and the model uses nine levels in the vertical. The model version used by Barron and Washington (1984) utilized mean annual insolation and an energy balance ocean. Sea surface temperatures were computed based on the surface energy balance and the ocean model lacks heat transport or heat storage. The Cretaceous geography experiment resulted in a globally averaged surface temperature warming of 4.8°K (Figure 3). The same two conclusions derived from the EBM studies above applied in this case, although the CCM had a greater sensitivity to geography than did the EBM experiment.

What explains the discrepancy between the climate models and the Cretaceous temperature data? From the perspective of climate models, the limitations of the above experiments are clear, lack of a realistic ocean in both cases, a seasonal cycle in the EBM but highly parameterized poleward heat transport and more realistic poleward heat transport but a lack of a seasonal cycle in the CCM experiment.

Cloud-climate feedback must also be considered as a potential limitation of any current climate model experiment. The efforts to resolve the role of model limitations in explaining the discrepancy between model results and Cretaceous temperature data were based on an investigation of the role of oceans and the seasonal cycle in CCM simulations.

## The Role of the Oceans in Explaining Cretaceous Temperatures

A greater role of the oceans in poleward heat transport as a mechanism for explaining Cretaceous polar warmth receives considerable qualitative support. First, in order to explain warmer poles with a reduced temperature gradient a greater portion of the heat transport must have been accomplished by the oceans, since with such conditions the role of the atmosphere must be reduced. Second, the presence or absence of land or land barriers, particularly at high latitudes should have a marked effect on oceanic poleward heat transport and therefore presumably on the total poleward heat transport of the ocean-atmosphere

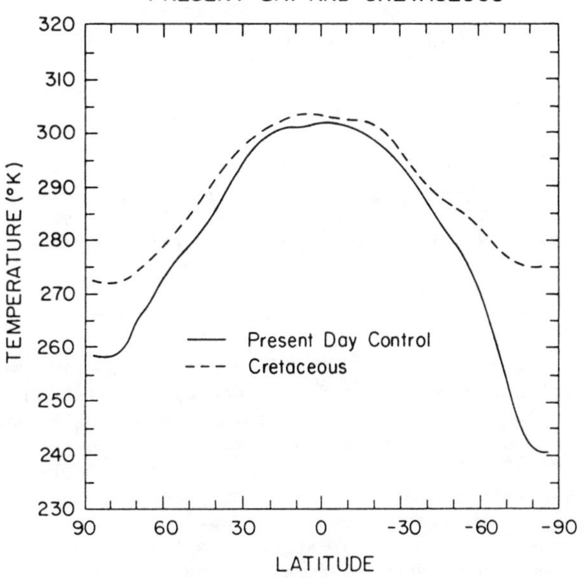


Fig. 3. Zonally-averaged surface temperatures (° Kelvin) derived from mean-annual simulations using the NCAR Community Climate Model (CCM), a spectral General Circulation Model of the Atmosphere. Present-day simulated surface temperatures are compared with values generated from a simulation with mid-Cretaceous geography (after Barron and Washington, 1984).

system. In fact, the role of ocean heat transport in climatic change is rather poorly known (see review by Covey and Barron, 1988). However, for the specific case of the Cretaceous a series of sensitivity experiments with the CCM were completed by Schneider et al. (1985) to evaluate whether a greater role by the ocean could explain the model-observation discrepancies cited above.

The Schneider et al. (1985) experiments did not utilize a dynamic ocean circulation model coupled to the CCM. Instead, extreme assumptions on the efficiency of ocean heat transport and how these assumptions could alter sea surface temperature were used to specify sea surface temperatures in CCM experiments. The experiments were not designed to be realistic, but rather to evaluate the impact of a greater role by the ocean and reduced equator-to-pole temperature gradients on the climate. In particular, a greater role by the oceans and warmer polar oceans (even if set at 20°C) did not prove sufficient to maintain high latitude continental regions above freezing in winter. This result is in apparent contrast to data of extreme warmth and lack of winter freezing at high latitudes. In fact, experiments with an assumed or implied very efficient ocean heat transport resulted in substantially colder continental interiors in winter. This result occurred because of a decreased vigor of the atmospheric circulation, which in turn provides insufficient ocean-to-land atmospheric heat transport to mitigate mid-winter continental radiative cooling.

If the above results are valid, a greater role by the oceans is unlikely to solve the problems presented by comparing Cretaceous climate model experiments with Cretaceous temperature data.

## The Role of the Seasonal Cycle in Explaining Cretaceous Temperatures

The role of the annual cycle of insolation was not addressed in the first general circulation model experiments to investigate geography as an explanation of Cretaceous warmth. Crowley et al. (1986) emphasize that, in particular, summer temperatures may be a critical parameter in initiating glaciation. Depressed summer temperatures on high latitude continents may allow winter snow to remain throughout the year. Of further importance, few geologic indicators respond to mean annual conditions. Seasonal information is essential to compare model results with observations and to ascertain the real nature of any discrepancy.

A full annual cycle simulation was performed with mid-Cretaceous geography using the version of the CCM described by Barron and Washington (1984) and discussed above, coupled with a mixed layer ocean. The ocean model makes use of a simple thermodynamic equation for heat storage in an ocean layer 50 meters thick (see Washington and Meehl, 1984 for additional discussion). The heat flux into the ocean is given by the surface energy balance.

The results of this experiment are in contrast with the results for the mean annual simulation with Cretaceous geography described by Barron and Washington (1984). First, the globally averaged surface temperature warming due to the change in geography and removal of permanent ice caps is near 3.0°C in the annual cycle experiment compared to 4.8°C in the mean annual experiment. Interestingly, using the same two models, the sensitivity to a doubling in atmospheric $CO_2$ (Washington and Meehl, 1984) was greater in the annual cycle experiment (3.5°C) than in the mean annual experiment (1.3°C). The greater sensitivity of the model to higher $CO_2$ levels evidently occurred because of the effect of incorporating the mixed layer on the latitude of the sea ice margin.

The primary question to be addressed is why the annual cycle experiment with Cretaceous geography illustrated less sensitivity than the mean annual model. The seasonal cycle experiment resulted in a greater discrepancy between model results and observations. A series of sensitivity experiments demonstrated which aspect of Cretaceous geography resulted in the warming in the mean annual experiments (Barron and Washington, 1984). Reduced northern hemisphere land area at high latitudes in comparison with the present day explained the majority of the Cretaceous northern hemisphere warming illustrated in Figure 3. Given mean annual insolation and an equilibrium simulation for an ocean without heat capacity, the differences between land and ocean surface characteristics dominate the model-derived northern hemisphere warming. Surface temperature isotherms are displaced poleward over land and ocean due to a warming related to the decrease in total land area.

The seasonal simulation included ocean heat storage. In this case the difference in land and ocean surface characteristics is much less significant than is the thermal inertia of the system. Summer heat storage and winter heat loss result in northern hemisphere ocean mixed layer temperatures which are not substantially different from the present day control experiment. The increased area of ocean resulted in a zonally averaged temperature at mid to higher latitudes in the northern hemisphere which is cooler in summer and warmer in winter. This model response is directly comparable to the change in land area (Figure 4) or the degree of continental flooding.

In fact, the majority of the globally-averaged surface temperature increase in the annual cycle experiment is in response to the removal of the Antarctic ice sheet. If this aspect is ignored,

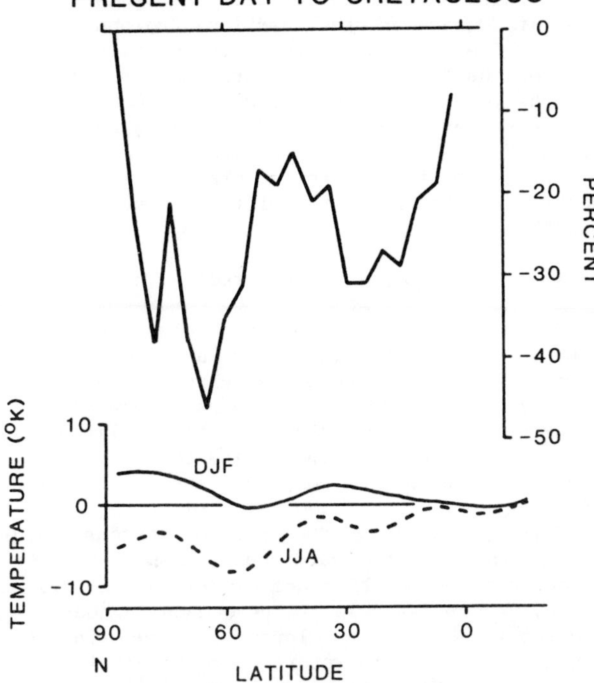


Fig. 4. Zonally-averaged surface temperature (° Kelvin) differences (Cretaceous minus the present day) for annual cycle simulations using the NCAR CCM coupled with a mixed layer ocean. DJF - December, January, February average. JJA - June, July, August average. The surface temperature differences are compared with the percentage change in land fraction for each 10° latitude belt from the Present day to the Cretaceous.

the change in geography resulted in a large change in the amplitude of the seasonal cycle, but little change in mean, global surface temperature. Evidently, the addition of the annual cycle of insolation does not solve the discrepancies between model simulations and Cretaceous observations. Further, this study continues to develop the theme noted by Rind (1986), that model sensitivity to geography may be quite different from model sensitivity to $CO_2$ variations.

### Other Model Factors

Potential model limitations which have a bearing on the problem of simulating Cretaceous warmth are not restricted to the ocean and the seasonal cycle. The discussions on the lack of agreement between model sensitivities to a $CO_2$ doubling described by Washington and Meehl (1986) and Schlesinger and Mitchell (1987) indicate a number of additional factors. The two most important factors are (1) cloud formulations and cloud-climate feedbacks and (2) snow and sea ice parameterizations and ice-albedo feedbacks of the control simulation. If the comparison of GCM studies for a $CO_2$ doubling are indicative, the GCM utilized by Barron and Washington (1984) can be characterized by a comparatively low sensitivity. In addition, this model has little negative lapse rate feedback at high latitudes in comparison with Manabe and Wetherald (1975). The model by Manabe and Wetherald (1975) had a stability-dependent vertical diffusion as do more recent versions of the Community Climate Model. This high latitude factor may be very significant for the degree of polar amplification of warming. As pointed out by Dickinson (1985) these factors, and the differences between climate models, suggest that potential model limitations cannot be eliminated as a solution to the discrepanies between model experiments and Cretaceous observations. Still the results of the annual cycle simulation and the lack of global temperature sensitivity in comparison with the same hierarchy of $CO_2$ experiments, indicates that other forcing factors, in addition to paleogeography, must also be considered as explanations of Cretaceous warmth.

### Are Other Climatic Forcing Factors Important

With the advent of plate tectonic theory, other climatic forcing factors which might explain the contrast between warm episodes and glacial time periods received little attention. Atmospheric $CO_2$ variations are the one significant exception. Variations in atmospheric carbon dioxide levels were suggested by Chamberlin (1899), Budyko and Ronov (1979) and Fischer (1982), but received little credible quantitative or observational support. Then, the geochemical models of Berner et al. (1983) and Lasaga et al. (1985), which are based on the carbonate-silicate geochemical cycle, provided much stronger support for the prospect of large variations in atmospheric $CO_2$ levels. Specifically these models linked plate tectonics (rapid sea floor spreading), continental area (flooding of continents is equated with rapid sea floor spreading), higher rates of volcanic degassing (with rapid sea floor spreading), geochemical weathering rates and fluxes, $CO_2$ levels and climate. The geochemical models predicted a several-fold increase in atmospheric $CO_2$ during the Cretaceous depending on a set of assumptions.

Much effort is now focussed on improving or examining limitations in the geochemical models (e.g. Berner and Barron, 1984; Kasting, 1984; Volk, 1987). The support for higher atmospheric $CO_2$ levels is not limited to model evidence, but may also be supported by observations of the sedimentary record (e.g. Sandberg, 1985) and independent evidence for higher rates of volcanism (Arthur et al., 1985). Consequently, higher $CO_2$ levels are a logical alternative to geography, or an additional climatic forcing factor.

However, the large number of assumptions and limitations in the geochemical cycle models prevent the $CO_2$ "forcing factor" from being specified in climate models in the same fashion as the geography. Consequently, the nature of the problem changes: what level of atmospheric $CO_2$ would be required to explain Cretaceous warmth given Cretaceous paleogeography, and is this level consistent with other sources of information.

The results of Barron and Washington (1985) indicate that a four-fold increase in atmospheric carbon dioxide in the mean annual version of the CCM produces a globally averaged surface temperature sufficient to reach the lower limit of Cretaceous temperature observations. Recognizing that this model is at the lower end of the GCM model spectrum in sensitivity to a $CO_2$-doubling, moderate and reasonable $CO_2$ level increases might well solve the discrepancy between model results and Cretaceous observations described earlier. The results (Figure 5) are not without problems, however. The degree of polar amplification of the warming seems insufficient, and tropical temperatures exceed paleoclimatic estimates. In fact, the temperatures in the model simulations are approaching some limits of tropical organisms. Even if $CO_2$ answers the question of global warmth, the results return us to potential model limitations (nature of poleward heat transport and the role of the oceans, cloud-climate feedbacks, and the nature of the lapse rate feedback at higher latitudes).

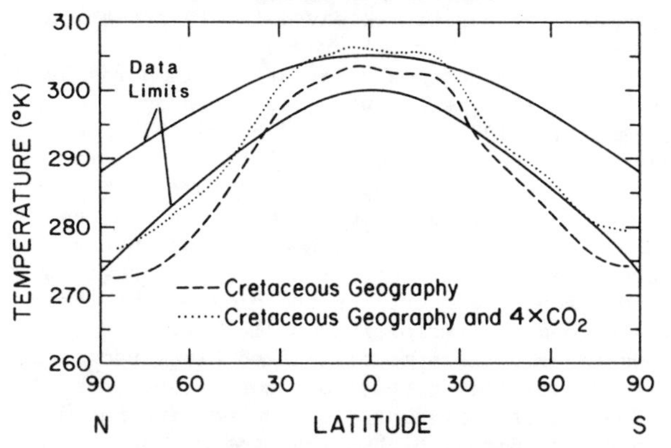


Fig. 5. A comparison of mid-Cretaceous temperature (° Kelvin) limits given in Figure 2 with mid-Cretaceous model results described in Figure 3 and a second simulation with Cretaceous geography and specified 4x present levels of atmospheric $CO_2$.

Finally, other forcing factors may also have contributed to the nature of the Cretaceous warmth.

## Is the Cretaceous Record Correctly Interpreted?

The differences between the Cretaceous model experiments and observations related to temperature could also be explained if the interpretations of the Cretaceous record are incorrect in two respects: (1) the amplitude of annual temperature in continental interiors and (2) the nature of polar climate, particularly in winter.

More specifically, the proposed increased role of the oceans in poleward heat transport was rejected based on the model experiments described above. A greater role by the oceans in heat transport resulted in a larger amplitude in the annual cycle of temperature in continental interiors at high latitudes. In addition, the constraint of continental equability and essentially frostless or very warm winters forces estimates of Cretaceous warmth to the upper limits illustrated in Figure 2.

The problems would be distinctly different if the Cretaceous record was strongly biased by maritime coastal climates, especially in the case of warmer polar oceans. In fact, the vast majority of data come from coastal regions, which is evident when the data are plotted on paleogeographic maps. There exists one important exception. Alligator-related reptiles, lake records and other indicators from interior Asia (Mongolia) support the interpretation of warmth and equability in Cretaceous continental interiors in the mid-latitudes. Such sites are rare, and much of the real weight of evidence for the amplitude of the seasonal cycle rests on this sparse data set.

The nature of polar climates is even more critical. Some of the better paleoclimatic evidence from fossil floras (e.g. Spicer and Parrish, 1986) indicates a north polar (~85 N) mean annual temperature near 10°C. Cool or cold winters are plausible. The interpretations for Antarctica are much less certain. All the evidence from Antarctica is associated with the continental margin or based on the lack of a record of Antarctic glaciation. The evidence for a lack of permanent ice is indirect. If Pleistocene arguments for the effect of polar ice on the oxygen isotopic record of the oceans (e.g. Emiliani and Shackleton, 1974) are applied to the

Cretaceous isotopic record, the occurrence of substantial ice is unlikely or can be rejected. If substantial ice enriched in 0-16 were present, the ice would be detectable in the oceanic oxygen isotopic record. Warmth at the Antarctic margin and the oxygen isotopic record support a case for the lack of permanent ice on Antarctica.

A number of cases can be presented which favor some permanent ice or at least seasonally subfreezing continental temperatures, either of which would substantially reduce the nature of the problems described here.

First, waxing and waning of ice sheets produces fast and large variations in sea level. We might expect that the character of the record of sea level variations for glacial and non-glacial climates would be different. However, differences are not noted in reconstructions of sea level by Haq et al., 1987 or Vail et al., 1977) throughout the last 100 million years.

Second, the Antarctic warmth is restricted to data from the continental margin. Numerous areas (e.g. New Zealand) can be cited as examples of coastal, maritime warmth adjacent to interior glaciers. Certainly, the condition of coastal warmth on a large polar continent is not necessarily indicative of interior conditions.

Third, the oxygen isotopic effect could be arguable if ice caps were small or the product of evaporation from adjacent warm oceans, with a short path length through the atmosphere (little distillation and fractionation of isotopes) and if Antarctica was characterized by little topographic expression (see Covey and Haagenson, 1984, for a discussion of variables which influence oxygen isotopic composition of snow and glacial ice). Certainly, the oxygen isotope record does not eliminate isolated glaciers, and might not eliminate small polar ice caps.

Even the condition of sub-freezing winter temperatures and summer warmth (a mean near zero) would substantially reduce the need for polar amplification of Cretaceous warming in model experiments. However, all the evidence described above is circumstantial and speaks only to possibilities. A substantial increase in knowledge from continental interiors and specifically from Antarctica is required to determine if the nature of Cretaceous warmth has been correctly interpreted.

## Geography as a Forcing Factor for Climatic Change

The role of geography in climatic change should be divided into two questions: (1) Is Cretaceous geography the explanation for Cretaceous warmth and (2) is geography an important factor in climatic change? The answer to the first question is probably negative and the answer to the second question is almost certainly positive.

The model studies summarized in this contribution suggest that Cretaceous geography is an insufficient forcing factor because (a) the magnitude of the global warming is insufficient and (b) the degree of polar amplification is insufficient. At least two of the major limitations in the model studies (the role of the oceans and of the annual cycle) were addressed without abrogating these deficiencies. However, the strength of the negative response to the first question is still constrained by model limitations. Three limitations have been identified: (1) surface-albedo feedbacks, (2) cloud-climate feedbacks, and (3) polar lapse rate feedbacks. In addition, the nature of the sensitivity experiments which attempted to examine the role of the oceans is far too crude to eliminate completely a different role by the oceans as a major causitive factor in explaining Cretaceous climate.

Cretaceous geography might also be the major explanation of Cretaceous climate if the nature of the Cretaceous warmth is mis-interpreted. Two key climatic characteristics have been identified in this regard: (1) the amplitude of the seasonal cycle in continental interiors and (2) the nature of polar climates especially in winter. In both cases, the data are either sparse or inferential. The Cretaceous was substantially warmer than the present climate, but the condition of cold continental interiors during winter at higher latitudes and permanent ice in the interior of Antarctica would substantially reduce the key differences between model experiments and observations. The model studies in this case have directed attention at the two weakest points in the Cretaceous climatic reconstructions.

If the stated model limitations and the weaknesses in the climatic record are not the answer to the problems associated with simulating sufficient global warmth and the degree of polar amplification of temperature then the alternative is additional or alternative climatic forcing factors. The series of sensitivity experiments indicate that alternatives must be considered and may be likely. The most plausible additional factor is the possibility of higher $CO_2$ levels. The level of $CO_2$ required to simulate sufficient global warmth (4-10x present day) is within the range suggested by geochemical models as reasonable or likely. Even in this case, however, the polar amplification of warmth may be problematic and the degree of tropical warming becomes a critical factor.

The weight of the arguments support a conclusion, with several caveats, that Cretaceous geography is insufficient to explain fully the observations of Cretaceous warmth. However, geography remains a substantial climatic forcing factor. A hierarchy of models yield an increase in globally averaged surface temperature of several degrees. The amplitude of the annual

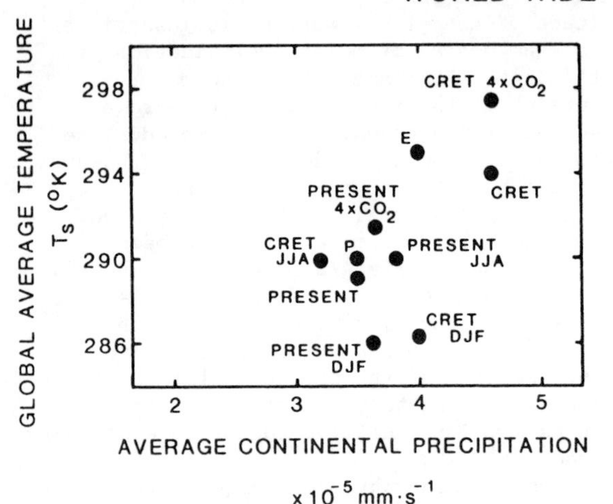


Fig. 6. A comparison of model simulated continental precipitation ( $x\ 10^{-5}$ $mm.s^{-1}$) with model simulated globally-averaged surface temperature (° Kelvin) for a series of NCAR Community Climate Model experiments. Cret - Cretaceous mean annual, Cret 4x - Cretaceous mean annual with 4x present day $CO_2$, Present - Present day mean annual, Pres 4x - Present day mean annual with 4x present day $CO_2$, P - hypothetical polar land mass, and E - hypothetical equatorial land mass. Present DJF, Present JJA, Cret DJF, and Cret JJA are annual cycle experiments for the Present day (Pres) and Cretaceous (Cret) for December, January and February (DJF) and June, July and August (JJA) averages.

cycle of temperature decreased by as much as 10°C in the annual cycle simulations in some latitude zones. Barron et al. (1984) estimate that the range in model simulated globally-averaged surface temperature due to variations in geography in the mean-annual version of the CCM is greater than 7°C. Each of these results is a substantial climatic sensitivity to geography.

In addition, the focus on temperature has hardly touched the degree to which geography influences climate. As one example, the range of CCM experiments described above alter continental precipitation (Figure 6) substantially, yielding almost an order of magnitude variation in continental run-off. Such differences have considerable implications for plants, plant productivity, continental weathering and geochemical cycles among other factors. Further, the role of geography in modifying the atmospheric circulation and the nature of the ocean circulation has received little attention. Consequently, the ways in which geography alter climate are only beginning to be identified and understood.

Acknowledgments. The author gratefully acknowledges the efforts of the convenors and editors, A. Berger, R. Rickinson and J. Kidson. This research was partially supported by Grant ATM-8715499 from the National Science Foundation.

References

Arthur, M. A., W. E. Dean, and S. O. Schlanger, Variations in the global carbon cycle during the Cretaceous related to climate, volcanism, and changes in atmospheric $CO_2$, in The Carbon Cycle and Atmospheric $CO_2$: Natural Variations Archean to Present, Geophysical Monograph Series: 32, edited by E. T. Sundquist and W. S. Broecker, pp. 504-530, AGU, Washington, DC, 1985.

Barron, E. J., A warm, equable Cretaceous: The nature of the problem, Earth Sci. Rev., 19, 309-338, 1983.

Barron, E. J., S. L. Thompson and S. H. Schneider, An Ice-Free Cretaceous? Results from Climate Model Simulations, Science, 212, 501-508, 1981a.

Barron, E. J., C. G. A. Harrison, J. L. Sloan, and W. W. Hay, Paleogeography, 180 million years ago to the present, Eclogae Geol. Helv, 74, 443-470, 1981b.

Barron, E. J., S. L. Thompson, and W. W. Hay, Continental distribution as a forcing factor for global-scale temperature, Nature, 310, 574-575, 1984.

Barron, E. J. and W. M. Washington, The role of geographic variables in explaining paleoclimates: Results from Cretaceous climate model sensitivity studies. J. Geophys. Res., 89, 1267-1279, 1984.

Barron, E. J. and W. M. Washington, Warm Cretaceous climates: high atmospheric $CO_2$ as a plausible mechanism, in The Carbon Cycle and Atmospheric $CO_2$: Natural Variations Archean to Present, Geophysical Monograph Series: 32, edited by E. T. Sundquist and W. S. Broecker, pp. 546-553, AGU, Washington, DC, 1985.

Beaty, C., The causes of glaciation, Am. J. Sci., 66, 452-459, 1978.

Berggren, W. A. and C. D. Hollister, Paleogeography, paleobiogeography and the history of circulation in the Atlantic Ocean, in Studies in Paleo-Oceanography, Soc. Econ. Paleotol. Min. Spec. Publ. 20, edited by W. W. Hay, pp. 126-186, 1974.

Berner, R. A., A. C. Lasaga, and R. M. Garrels, The carbonate-silicate geochemical cycle and its effect on atmospheric carbon dioxide over the past 100 million years, Am. J. Sci., 283, 641-683, 1983.

Berner, R. A. and E. J. Barron, Comments on the BLAG model: factors affecting atmospheric $CO_2$ and temperature over the past 100 million years, Am. J. Sci., 284, 1183-1192, 1984.

Budyko, M. and A. Ronov, Chemical evolution of the atmosphere in the Phanerozoic, Geokhimiya, 5, 643-653, 1979.

Chamberlin, T. C., An attempt to frame a working hypothesis of the cause of glacial periods on an atmospheric basis, J. Geol., 7, 545-585, 1899.

Covey, C. and E. Barron, The role of ocean heat transport in climatic change, Earth Sci. Rev., 24, 429-445, 1988.

Covey, C. and P. L. Haagenson, A model of oxygen isotope composition of precipitation: implications for paleoclimate, J. Geophys. Res., 89, 4647-4656, 1984.

Crowell, J. and L. A. Frakes, Phanerozoic glaciation and the causes of the Ice Ages, Am. J. Sci., 268, 193-224, 1970.

Crowley, T. J., D. A. Short, J. G. Mengel, and G. R. North, Role of seasonality in the evolution of climate during the last 100 million years, Science, 231, 579-584, 1986.

Dickinson, R. E., Climate sensitivity, Advances Geophys., 28A, 99-129, 1985.

Donn, W. and D. Shaw, Model of climate evolution based on continental drift and polar wandering, Geol. Soc. Am. Bull., 88, 390-396, 1977.

Emiliani, C. and N. J. Shackleton, The Brunhes Epoch: Isotopic paleotemperatures and geochronology, Science, 183, 511-514, 1974.

Fischer, A. G., Long-term climatic oscillations recorded in stratigraphy, in Climate in Earth History, edited by W. H. Berger and J. C. Crowell, pp. 97-104, National Academy Press, Washington, DC, 1982.

Frakes, L. A. and E. Kemp, Paleogene continental positions and evolution of climate, in Implications of Continental Drift to the Earth Sciences, edited by D. Tarling and S. Runcorn, pp. 535-550, Academic Press, New York, NY, 1973.

Frakes, L. A. and E. Kemp, Influence of continental positions on Early Tertiary climates, Nature, 240, 97-100, 1972.

Hag, B. U., J. Hardenbol, and P. R. Vail, Chronology of fluctuating sea levels since the Triassic, Science, 235, 1156-1167, 1987.

Hays, J. and W. Pitman, Lithospheric plate motion, sea level changes, and climatic and ecological consequences, Nature, 246, 18-22, 1973.

Kasting, J., Comments on the BLAG model: the carbonate-silicate geochemical cycle and its effect on atmospheric carbon dioxide over the past 100 million years, Am. J. Sci., 284, 1175-1182, 1984.

Kennett, J. P., Cenozoic evolution of Antarctic glaciation, the circum-Antarctic Ocean, and their impact on global paleoceanography, J. Geophys. Res., 82, 3843-3860, 1977.

Lasaga, A. C., R. A. Berner, and R. M. Garrels, An improved geochemical model of atmospheric $CO_2$ fluctuations over the past 100 million years, in The Carbon Cycle and Atmospheric $CO_2$: Natural Variations Archean to Present, Geophysical Monograph Series: 32, edited by E. T. Sundquist and W. S. Broecker, pp. 397-411, AGU, Washington, DC, 1985.

Luyendyk, B., D. Forsyth, and J. Phillips, An experimental approach to the paleocirculation of oceanic surface waters, Geol. Soc. Am. Bull., 83, 2649-2664, 1972.

Manabe, S. and R. T. Wetherald, The effects of doubling the $CO_2$ concentration on the climate of a general circulation model, J. Atmos. Sci., 32, 3-15, 1975.

Rind, D., The dynamics of warm and cold climates, J. Atmos. Sci., 43, 3-24, 1986.

Sandberg, P. A., Nonskeletal aragonite and $pCO_2$ in the Phanerozoic and Proterozoic, in The Carbon Cycle and Atmospheric $CO_2$: Natural Variations Archean to Present, Geophysical Monograph Series: 32, edited by E. T. Sundquist and W. S. Broecker, pp. 585-594, AGU, Washington, DC, 1985.

Schlesinger, M. E. and J. F. B. Mitchell, Climate model simulations of the equilibrium climate response to increased carbon dioxide, Rev. Geophys., 24, 760-798, 1987.

Schneider, S. H., S. L. Thompson, and E. J. Barron, Mid-Cretaceous continental surface temperatures: Are high $CO_2$ concentrations needed to simulate above freezing winter conditions?, in The Carbon Cycle and Atmospheric $CO_2$: Natural Variations Archean to Present, Geophys. Monogr. Ser. 32, edited by E. T. Sundquist and W. S. Broecker, pp. 554-560, AGU, Washington, DC, 1985.

Spicer, R. A. and J. T. Parrish, Paleobotanical evidence for cool north polar climates in middle Cretaceous (Albian-Cenomanian) time, Geology, 14, 703-706, 1986.

Tarling, D. H., The geological-geophysical framework of ice ages, in Climatic Change, edited by J. Gribben, pp. 3-24, Cambridge Univ. Press, Cambridge, 1978.

Vail, R. R., R. M. Mitchum, Jr., and S. Thompson, III, Seismic stratigraphy and global changes of sea level, in Seismic Stratigraphy and Global Changes of Sea Level, AAPG Memoir 26, edited by C. E. Payton, pp. 49-212, AAPG, Tulsa, Oklahoma, 1977.

Volk, T., Feedbacks between weathering and atmospheric $CO_2$ over the last 100 million years, Am. J. Sci., 287, 763-779, 1987.

Washington, W. M. and G. A. Meehl, General circulation model $CO_2$ sensitivity experiments: snow-sea ice albedo parameterizations and globally averaged surface air temperature, Climatic Change, 8, 231-241, 1986.

Washington, W. M. and G. A. Meehl, Seasonal cycle experiment on the climatic sensitivity due to a doubling of $CO_2$ with an atmospheric general circulation model coupled to a simple mixed-layer ocean model, J. Geophys. Res., 89, 9475-9503, 1984.

# SIMULATIONS OF THE LAST GLACIAL MAXIMUM WITH AN ATMOSPHERIC GENERAL CIRCULATION MODEL INCLUDING PALEOCLIMATIC TRACER CYCLES

Sylvie Joussaume[1], Jean Jouzel[2], and Robert Sadourny[1]

Abstract. Simulations of the Last Glacial Maximum have been performed with an atmospheric general circulation model. We focus on a new approach of the problem with the modeling of important climatic tracers: water isotopes and desert dust particles. The mean dependency of the water isotope content of precipitation with temperature is similar to the present-day one, but can depart locally. The global amount of dust is little changed for the ice age, but important changes are simulated over some regions.

## Introduction

Atmospheric general circulation models (AGCMs) can be useful tools to reconstruct past climates. This approach has been extensively applied to the Last Glacial Maximum (LGM 18000 years Before Present), prescribing boundary conditions (sea surface temperature, sea ice, topography, surface albedo) given by CLIMAP [1981]. To validate these simulations it is then important to compare model results with paleodata generally obtained through climatic tracers (e.g. water isotopes, dust, pollens). We propose, as a complementary approach, to introduce a modeling of important climatic tracer cycles within the AGCM. On one hand, this approach allows us to obtain a more direct comparison with observations. On the other hand, the knowledge of both the climatic response of tracers and of atmospheric parameters should help us in translating paleodata into atmospheric parameters. This approach has been applied to water isotopes (HDO,$H_2{}^{18}O$) and desert dust particles.

Present-day observations of water isotopes display a good correlation between isotope ratios of precipitation and surface air temperature, well understood by simple isotope models. However, the isotope atmospheric cycle is quite complex and is sensitive to the entire atmospheric circulation. To investigate the sensitivity to a glacial/interglacial oscillation, AGCMs are the only appropriate tools.

As regards desert dust particles, a large increase in the amount of dust in ice cores has been observed for the Last Glacial Maximum [Petit et al., 1984]. Thus AGCMs can help to investigate the link between climate (aridity, atmospheric circulation) and dust deposits.

## Modeling

Evolution in time of a tracer is governed by source and removal processes, which may be highly dependent on the nature and type of tracers, large scale transport and diffusion processes - turbulent vertical diffusion within the planetary boundary layer and convective diffusion within convective clouds. The transport and diffusion processes of tracers have been introduced using modeling schemes, which are similar to those found in our AGCM for the transport and diffusion of water vapor. The AGCM of the Laboratoire de Météorologie Dynamique is described in Sadourny and Laval [1984].

### Water isotopes

The modeling of the water isotope cycles in AGCMs follows the basic water cycle: surface evaporation and condensation, water vapor transport, atmospheric condensation processes, subsequent evaporation of precipitation and ground hydrology. At each phase change, differentiation occurs between $H_2{}^{16}O$ and the isotopes $H_2{}^{18}O$ and HDO because of the slight differences in their vapor pressures and their molecular diffusivities. The liquid phase is enriched in heavier isotopes relatively to the vapor phase. Advection is a critical process in isotope modeling because it is necessary to avoid the production of negative values and to transport isotopic ratios. A description of the modeling and preliminary results can be found in Joussaume et al. [1984].

---

[1]Laboratoire de Météorologie Dynamique
24 rue Lhomond, 75231 Paris Cedex 05, France
[2]Laboratoire de Géochimie Isotopique/LODYC, CEN/Saclay, 91191 Gif sur Yvette, France

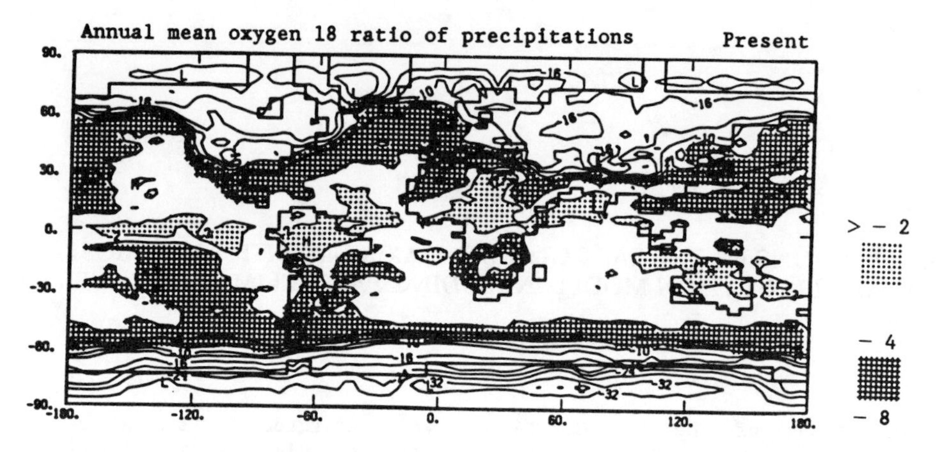


Fig. 1. Simulated oxygen 18 ratio of precipitations for the estimated annual mean (February + August) present day climate. Isolines every 2°/oo down to -12°/oo, every 4°/oo for lower values. Dotted areas above -2°/oo, hatched areas between -4°/oo and -8°/oo.

Desert dust

In this first approach, only one size range of particles has been modeled, in the order of 1 μm, to deal with long range transport. Morever, particles are considered here as passive scalars, which means that their eventual impact on climate, either as condensation nuclei and through their radiative properties, is not included.

Dust is raised by winds over desert zone areas. The extent of source regions is generated by the model itself as regions that are dry both in February an August. The surface flux of dust over source regions depends on the surface wind speed and on the vertical gradient of dust near the surface. A constant dust concentration is assumed at the soil-atmosphere interface arbitrarily taken equal to 1, which defines all dust quantities in arbitary units.

Removal by the rainout process has been included, assuming the same efficiency for particles and water vapor. Gravitational settling, although weak, has been considered. Dry deposition near the ground, for example by interception or impaction by obstacles, has been introduced depending on surface wind speed and dust concentration. For a more detailed description see Joussaume [1985].

Results

Simulations of February and August present-day and LGM climates (200 days each with tracers introduced over the last 100 days) have been performed using the LMD AGCM with the standard spatial resolution of 64 points in longitude, 50 points in latitude, and 11 levels, that corresponds to a horizontal mesh size of 400 x 400 km$^2$ at 50° of latitude. The results presented here are averaged over the last 60 days of the simulations, as 20-30 days only are necessary to reach an equilibrium regime for tracers.

Climate

Simulated results are in general agreement with other model results [e.g. Kutzbach and Guetter, 1986; Rind, 1987], for example a decrease in surface air temperature of 3.1° (February) and 4.1° (August), a southward shift of storm tracks in Northern Atlantic, a splitting of the Jet Stream around the Laurentide ice sheet (February) and a weaker summer Indian monsoon. This last feature is corroborated by observations. Globally, the hydrological cycle exhibits nearly no change in February and a small decrease in August (3%). However, the latitudinal distribution of evaporation and precipitation is modified. Note that simulations are short and that the statistical significance of results cannot thus be obtained.

Water isotopes

The simulated charts of the oxygen 18 ratio ($H_2{}^{18}O/H_2{}^{16}O$) of precipitation for present-day (Figure 1) and LGM climates exhibit a good dependency with temperature in middle to high latitudes. In tropical regions, the oxygen 18 ratio is modulated by the amount of precipitation. In high latitudes, the simulated difference (LGM-Present-day) in the oxygen 18 ratio of precipitation is in good agreement with ice core data (of the order of -5 °/oo in Antarctica). Considering the statistical dependency between $^{18}O$ in precipitation and surface air temperature (Figure 2), no clear difference is simulated between present-day and the LGM climates. However, the variability displayed in this relationship imp-

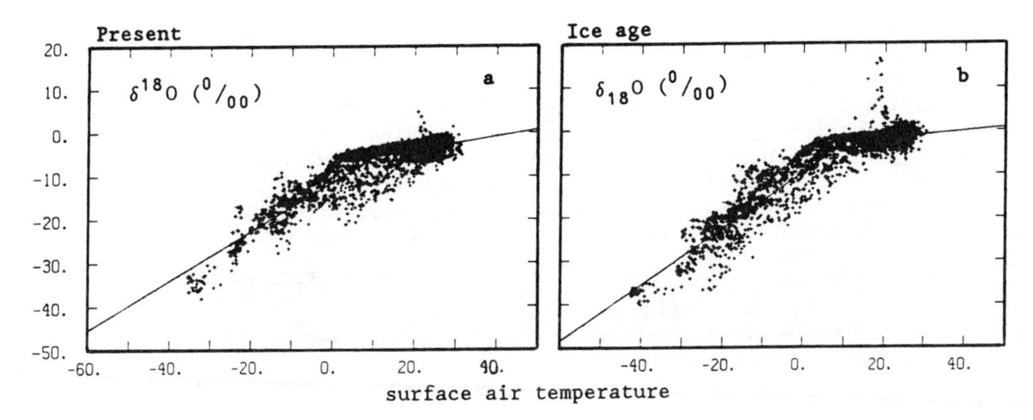


Fig. 2. Simulated relationship between the oxygen 18 ratio of precipitations and surface air temperature for the estimated annual mean (February + August) : a) present day climate, b) Last Glacial Maximum, and their associated regression lines (below 15°C, slopes a).57 and b).65).

lies some caution when interpreting regional results.

## Desert dust

The simulated extent of arid regions, considered as source areas of dust, is reasonable, except for an underestimation of the Australian desert. Simulated modern desert dust results show a clear seasonal cycle, with a double amount of the atmospheric content of dust in August compared to February [Joussaume et Sadourny, in press]. During the Last Glacial Maximum, a small increase is simulated by the model both for the source regions (+18%) and the atmospheric dust content (+8%), with a stronger increase in February, leading to a weaker seasonal contrast. However, important changes are simulated in some regions, for example over the tropical Atlantic Ocean, and over Europe (Figure 3) and are mainly associated with changes in the atmospheric circulation. The simulated change in dust deposits over East Antarctica is weak compared to observations [Petit et al., 1981]; this discrepancy is more likely to be due to an underestimation in the change of source regions for the LGM.

Acknowledgments. This work was supported by the Programme National d'Etude de la Dynamique du Climat. The computer time was contributed by the Centre de Calcul Vectoriel pour la Recherche.

## References

Joussaume S., Modélisation des cycles des espèces isotopiques de l'eau et des aérosols d'origine désertique dans un modèle de circulation générale de l'atmosphère, Thèse de 3ème cycle, Université de Paris VI, 1983.

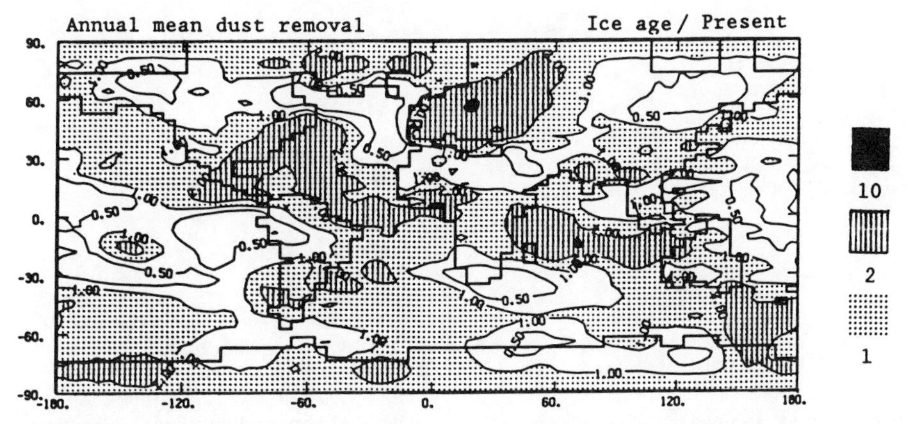


Fig. 3. Simulated total removal of dust (dry and wet) change : ice age/present day ratio for the estimated annual means. Isolines 1/2, 1, 2, 5 (shading for values above 1.)

Joussaume S., J. Jouzel and R. Sadourny, A general circulation model of water isotope cycles in the atmosphere, Nature, 311, 24-29, 1984.

Joussaume S., Simulation of airborne impurity cycles using atmospheric circulation models, Annals of Glaciology, 7, 131-137, 1985.

Joussaume S. and R. Sadourny, Simulation of the desert dust cycle using an atmospheric general circulation model, IAMAP Conference on "Aerosols and Climate", Vancouver 1987, published by A. Deepak, in press.

Kutzbach J.E. and P.J. Guetter, The influence of changing orbital parameters and surface boundary conditions on climate simulations for the past 18,000 years, J. Atmos. Sciences, 43, 1726-1759, 1986.

Petit J.R., M. Briat and A. Royer, Ice age aerosol content from East Antarctic ice core samples and past wind strength, Nature, 293, 391-394, 1981.

Rind D., Components of the ice age circulation, J. Geophys. Res., 92, 4241-4281, 1987.

Sadourny R. and K. Laval, January and July performance of the LMD general circulation model. In A. Berger and C. Nicolis (eds) New perspectives in Climate Modelling, Developments in Atmospheric Sciences, 16, Elsevier, 173-198 1984.

# PROGRESS AND FUTURE DEVELOPMENTS IN MODELLING THE CLIMATE SYSTEM WITH GENERAL CIRCULATION MODELS

P. R. Rowntree

Dynamical Climatology Branch, Meteorological Office
Bracknell, England, RG12 2SZ

## Introduction

The objective of this paper is to review the present status of climate models as a tool in studies of climate change, and to consider the needs for future developments. Progress in development of general circulation models (GCMs) for use in climate studies is a worldwide effort, with many of the major contributions coming from the USA modelling groups. For purposes of illustration here I have mainly used results from the UK Meteorological Office GCM, and must record my debt to those who did most of the actual work of running the models and extracting the results.

## Definition of Climate Models

Firstly, I must say what is generally meant by a climate model in the GCM context. It contains models of the atmosphere, ocean, sea-ice and land; since the pioneering GFDL models constructed by Smagorinsky, Manabe and Bryan in the 1960s the variables have been generally at least surface pressure, wind, temperature and moisture in the atmosphere, current, temperature and salinity in the ocean and snow, surface temperature and soil moisture on land. There have been increases in resolution and complexity. For example, the present UK Meteorological Office climate model has 11 atmosphere layers and 17 ocean layers with a 2 $^1/_2$° latitude by 3 $^3/_4$° longitude horizontal resolution, and a much more complete representation of the land surface is being incorporated.

## Uses of Climate Models

Let us now consider the uses of climate models. An important area is the investigation of natural, internal variability of the climate system; one example is the irregular 3-5 year time scale tropical Pacific ('El Nino') ocean surface temperature variation. Another major use is for the study of the effects of naturally forced perturbations of climate. These fall into 3 groups (Table 1). Firstly, there are astronomic variations:- in solar output these range from long trends over billions of years to shorter term variations, 22 or 80 year cycles perhaps; then there are the perturbations of the earth's orbit which have been discussed in this symposium by Peltier, and in a union lecture by Berger. Secondly, there are variations in atmospheric composition - in volcanic aerosols as discussed by Sigurdsson in this symposium, trace gases such as the $CO_2$ variations associated with glacial cycles, and tropospheric aerosols in which variations may arise due to changes in vegetation or surface winds -these may also be associated with glacial cycles. Thirdly, there are changes in surface boundary conditions - those due to continental drift as Barron has discussed in this symposium and those in vegetation, ice caps etc. associated with glacial cycles.

Turning now to man-made perturbations (Table 2) there are again three groups. Firstly, atmospheric composition - study of the effects of increases in trace gases is probably the most important use of climate models and is discussed elsewhere in these symposium proceedings. Secondly, man's tendency to change the land surface, like his effect on $CO_2$, imposes on climatologists the responsibility of assessing the effects of such changes (see Nicholson, this symposium). Thirdly, there are changes in water management - diversion of rivers, construction of reservoirs, drainage of wet lands. I have indicated under each item in Table 2, the processes in a climate model affected by the changes. This is relevant to the next topic -what requirements these uses of climate models place on the models. Table 3 shows these for some applications. There are two types. Firstly, to be able to study the effects of these changes it is necessary to be able to represent the perturbation. Thus, we cannot study the effects

Published 1989 by the American Geophysical Union.

TABLE 1. Natural Perturbations of Climate

(a) Astronomic variations:-
Solar output
(long term trends, oscillations)
Earth's orbit
(b) Atmospheric composition variations
Volcanic stratospheric aerosols
Trace gases, tropospheric aerosols
(glacial cycle variations in vegetation etc)
(c) Surface boundary conditions
Continental drift
Glacial cycles
(icecaps, sea level, lakes, vegetation)

of aerosols on precipitation formation without a sufficiently realistic representation of cloud microphysics, in which the effect of aerosol size distribution and other factors is taken into account. Deforestation cannot be properly simulated if the model's hydrology parametrization cannot represent the effects of a change from forest to, say, pasture.

The other requirement is that the effects be properly described. Here the need is not only for adequate model diagnostics but also sufficiently realistic simulations of the required model parameters. Just taking one example, agriculture (Table 3), you see that the list is quite extensive. Note also how what is needed is nearly always surface data - not upper winds and temperatures; indeed, of the non-surface variables, only clouds will occur frequently in the list of requirements and then mainly because of the impact on surface radiation. Also, it is often frequency distributions and interannual variability that are required, not means; and the frequency distributions may be complex - e.g., occurrence of low temperatures together with

TABLE 2. Manmade Perturbations of Climate

(a) Atmospheric composition
Trace gases
(fossil fuel, CFCs)
- Radiation
Aerosols
(land use, ocean biota)
- Radiation, cloud physics
(b) Land surface
(deforestation, afforestation, desertification, irrigation)
- Radiation, turbulence, hydrology
(c) Water management
(river diversions, reservoirs)
- Runoff, salinity, evaporation

TABLE 3. Requirements on Climate Models

(a) Represent perturbations
e.g. deforestation, trace gas radiation effects.
(b) Describe effects
e.g. Agriculture -
radiation
temperature
soil moisture (precipitation, evaporation)
snow cover, frost
sea level
similarly for:
Fisheries, Water, Power, Transport, Construction, etc.

strong winds for ship icing. However, today I shall restrict my discussion to time averaged results.

## Some Results

Let me now take you fairly quickly through some recent results, most of which provide some illustration of progress in modelling. One quite instructive model field is that of sea level pressure. The two top frames in Figure 1 show a 4 year observed northern winter normal - very like longer periods - and an 8 year mean from our model as it was around 1984. Note the excessive westerly flow, a problem common to relatively high resolution models at that time. This was not a problem in the Southern Hemisphere - not shown here (see Slingo and Pearson [1987]) but the simulation there is very good with mean sub-Antarctic pressures of about 980 mb and realistic sub-tropical ridges. This suggests a problem with orography, as this is relatively unimportant in the Southern Hemisphere. The first solution proposed was to enhance the orography (e.g. Hills [1979]) in order to represent the barrier effects of mountains better. Later, the envelope orography technique [Wallace et al, 1983] was tested [Slingo and Pearson, 1987]. This worked quite well in winter but the northern summer circulation was degraded, probably because the elevated heat sources were too elevated.

The second solution was to incorporate a representation of the vertical transfer of momentum by gravity waves, in other words gravity wave drag [Palmer et al., 1986, Slingo and Pearson, 1987]. This also had a dramatic effect (Figure 1c) on the simulation as easterly stress or momentum flux convergence, applied at the earth's surface over rough (i.e., mountainous) areas, was transferred to higher levels, where it decelerated the westerly flow. With this change, how realistic are model winds? The results in Figure 2 are for a 4 winter mean from our latest

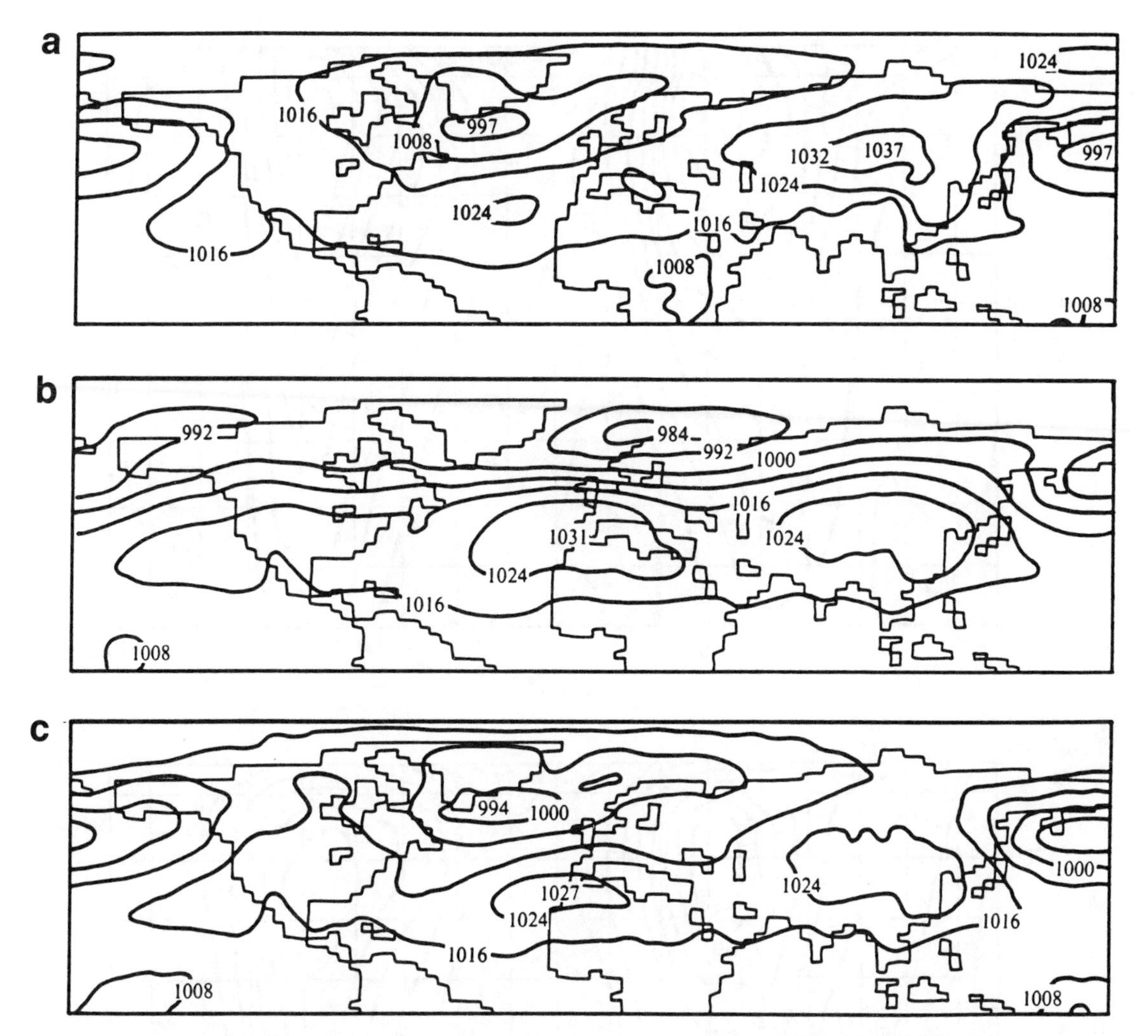


Fig. 1. December-February sea level pressure: (a) observed, based on Meteorological Office (MO) operational analyses for 1983-86; (b) modelled, average for 8 winters without gravity wave drag; (c) modelled, average for 4 winters with gravity wave drag.

long run of the model. The pattern of zonal mean winds is quite close to observations, though the jets are too strong. Differences greater than 5 m/s are shown in Figure 3 -mostly they are confined to the stratosphere. The other contours here are for temperature errors - mostly these are less than 4 K except in the polar stratospheres. A similar result is obtained in June to August (Figure 4). This stratospheric coolness is not a problem peculiar to our model - all three of the models used for $CO_2$ experiments reviewed by Schlesinger and Mitchell [1985] suffered from such errors, especially in the Southern Hemisphere. A possible cause is the rather poor resolution of the stratosphere - we have only three layers above 200 mb - in view of the important role of vertical motions in the stratospheric heat balance. However, Mahlman and Umscheid [1984] have errors approaching 10 K in the January of their simulation with a much higher resolution (40 level) model. One might also suspect radiation as a cause. The ICRCCM (Intercomparison of Radiation Codes for Climate Models) organised a valuable comparison of line-by-line models and parametrizations for climate models; Figure 5 compares two line-by-line comparisons of long wave cooling - GFDL and LMD. Though this shows generally good agreement, there are differences. Apart from the coarse resolution, the MO model's differences from the line-by-line calculations are of comparable magnitude; however there is too much cooling from 60 to 370 mb. Note that these schemes omit the water vapour continuum. Though this is a poorly understood area of radiation theory, a better representation would not affect the stratospheric cooling.

Further intercomparison programmes are underway - for boundary and surface fluxes, as discussed by

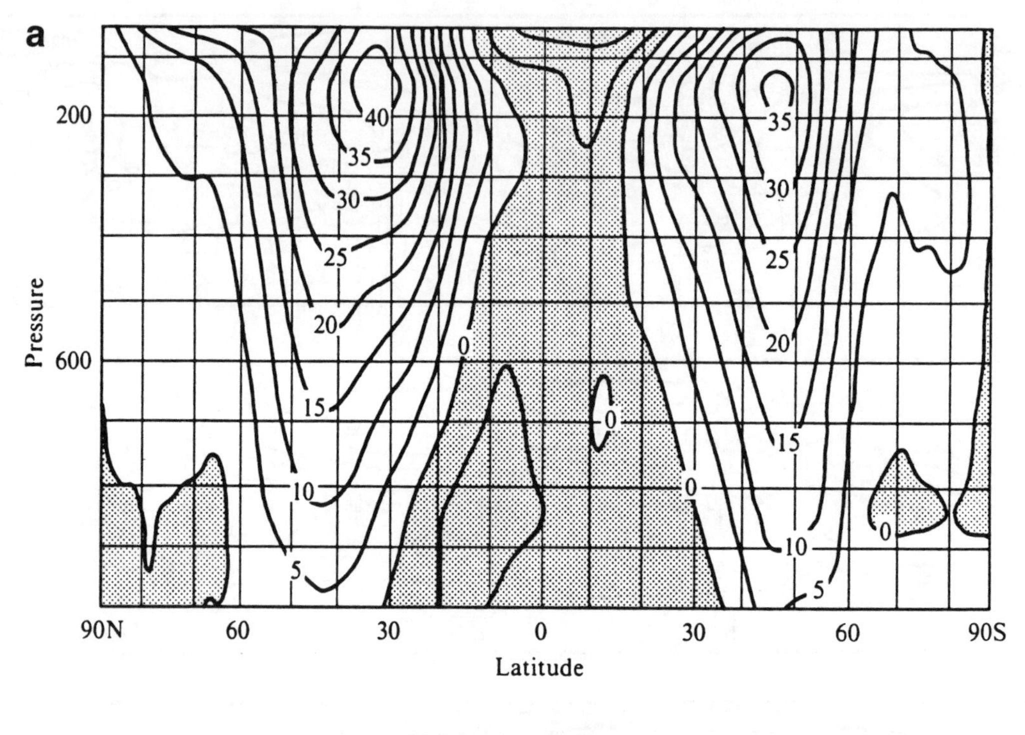


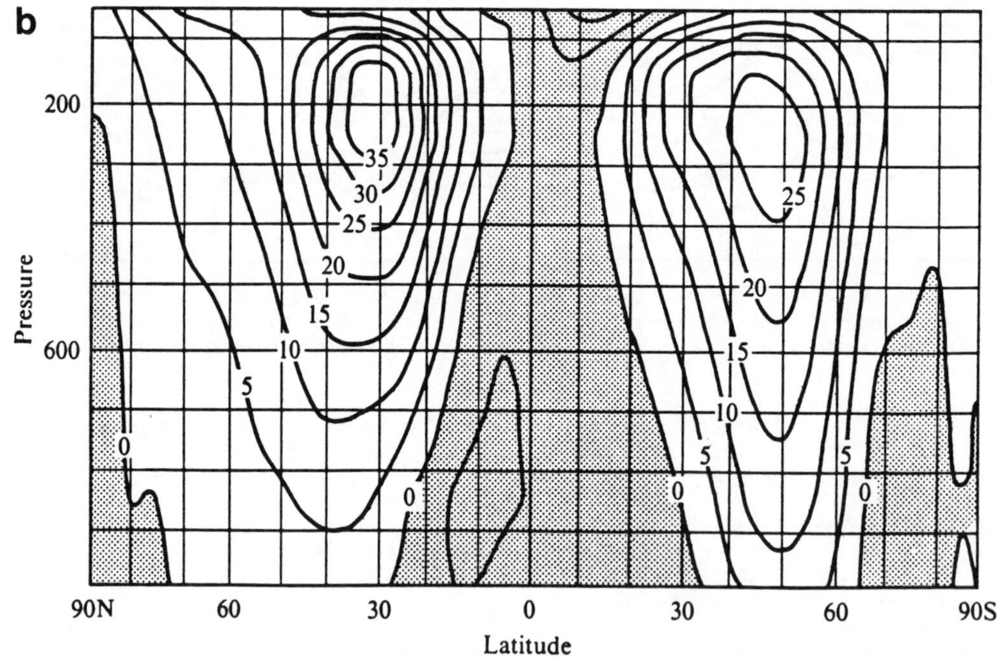


Fig. 2. December-February zonally averaged zonal winds: (a) modelled 4 year average; (b) observed 3 year mean (1983/4-1985/6). Contour interval 5 m/s; easterlies shaded.

McBean (this symposium). In this context, it is relevant to examine the surface winds which are of such importance for driving an ocean model. Figure 6 shows the zonal mean of the eastward surface stress on the ocean surface, averaged over the year, compared with Hellerman and Rosenstein's estimate from observed data. There are differences, but mostly these probably owe as much

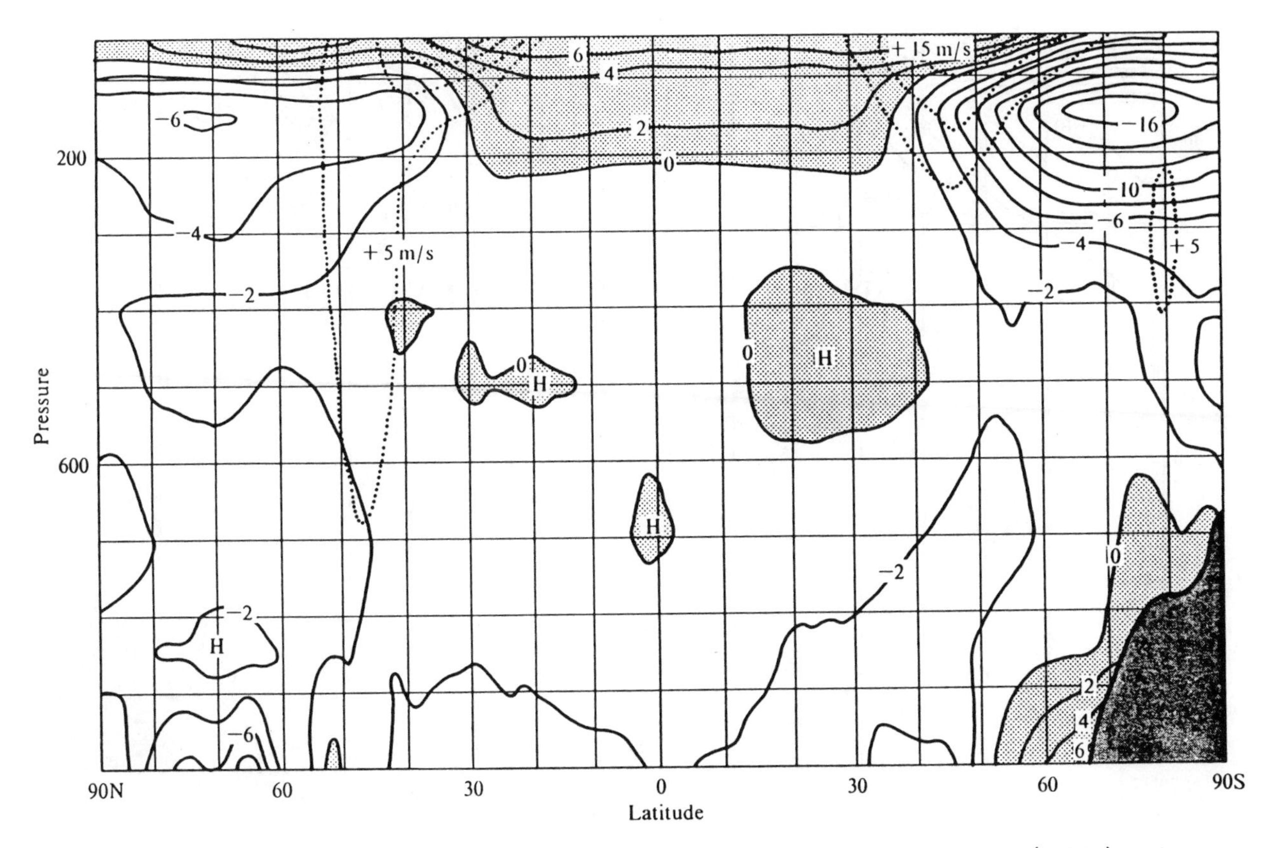


Fig. 3. December-February zonally averaged model errors for temperatures (solid) and zonal winds (dashed). Contour interval 2K for temperature, 5 m/s for winds.

to uncertainties in the data and the differences in the stress formulation as to errors in the model winds. An exception is probably near 50-60° S, where the model's peak westerlies are a few degrees too far from the pole, and north of 60° N, where some easterly bias is present.

Rainfall (Figure 7) is important for agriculture over land and over the ocean because of salinity's role in driving ocean currents. This simulation, though not as good in several respects as some of our earlier ones, nevertheless serves to demonstrate that broadly the patterns are realistic. The 'desert' areas with less than 1 mm/day and the rainy areas with more than 5 mm/day are broadly similar in the simulated and observed maps. Features which were worse here than in most of our experiments included several parts of the tropics, for example the excessive rain over South America and northern Australia and the deficient rain over the tropical Atlantic and southern tropical Africa. For many purposes, it is also important to simulate the runoff well - we have recently started to calculate runoff for selected basins, simply summing the runoff of grid boxes within the catchments. One basin we considered was the Mackenzie River which runs through Arctic Canada. The result (Figure 8) was disappointing with too strong a summer snow melt peak and too little the rest of the year. However, the model is not necessarily so wrong in high latitudes. Another comparison (Figure 9) we made was for the region of Siberia from 60-105° E north of 50° N, which closely approximates the basins of the Ob'Irtysh and Yenisei. Here, the model had nearly 50% too much run off, but the apportionment through the year is close to observed.

An important need in climate change studies, especially those involving $CO_2$, may be sea level information. To obtain this, it is essential to have an ice sheet model and a prerequisite for that is realistic snow accumulation. This is obviously very dependent on getting precipitation right but other factors are also important, as shown by the ice budget analysis for the North American ice sheet at the last glacial maximum in Manabe and Broccoli [1985]. Their results show the sublimation to be nearly half as large as snowfall. We have calculated snow budget data for Antarctica and compared it to the estimates by Schwerdtfeger [1970] (Figure 10). The broad pattern of a minimum of less than 5 g $cm^{-2}$ $yr^{-1}$

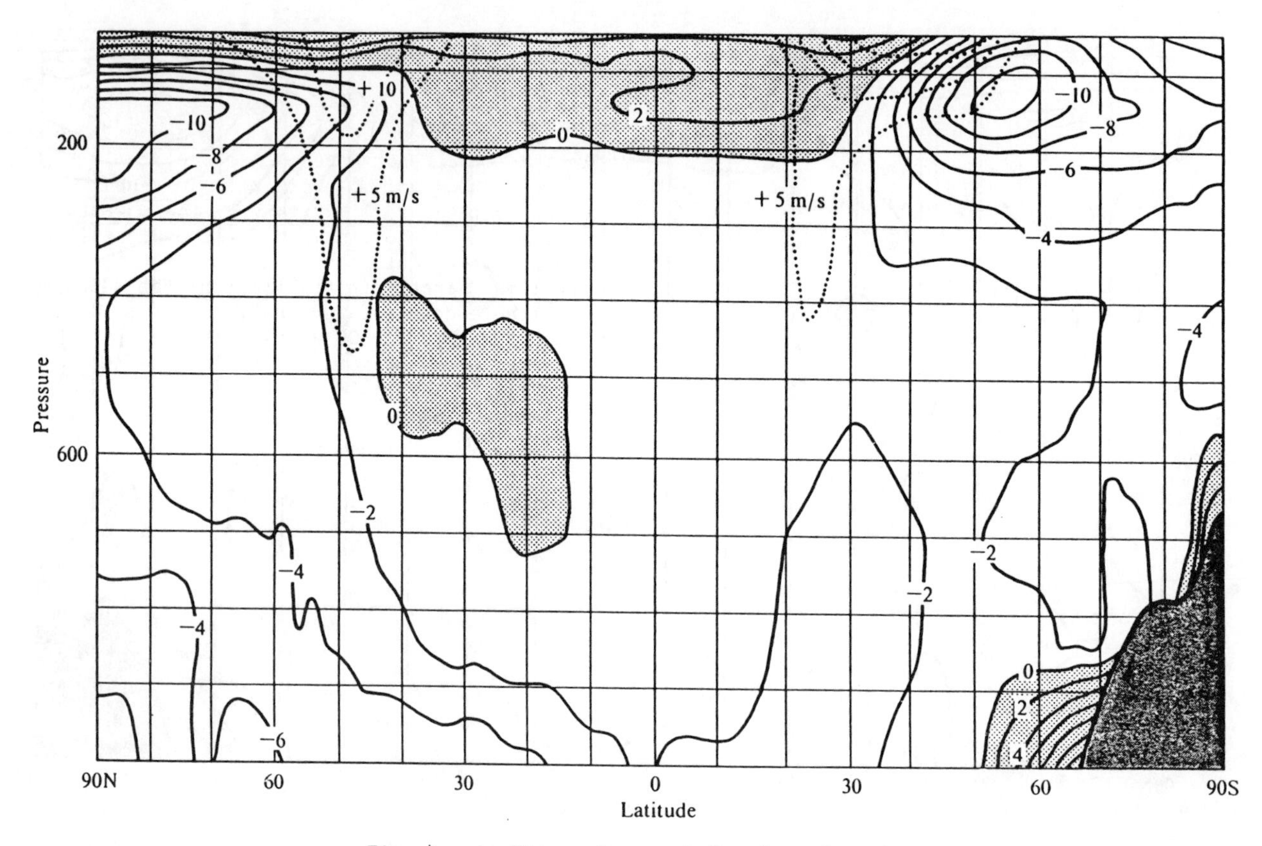


Fig. 4. As Figure 3 except for June-August.

over most of eastern Antarctica, with peak values near the coast of over 40 g $cm^{-2}$ $yr^{-1}$ in places, is simulated quite well. The South Pole apparently has too little - there is a maximum on the south edge of the Ross Ice Shelf, but like much of western Antarctica, magnitudes are too small.

The results discussed so far are from atmosphere models. It is obvious that the simulation of the surface climate from coupled models cannot, except through chance cancellation of errors, be expected to be as good as that for atmospheric GCMs, which use climatologically correct sea surface temperatures (SSTs) and sea ice. So the crucial question is how good are the SST and ice simulations in coupled models - I will discuss only SST, referring to Han et al. [1985]'s useful comparison of annually averaged SST errors. Their maps show rather large areas with errors exceeding 4 K -i.e., greater than the largest observed anomalies - and there is much in common between the simulations which is only partly disguised by the range of up to 2 K in the mean errors. These common features include a warm Antarctic Ocean and eastern southern hemisphere oceans, cool water in the northwestern tropical and sub-tropical oceans, warm north Pacific and northwest Atlantic and cool northeast Atlantic. Most of these errors are, I think, not well understood.

## Future

Now I want to touch on a few of the many challenges and problems in modelling climate. Table 4 indicates some likely future developments

TABLE 4. Future Developments, Problems

(a) Ocean: - eddies
    Tropical (El Nino)
    Extratropical (glacial cycles)
(b) Cloud: - water content -> albedo
(c) Mountains: - elevation
    (barrier or heating)
    gravity waves
(d) Land surface: - soil, etc datasets
(e) Other possible developments:
    vegetation models
    ice sheet models.

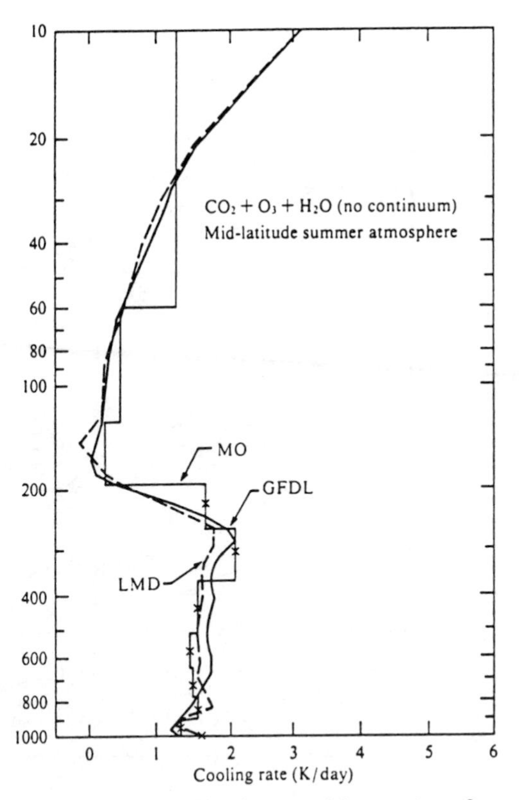


Fig. 5. Longwave radiative cooling rates from line-by-line calculations with GFDL (solid curve) and LMD (dashed) schemes (Luther, 1984) and the MO GCM broadband scheme of Slingo and Wilderspin (1986) (A. Slingo, personal communication) (thin solid lines).

and problems. One of the most important may be ocean eddies - how important for modelling climate is it to get the eddy transports right? And if they are crucial, how can they be simulated without the enormous increase in computing power needed to represent the eddies explicitly - about 3 orders of magnitude with a 10 km mesh instead of a 300 km one? Possibly they can be parametrized -though the methods have not yet been developed. Table 4 also lists some other ocean modelling problems. Firstly initialization: can WOCE provide a suitable data set for, say, initializing the ocean for 100 year $CO_2$ runs starting in 1995? Secondly, the tropical oceans - do we need to simulate the delicate balance illustrated in real data by the existence of two quasi-steady states in the tropical Pacific? Thirdly, the high latitude oceans - the large and quite rapid variations of North Atlantic circulation, for example around the Younger Dryas period, suggest another delicate balance.

Another major problem area is clouds. As discussed by Schlesinger in this symposium, current GCMs mostly omit any dependence of cloud radiative properties on water or ice content, and so omit a possibly important negative feedback if, as suggested by Somerville and Remer [1984], albedo tends to increase with temperature. The problem is not with calculating the water or ice content. There are ways of doing this already. Figure 11 shows a calculation of zonally averaged cloud water content for September (R N B Smith, personal communication) with a scheme developed from that of Sundqvist's. It verifies quite well against SMMR data for ocean only, except that the falloff towards high latitudes is too early and

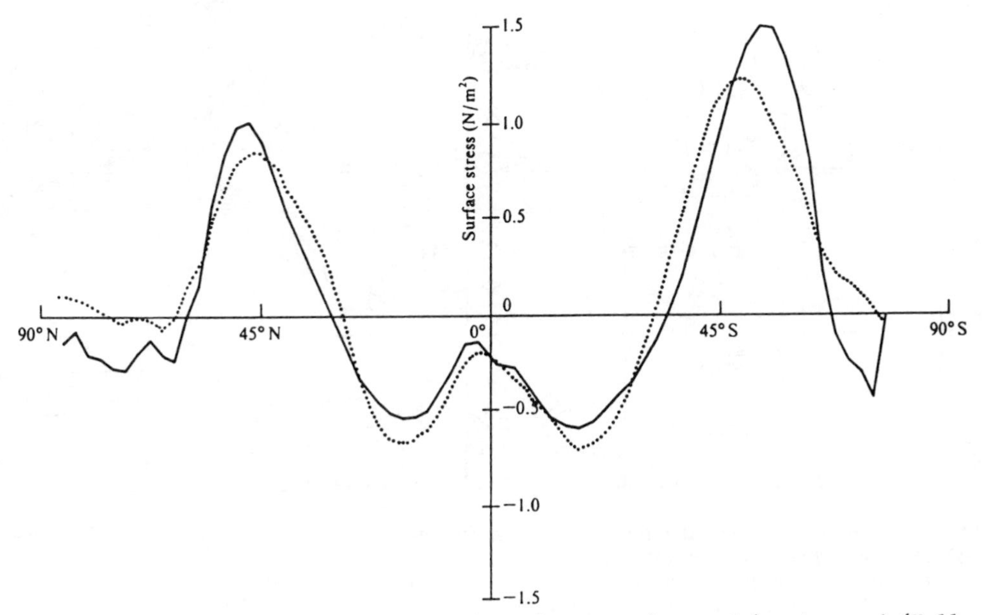


Fig. 6. Zonally averaged annual mean zonal wind stress (sea only), observed (Hellerman and Rosenstein, 1983) (dotted), and modelled (4 year mean for MO model) (solid).

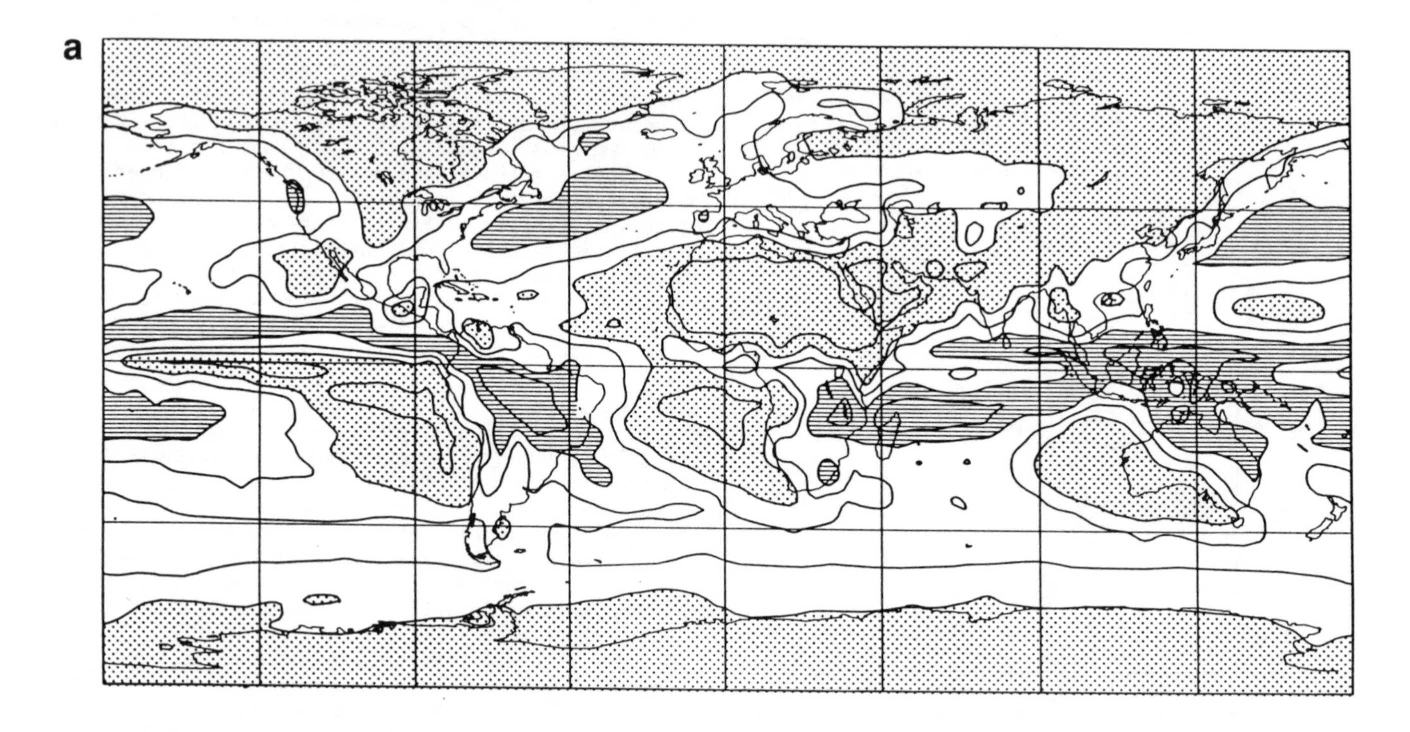


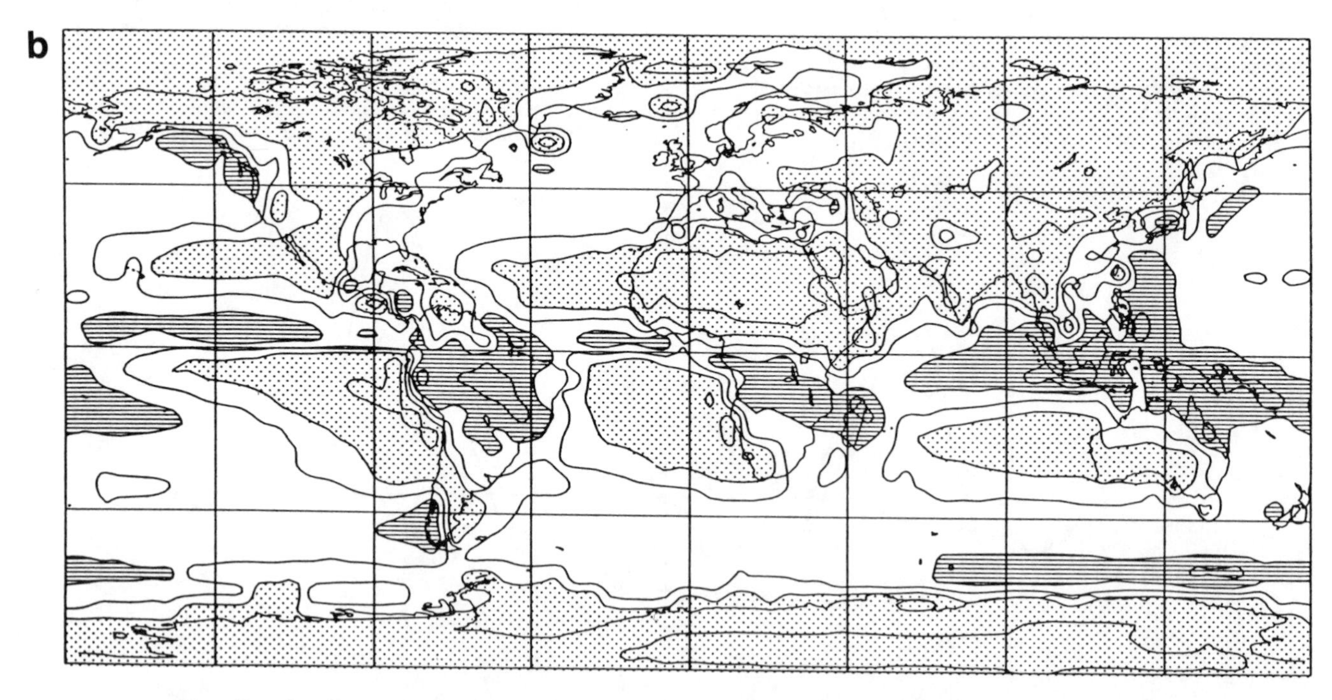


Fig. 7. December-February precipitation (mm/day): (a) 4-year mean from MO model; (b) observed from Jaeger (1976).

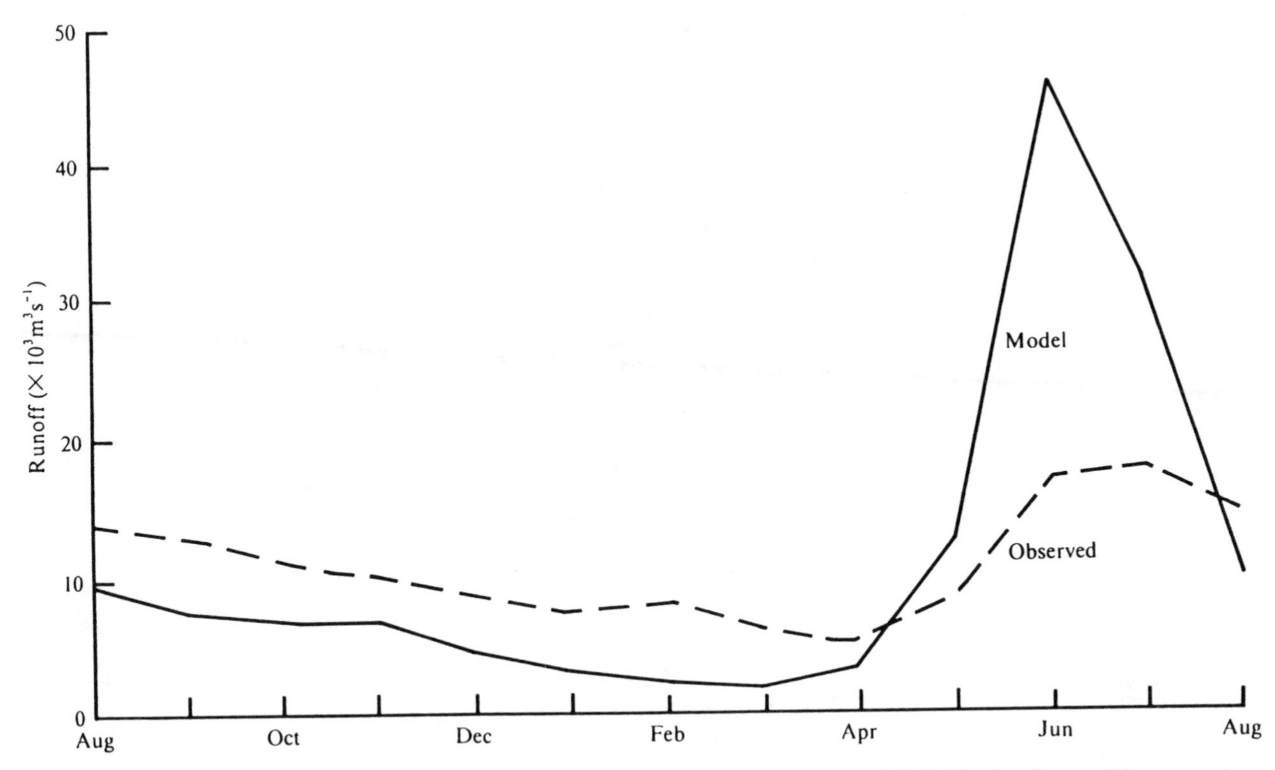


Fig. 8. Modelled (solid) and observed (Mackay and Loken, 1974) (dashed) runoff for the Mackenzie basin ($10^3 m^3 s^{-1}$).

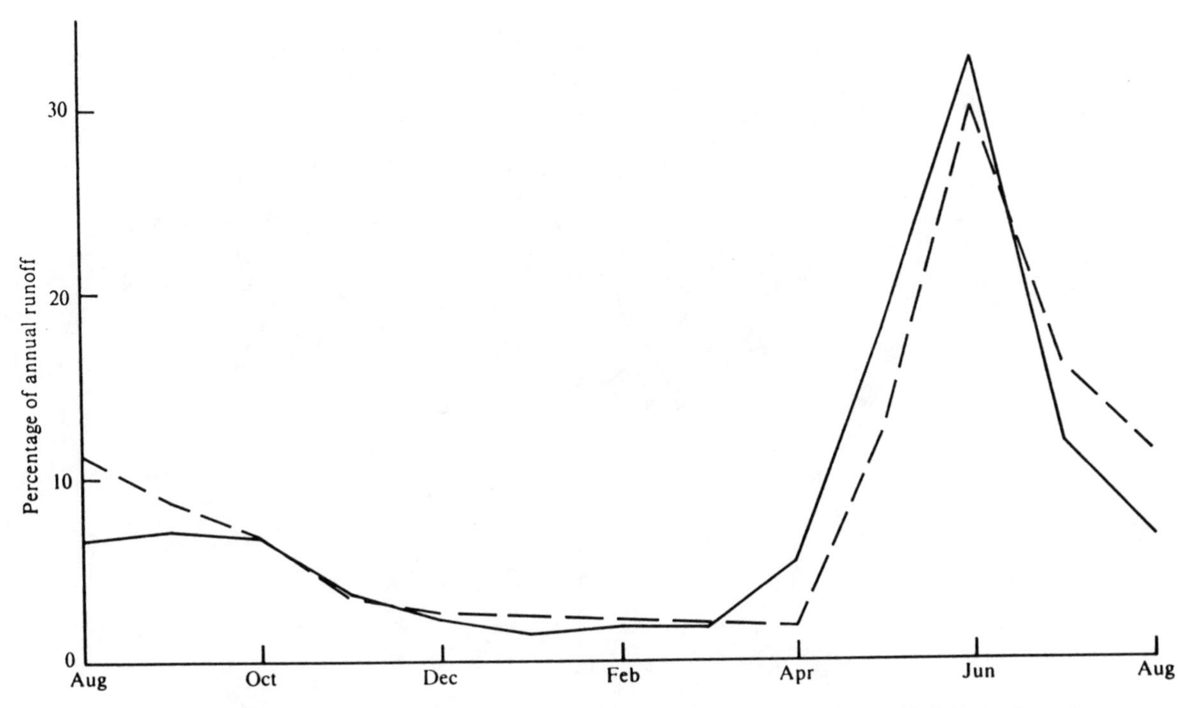


Fig. 9. As Figure 8 but as percentage of annual total and for Ob'-Irtysh and Yenisei basins (observed) and 60-105 E north of 50 N (modelled).

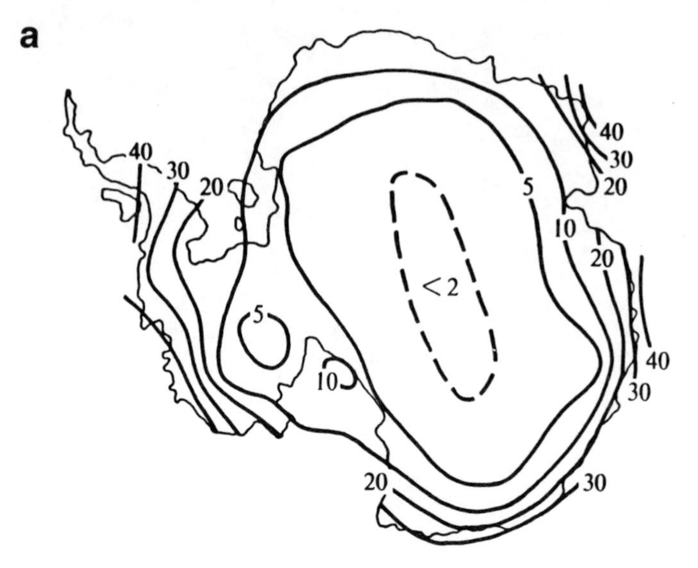


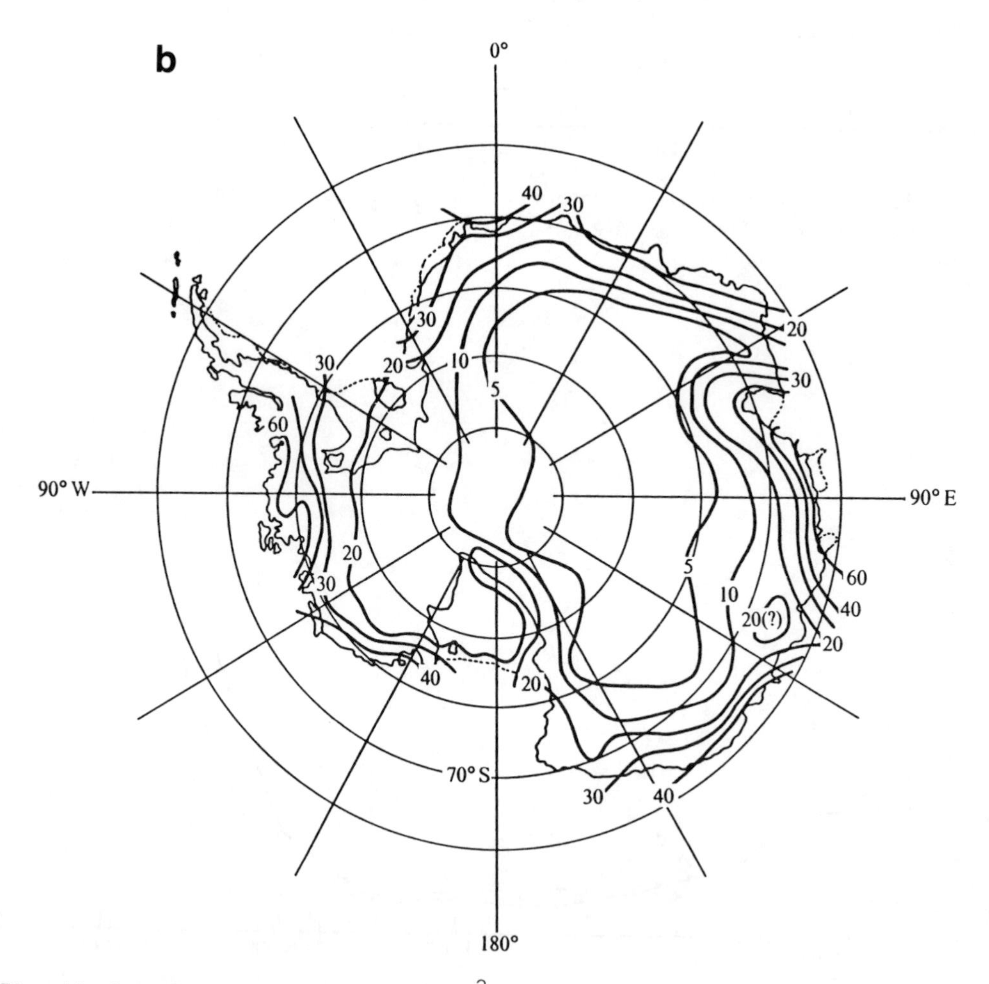


Fig. 10. Annual snow accumulation (g/cm$^2$/yr) for Antarctica: (a) modelled from MO model; (b) observed according to Giovinetto from Schwerdtfeger (1970)

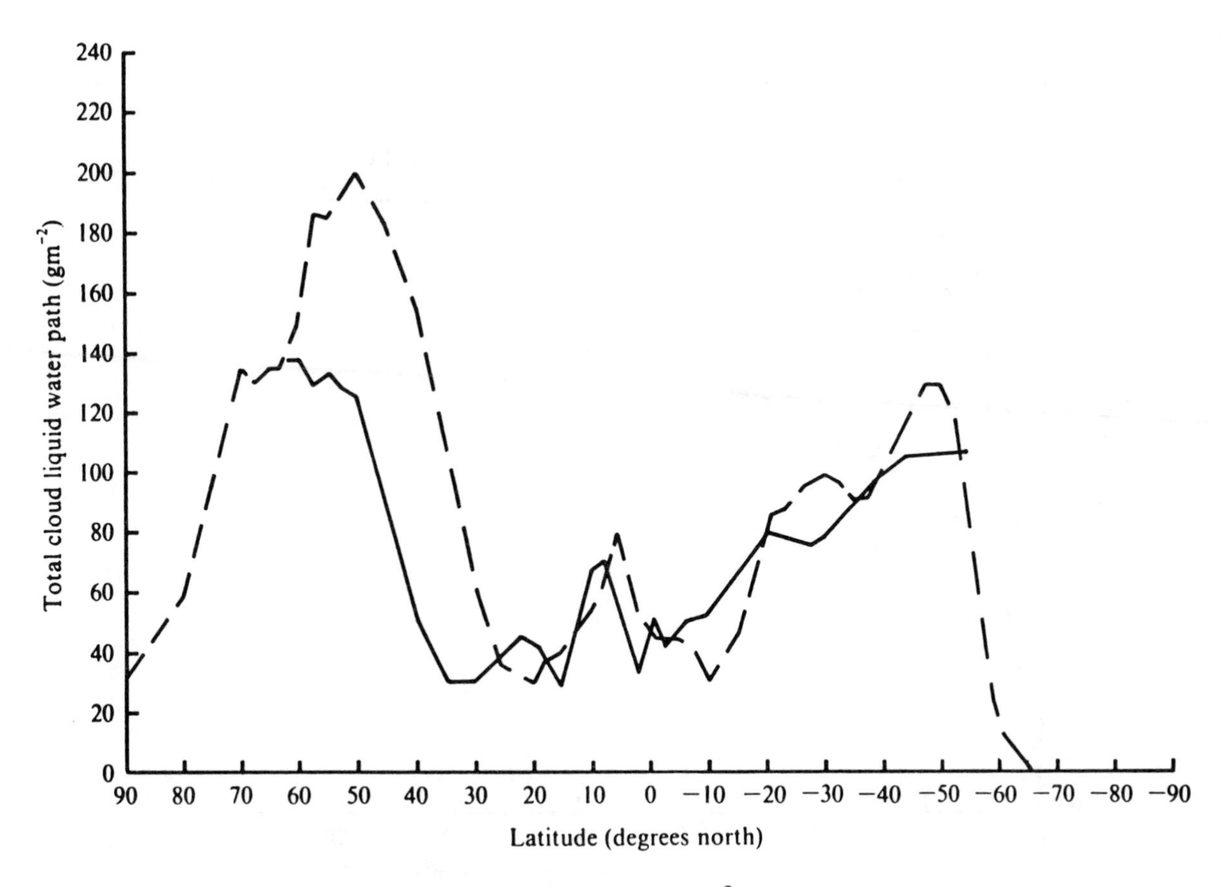


Fig. 11. Zonally averaged cloud liquid water (g $m^{-2}$) as observed (ocean only) from Njoku and Swanson (1983) for July 11-August 10 1978 (solid) and modelled for June-August from MO model (all longitudes) (dashed).

too fast compared with the satellite data. However, the water path must still be interpreted in terms of cloud radiative properties. This is not straightforward as Figure 12 (W. Ingram, personal communication), showing several proposed relationships between reflectivity and water content, illustrates. Clearly, estimates of drop size effective radii are needed if uncertainties of up to 10 per cent in albedo are to be avoided.

Land surface parametrizations are becoming much more comprehensive. However, the land surface poses some other interesting problems. Let me illustrate these with the simulation of the climate 9000 years ago when the solar radiation at the top of the atmosphere differed from that of the present day by about as much as it has in the last few tens of thousands of years. There were increases of 30-40 $Wm^{-2}$ over parts of the Northern Hemisphere in summer. This was due mainly to the different time of perihelion - in July compared to January at present. Kutzbach and Guetter [1986] simulated the effects of this in the NCAR CCM and obtained precipitation increases in summer across the Sahara and Arabia, as well as southern Asia. These results are consistent with paleoclimatic data indicating high lake levels over the Saharan region at about 9000-6500 years BP. Mitchell et al. [1988] used the UK Meteorological Office model with a 50 m slab mixed layer ocean to simulate the effects, but although rainfall increased over southern Asia and eastern Africa, they obtained little change, even some decreases over the central and western Sahara. The explanation of this difference between the models was traced to the formulations in evaporation: the NCAR model used one quarter of the potential evaporation, however little soil moisture there was, so that evaporation was able to respond to increases of radiation and allow the model to develop a moist climate. In the Meteorological Office scheme, any increase in evaporation had to be met from soil moisture and, as there usually is little over the model Sahara, no evaporation increase occurred. However, if the albedo was decreased to values typical of grassland, precipitation did increase fairly widely as a result of the Charney albedo feedback mechanism. Now, such a vegetation change could occur

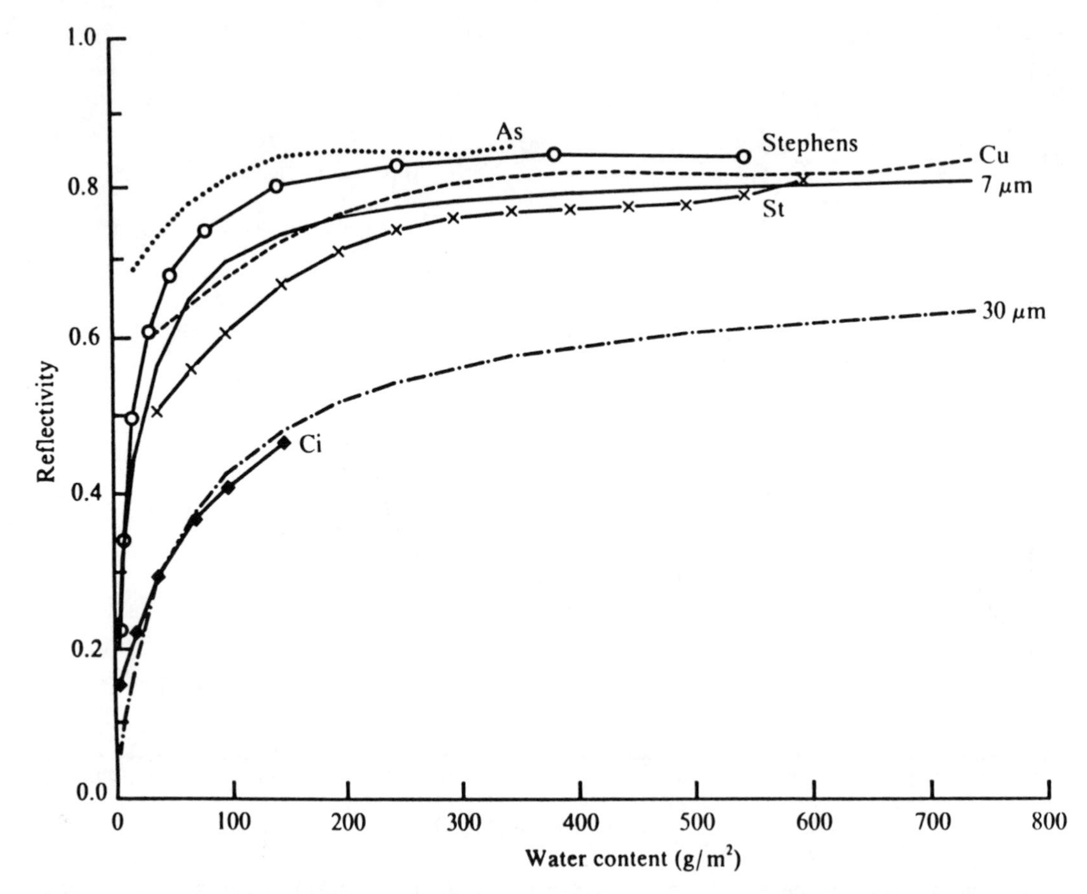


Fig. 12. Estimates of the relation between reflectivity and cloud water content for different clouds (stratus, cumulus, altostratus and cirrus from Liou and Wittman (1979), Charlock and Ramanathan (1985) 7 and 30μm effective drop radius and Stephens' (1984) spectral average.

gradually as a result of a slow increase in moisture availability but, without interactive vegetation, the climate model cannot represent it.

This brings me to my final point, that we must expect to expand climate models. Such an expansion to include modelling of vegetation appears necessary. Eagleson's talk in this symposium suggests possible directions for such a development. However, note that the role of man in determining vegetation implies a need for models of society. The same applies to other expansions - not perhaps to ice sheet models, but certainly carbon cycle models (see Bolin's and Toggweiler's papers), and ozone chemistry models needed for experiments exploring the effects of chlorofluorocarbons.

## Summary and Conclusions

(1) **There are many possible uses for climate models.** These include the study of the climatic effects of natural and manmade perturbations of the forcing. Such applications mostly require diagnostic output relevant to climate at the surface, rather than higher in the atmosphere.

(2) Model characteristics are obviously determined in part by the requirement for a realistic simulation. However, the need to represent perturbations of the forcing and the nature of the guidance required are also important constraints.

(3) Atmospheric GCMs are generally adequate for climate simulation purposes though developments are needed in the representation of cloud radiative properties and mountains and in the simulation of regional variations. Important problems lie ahead in ocean modelling, at least if eddies need to be resolved, and in the incorporation of vegetation and ice sheet models and the representation of interactions between climate and society.

Acknowledgments. The author acknowledges the considerable help from staff of the Dynamical Climatology Branch who developed and ran the model and diagnostic programs used for many of the results in this paper, and also the constructive suggestions made by Dr Howard Cattle.

References

Charlock, T.P., and Ramanathan, V., The albedo field and cloud radiative forcing produced by a general circulation model with internally generated cloud optics, J. Atmos. Sci., 42, 1408-1429, 1985.

Han, Y.-J., Schlesinger, M.E., and Gates, W.L., An analysis of the air-sea-ice interaction simulated by the OSU coupled atmosphere-ocean general circulation model, Coupled Ocean-Atmosphere Models (ed. J. C. J. Nihoul), Elsevier, pp 167-182, 1985.

Hellerman, S., and Rosenstein, M., Normal monthly wind stress over the world ocean with error estimates, J. Phys. Oc., 13, 1093-1104, 1983.

Hills, T.S., Sensitivity of numerical models to mountain representation, ECMWF Workshop on mountains and numerical weather prediction, pp 139-161, 1979.

Jaeger, L., Monthly maps of precipitation for the whole world, Berichte des Deutschen Wetterdienstes, 18, No. 139, 1976.

Kutzbach, J.E., and Guetter, P.J., The influence of changing orbital parameters and surface boundary conditions on climate simulations for the past 18000 years, J. Atmos. Sci., 43, 1726-1759, 1986.

Liou, K.-N., and Wittmann, G.D., Parametrization of the radiative properties of clouds, J. Atmos. Sci., 36, 1261-1273, 1979.

Luther, F., Intercomparison of radiation codes in climate models (ICRCCM): Longwave clear sky calculations, WMO/ICSU WCP-93, 1984.

Mackay, D.K., and Loken, O.H., Arctic hydrology, Arctic and alpine environments (ed. J.D. Ives and R.G. Barry), Methuen, pp. 111-132, 1974

Mahlman, J.D., and Umscheid, L.J., Dynamics of the middle atmosphere; Successes and problems of the GFDL 'SKYHI' general circulation model, Dynamics of the middle atmosphere (ed. J.R. Holton and T. Matsuno), (Terra Sci./Reidel) pp. 501-525, 1984.

Manabe, S., and Broccoli, A.J., The influence of continental ice sheets on the climate of an ice age, J. Geophys. Res., 90, 2167-2190, 1985.

Mitchell, J.F.B., Grahame, N.S., and Needham, K.J., Climate simulations for 9000 years before present; seasonal variations and the effect of the Laurentide ice sheet, J. Geophys. Res., in press, 1988.

Njoku, E.G., and Swanson, L., Global measurements of sea surface temperature, wind speed and atmospheric water content from satellite microwave radiometry, Mon. Weath. Rev., 111, 1977-1987, 1983.

Palmer, T.N., Shutts, G.J., and Swinbank, R., Alleviation of a systematic westerly bias in general circulation and numerical weather prediction models through an orographic gravity wave drag parametrization, Quart. J.R. Met. Soc., 112, 1001-1039, 1986.

Schlesinger, M.E., and Mitchell, J.F.B., Model projections of the equilibrium climatic response to increased carbon dioxide, Projecting the climatic effects of increasing carbon dioxide, US Dept. of Energy, pp 81-147, 1985.

Schwerdtfeger, W., The Climate of the Antarctic, Climate of the polar regions, World Survey of Climatology, Elsevier, Vol. 14, pp 253-355, 1970.

Slingo, A., and Pearson, D.W., A comparison of the impact of an envelope orography and of a parametrization of orographic gravity-wave drag on model simulations. Quart. J. R. Met. Soc., 113, 847-870, 1987.

Slingo, A., and Wilderspin, R.C., Development of a revised longwave radiation scheme for an atmospheric general circulation model, Quart. J. R. Met. Soc., 112, 371-386, 1986.

Somerville, R.C.J., and Remer, L.A., Cloud optical thickness feedbacks in the $CO_2$ climate problem, J. Geophys. Res., 89, 9668-9672, 1984.

Stephens, G.L., The parametrization of radiation for numerical weather prediction and climate models, Mon. Weath. Rev., 112, 826-867, 1984.

Wallace, J.M., Tibaldi, S., and Simmons, A.J., Reduction of systematic errors in the ECMWF model through the introduction of an envelope orography, Quart. J. R. Met. Soc., 109, 683-718, 1983.

# QUANTITATIVE ANALYSIS OF FEEDBACKS IN CLIMATE MODEL SIMULATIONS

Michael E. Schlesinger

Department of Atmospheric Sciences and Climatic Research Institute
Oregon State University, Corvallis, Oregon 97331

*Abstract.* Why do climate models, even within the same category of the climate model hierarchy, simulate different climatic changes for the same forcing? The first step toward answering this question is to calculate quantitatively the feedbacks of the physical processes in the models. This paper therefore presents the concept and terminology of classical feedback analysis; three different feedback analysis methods, one for radiative-convective models and two others for general circulation models (GCMs); and the application of these feedback analysis methods to the simulations of $CO_2$-induced climatic change. A rigorous intercomparison of the quantitative values of the feedbacks for the GCMs is made difficult by the fact that these feedbacks have been determined by two different methods. This notwithstanding, the intercomparison indicates that the contribution of the individual feedback processes to the simulated climatic changes is not the same for different models, even when they are within the same category of the climate model hierarchy. Consequently, based on these and future feedback analyses, the second step toward answering the above question is to intercompare the parameterizations of the most important feedback processes among themselves, with more-detailed models, and with observations.



## Introduction

A hierarchy of climate models has been and is being used to simulate past, present and potential future climates. For example, the potential climatic changes induced by projected future levels of carbon dioxide ($CO_2$) have been simulated by: 1) energy balance models (EBMs), which calculate only the temperature at the Earth's surface, generally in terms of what is identified as the global mean; 2) radiative-convective models (RCMs), which calculate the vertical profile of temperature at a point, again in terms of what is identified as the global mean; and 3) general circulation models (GCMs), which calculate the global geographical distributions of a wide variety of climatic quantities. Yet these models generally do not depict the same climatic changes for a given forcing such as a $CO_2$ doubling, even for the models within each of the above categories. Why is this?

One way to answer this question is to identify quantitatively the contribution to the climatic changes by each of the physical processes in the models, and then rank these physical processes in decreasing order in terms of their contributions. Doing this will identify which of the processes in each model is important for the simulated climatic changes and, thereby, permit a systematic intercomparison of the models. Insofar as any physical process contributes differently to the models' simulated climatic changes, further investigation of the treatment of that process will be warranted. In particular, such further investigation should intercompare the parameterizations of the contributory physical processes among themselves, with more-detailed models, and with observations. It is only through such a systematic approach that the causes for the models' different simulated climatic changes can be identified, understood and ameliorated. The purpose of this paper is to present a method whereby the contributions of the individual physical processes to a simulated climatic change can be determined quantitatively, namely, by the evaluation of their feedbacks.

The concept of feedback had its origins in Electrical Engineering [e.g., Bode, 1975] and was concerned with the design of electronic amplifiers such as those employed in radio. However, a somewhat different definition of feedback has been used in the study of climate [Dickinson, 1981]. In this climatic context, a feedback parameter $\lambda$ has been defined which decreases as the sensitivity of the climate system increases and increases as the sensitivity of the climate system decreases. More recently, the classical definition of feedback has been introduced into the study of climate [Hansen et al., 1984], however, with a terminology that is the reverse of the classical terminology. Therefore, the second objective of this paper is to introduce both the concepts and terminology of classical feedback analysis. Earlier expositions of classical feedback analysis in the climatic context have been presented by Schlesinger [1985, 1986 and 1988a]. Here we present the classical concepts and terminology of feedback and develop a method whereby feedbacks can be quantitatively analyzed. This method is then applied to analyze the feedbacks in RCM and GCM simulations of climatic change. Finally, the conclusions of this study are presented.

## Feedback: Concept and Terminology

The concept of feedback can be most easily introduced with the aid of the block diagram for the climate system shown in Fig. 1. In this figure $\Delta Q$ represents the "input" to the climate system in the form of a radiative forcing due to an external perturbation such as a doubling of $CO_2$, and $\Delta T_*$ is

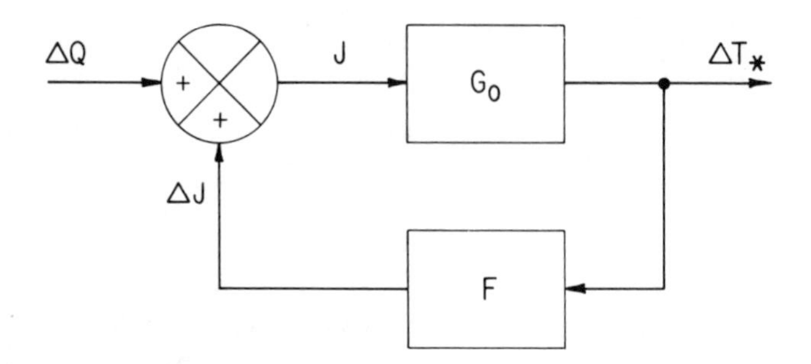


Fig. 1. Block diagram for the climate system. $\Delta Q$ and $\Delta T_* = G_0\ J$ are the forcing and surface-temperature response of the climate system, respectively, with $G_0$ the gain of the system in the absence of feedback and $J = \Delta Q + \Delta J$, with $\Delta J = F\ \Delta T_*$. The feedback of the climate system is $f = G_0\ F$.

the "output" from the climate system in the form of a change in the global-mean surface temperature. The quantity $G_0$ is the gain of the climate system, defined as the "output"/"input", when there is no feedback, and F is a measure of the feedback in the climate system. If there is no feedback, $F = 0$, and the input is transferred directly to the output so that $J = \Delta Q$. Then

$$\Delta T_* = (\Delta T_*)_0 = G_0 \Delta Q \ , \tag{1}$$

where $(\Delta T_*)_0$ is the surface temperature change in the absence of feedback. If there is feedback, $F \neq 0$, and part of the output is transferred through the feedback loop back to the input as $\Delta J$. Then $J = \Delta Q + \Delta J$, where $\Delta J = F\ \Delta T_*$, hence

$$\Delta T_* = G_0 J = G_0 (\Delta Q + F \Delta T_*) \ . \tag{2}$$

Solving for $\Delta T_*$ then gives

$$\Delta T_* = \frac{G_0}{1 - f} \Delta Q \ , \tag{3}$$

where $f = G_0F$ is the feedback factor [Bode, 1975, p. 32] or, more simply, the feedback. Equation (3) can also be written as

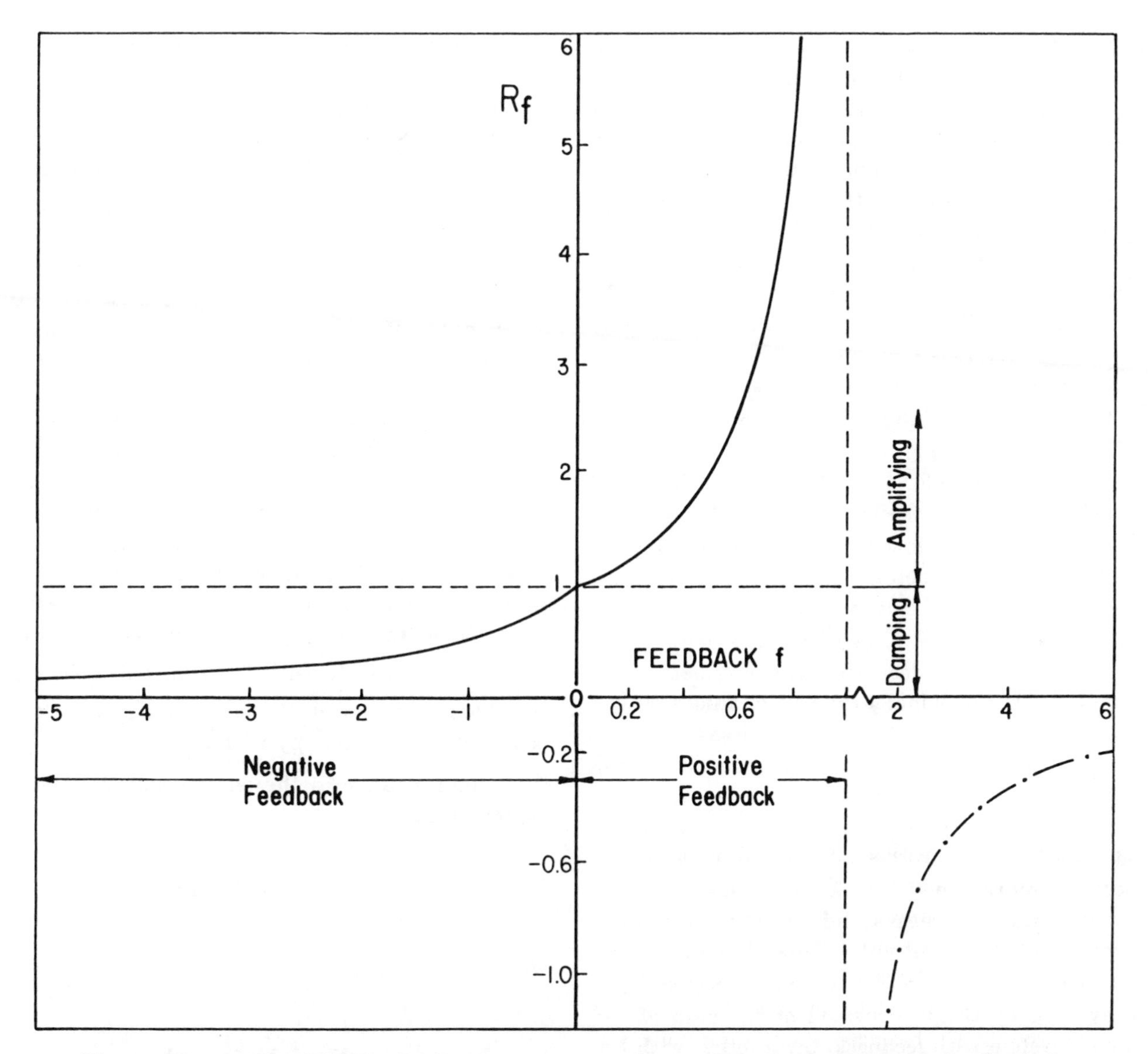


Fig. 2. The feedback/no-feedback response ratio $R_f = \Delta T_*/(\Delta T_*)_0$ versus the feedback f, where $\Delta T_*$ is the surface-temperature response of the system with feedback and $(\Delta T_*)_0$ is surface-temperature response without feedback. There is a change in the scales of both axes at the origin.

$$\Delta T_* = G_f \Delta Q \; , \tag{4}$$

where

$$G_f = \frac{G_0}{1 - f} \tag{5}$$

is the gain of the climate system with feedback.

The effect of the feedback can be characterized on the basis of the ratio of the temperature change with feedback to the temperature change without feedback,

$$R_f = \frac{\Delta T_*}{(\Delta T_*)_0} = \frac{G_f}{G_0} = \frac{1}{1 - f} \; , \tag{6}$$

the latter by Eqs. (1), (4) and (5). This ratio is shown in Fig. 2. For f = 0 there is no feedback and $R_f = 1$.

For $f < 0$ the feedback is negative and $0 \le R_f < 1$. Consequently, when there is negative feedback, regardless of its magnitude, the sign of the response is the same as the sign of the forcing. This is in contrast to what has sometimes been erroneously inferred for the outcome of negative feedback. For $0 < f < 1$, the feedback is positive and $R_f > 1$. However, for positive feedback with $f > 1$, $R_f < 0$ and the sign of the response is opposite to the sign of the forcing. This outcome, while mathematically possible (and actually obtained by one improperly formulated EBM; see Schlesinger [1985, 1988a]), is not physically consistent and must therefore be rejected.

The classical expression for feedback given by Eq. (4) can be contrasted with the definition of the "feedback parameter $\lambda$" used by Dickinson [1981], namely,

$$\Delta T_* = \frac{1}{\lambda} \Delta Q \; . \tag{7}$$

From Eqs. (4), (5) and (7) it can be seen that

$$\lambda = \frac{1}{G_f} = \frac{1 - f}{G_o} \; . \tag{8}$$

Consequently, as f increases and the climate system becomes more sensitive, $\lambda$ decreases, while $\lambda$ increases as f decreases and the climate system becomes less sensitive. This "upside-down" behavior of $\lambda$ is unnecessarily confounding and can be avoided by the use instead of the gain of the climate system with feedback, $G_f$, together with Eq. (4). In fact, Cess and Potter [1984] employed the relation given by Eq. (4), but instead of $G_f$ they used the symbol $\lambda$, identified as the "surface response function," which was therefore easily confused with the same symbol used by Dickinson [1981]. Moreover, Hansen et al. [1984] used an expression equivalent to Eq. (6), but with the equivalent of f (their g) identified as the "gain" of the climate system, and $R_f$ (their f) as the "net feedback factor."

In this paper we will use the classical terminology in which f is the feedback and G is the gain. Also, by virtue of Eq. (6), we will call $R_f$ the feedback/no-feedback response ratio. In the following we will derive expressions for f and $G_o$ for the climate system.

*The Feedback f For The Climate System*

The net radiation at the top of the Earth's atmosphere N can be expressed from the viewpoint of a planetary energy balance model as $N = N(\mathbf{E}, T_*, \mathbf{I})$. Here **E** is a vector of quantities that can be regarded as external to the climate system, that is, quantities whose change can lead to a change in climate, but which are independent of climate. **I** is a vector of quantities that are internal to the climate system, that is, quantities that can change as the climate changes and, in so doing, feed back to modify the climatic change. The external quantities include, for example, the solar constant, the optically-active ejecta from volcanic eruptions and the $CO_2$ concentration (although eventually it may change as a result of climatic change). The internal quantities include all the variables of the climate system other than the surface temperature $T_*$ such as the atmospheric temperature, water vapor and clouds. Because $T_*$ is the only dependent variable in this model, the internal quantities must be represented therein by $\mathbf{I} = \mathbf{I}(T_*)$.

A small change in the energy flux, $\Delta N$, can be expressed as

$$\Delta N = \Delta Q - (G_o^{-1} - F)\,\Delta T_* \; , \tag{9}$$

where

$$\Delta Q = \sum_i \frac{\partial N}{\partial E_i} \Delta E_i \tag{10}$$

is the change in N due to a change in one or more external quantity, $\Delta E_i$,

$$- G_o^{-1} \Delta T_* = \frac{\partial N}{\partial T_*} \Delta T_* \tag{11}$$

is the change in N due to the change in $T_*$ alone, and

$$F\,\Delta T_* = \sum_j \frac{\partial N}{\partial I_j} \frac{dI_j}{dT_*} \Delta T_* \tag{12}$$

is the change in N due to the change in the internal variables **I** through their dependence on $T_*$. When the equilibrium $\Delta T_*$ is reached in response to the forcing $\Delta Q$, $\Delta N = 0$ and

$$\Delta T_* = G_f \Delta Q = \frac{G_o}{1 - \sum_j f_j} \Delta Q \,, \quad (13)$$

where

$$f_j = G_o \frac{\partial N}{\partial I_j} \frac{dI_j}{dT_*} \,. \quad (14)$$

Equation (13) with

$$f = \sum_j f_j \quad (15)$$

is identical to Eqs. (4) and (5) which were obtained solely from our consideration of the feedback block diagram of Fig. 1.

It should be noted that this classical feedback analysis, together with its representation by Fig. 1, is linear because it neglects the second- and higher-order derivatives of N. Because this is equivalent to the assumption that $\partial^2 N/\partial I_j \partial I_k = 0$, where j and k are any two physical processes, the effects on N of the individual physical processes are considered to be independent.

As a result of the classical linear feedback analysis, Eq. (14) shows that the feedback of a physical process j depends on three quantities: 1) the sensitivity of the net flux N to the process as measured by $\partial N/\partial I_j$, 2) the sensitivity of the process to the surface temperature as measured by $dI_j/dT_*$, and 3) the zero-feedback gain of the climate system $G_o$. Consequently, the feedback of any physical process is zero if either the net flux N is independent of that physical process or the physical process is independent of the surface temperature.

As shown by Eq. (13) and Fig. 2, the influence of any particular feedback $f_j$ on the response of the climate system depends nonlinearly on the sum of the other feedbacks. For example, the addition of a physical process with $f_j = 0.2$ would increase $\Delta T_*$ by 1.6°C if added to a system with an existing feedback of 0.5, but would increase $\Delta T_*$ by only 0.5°C if added to a system with no existing feedback.

*Zero-Feedback Gain Of The Climate System, $G_o$*

The planetary radiative energy budget can be written as

$$N = \frac{1 - \alpha_p}{4} S - \varepsilon \sigma T_*^4 \,, \quad (16)$$

where S is the solar constant, $\alpha_p$ the planetary albedo, $\varepsilon$ the effective emissivity of the Earth-atmosphere system, and $\sigma$ the Stefan-Boltzmann constant. From this equation and Eq. (11) we can calculate the zero-feedback gain as

$$G_o = -\left(\frac{\partial N}{\partial T_*}\right)^{-1} = \frac{T_*}{(1 - \alpha_p) S} \,. \quad (17)$$

Taking $S = 1370$ Wm$^{-2}$, $\alpha_p = 0.3$ and $T_* = T_s = 288$ K, where $T_s$ is the observed global-mean surface air temperature, Eq. (17) gives $G_o = 0.3$°C/(Wm$^{-2}$). Thus, for $\Delta Q = \Delta R_T = 4$ Wm$^{-2}$ for a doubling of the $CO_2$ concentration, where $\Delta R_T$ is the change in the net longwave radiation at the tropopause, Eq. (1) gives $(\Delta T_*)_o = 1.2$°C. (The reasons for evaluating $\Delta Q$ at the tropopause rather than at the top of the atmosphere are described subsequently.) This value of $(\Delta T_*)_o$ for a $CO_2$ doubling is in agreement with what has been obtained by RCMs without feedbacks (see below and Schlesinger [1985, 1988a], and Schlesinger and Mitchell [1985]).

## Methods Of Feedback Analysis For Climate Model Simulations

Three methods have been used to evaluate the feedbacks in climate model simulations, one method for simulations by radiative-convective models, and two other methods for simulations by general circulation models. In the following we describe each of these feedback analysis methods.

*Radiative-Convective Models*

Schlesinger [1985, 1988a] evaluated the feedbacks in radiative-convective model simulations of $CO_2$-induced climatic change on the basis of Eq. (13) which may be written as

$$(\Delta T_s)_j = \frac{G_o}{1 - \sum_{i=1}^{j-1} f_i - f_j} \Delta Q \ , \quad (18)$$

and

$$(\Delta T_s)_{j-1} = \frac{G_o}{1 - \sum_{i=1}^{j-1} f_i} \Delta Q \ , \quad (19)$$

where $(\Delta T_s)_j$ and $(\Delta T_s)_{j-1}$ are the surface air temperature changes from two $1xCO_2$-$2xCO_2$ simulation pairs, the first with physical process j and the second without. (Here and in the following we replace the surface temperature change $\Delta T_*$ with the surface air temperature change $\Delta T_s$. We do this because the results for RCMs and GCMs are given for $\Delta T_s$ and not $\Delta T_*$, and because the difference between the two temperature changes is small.) Solving for $f_j$ from Eq. (18), substituting for the sum of the feedbacks from Eq. (19), and use of Eq. (1) then yields

$$f_j = \left[ \frac{1}{(\Delta T_s)_{j-1}} - \frac{1}{(\Delta T_s)_j} \right] (\Delta T_s)_o \ , \quad j = 1,...,J \ , \quad (20)$$

where J is the total number of feedback processes. By defining $\Delta T_s = T_s(\text{perturbation}) - T_s(\text{control})$, this feedback-analysis method can be used for a climatic-change simulation other than that for a $CO_2$ doubling.

*General Circulation Models*

The feedback analysis method defined by Eq. (20) requires that J+1 pairs of control-perturbation simulations be performed to generate the values of the $(\Delta T_s)_j$, one pair for the case of zero feedback (j = 0), and J pairs (j = 1 ,..., J) in which the feedback processes are sequentially added one at a time. Although this feedback analysis method is practicable for radiative-convective models because of their computational economy, it is not viable for general circulation models because of the large amount of computer time that would be required. Therefore, two methods different from that of Eq. (20) have been used to analyze the feedbacks in climatic-change simulations performed by general circulation models. In one method use is made of a radiative-convective model as a surrogate for the general circulation model. In the other method, only the radiative transfer model of the GCM itself is used. These two methods are described below.

*Use of a radiative-convective model.* Hansen et al. [1984] used a radiative-convective model to analyze the feedbacks in the GISS (NASA Goddard Institute for Space Studies, New York, New York) general circulation model simulation of the global-mean surface air temperature change induced by a doubling of the $CO_2$ concentration. The basis for this analysis is Eq. (6) which can be written as

$$f = \frac{\Delta T_s - (\Delta T_s)_o}{\Delta T_s} = \frac{\sum_{j=1}^{J} \delta_j(\Delta T_s)}{\Delta T_s} \ , \quad (21)$$

where $\delta_j(\Delta T_s)$ is the contribution to the total temperature change $\Delta T_s$ by feedback process j alone. Substituting Eq. (15) for f then gives

$$f_j = \frac{\delta_j(\Delta T_s)}{\Delta T_s} \ , \quad j = 1,...,J \ . \quad (22)$$

To use the feedback analysis method given by Eq. (22), Hansen et al. [1984] employed a radiative-convective model which had the same radiation code as the GISS GCM, and first ran a $1xCO_2$-$2xCO_2$ simulation pair with this RCM without any feedbacks. The result of this was $(\Delta T_s)_o = 1.2°C$, which is in agreement with the zero-feedback estimate given previously. The effect of each feedback was then determined by adding the GCM-simulated global-mean $2xCO_2$-$1xCO_2$ change in the appropriate quantity to the same quantity of the RCM $1xCO_2$ simulation without feedback, and rerunning the RCM $1xCO_2$ simulation with this fixed modified value. (During this rerun of the RCM $1xCO_2$ simulation, the temperature lapse rate in the troposphere was adjusted to the global-mean value of the GCM $1xCO_2$ simulation if it exceeded that value.) The difference between this RCM $1xCO_2$ simulation with the prescribed change in the appropriate quantity and the initial RCM $1xCO_2$ simulation without feedback gives $\delta_j(\Delta T_s)$.

*Use of a radiative transfer model.* The principal shortcoming of analyzing the feedbacks of a GCM simulation with an RCM is the need to use the global-mean values of the GCM as input to the RCM. This would be of no importance if the climate system were linear because then $[\mathcal{F}(x)]_{ave} = \mathcal{F}(x_{ave})$, where x and $\mathcal{F}$ represent the input to and output from the RCM, respectively, and "ave" denotes the global average. However, because radiative transfer is a nonlinear process, $[\mathcal{F}(x)]_{ave} \neq \mathcal{F}(x_{ave})$. Thus, the RCM analysis of the feedbacks of a GCM simulation is likely to produce erroneous results. Another method which does not suffer this shortcoming was introduced by Manabe and Wetherald [1980]. This method, which is based on Eq. (14), can be written using our nomenclature as

$$f_j = G_o \frac{\partial N}{\partial I_j} \frac{dI_j}{dT_s} = G_o \frac{[\delta_j N]_{ave}}{[\Delta T_s]_{ave}}, \qquad (23)$$

where $[\delta_j N]_{ave}$ is the global-mean change in the net radiation at the tropopause due to feedback process j, and $[\Delta T_s]_{ave}$ is the total $2xCO_2$-$1xCO_2$ change in the global-mean surface air temperature.

To use the feedback analysis method given by Eq. (23), Wetherald and Manabe [1988] employed the radiative transfer code of their GFDL (Geophysical Fluid Dynamics Laboratory, Princeton University, Princeton, NJ) GCM together with the monthly-mean quantities over the period of one annual cycle for both their $1xCO_2$ and $2xCO_2$ GCM simulations. For each calendar month the monthly-mean geographical distributions of the relevant climatic quantities for the $1xCO_2$ simulation were input to the radiation code, the geographical distribution of N was calculated, and the resultant 12 "monthly-means" of N were averaged both annually and globally. Then this radiative computation was repeated individually for each feedback process by replacing the appropriate monthly-mean quantity from the GCM $1xCO_2$ simulation with the corresponding monthly-mean quantity from the GCM $2xCO_2$ simulation. The difference between the latter and former calculations then gives $[\delta_j N]_{ave}$.

## Feedbacks For A $CO_2$ Doubling

In this section we present results, obtained by the above analysis methods, of the feedbacks in radiative-convective and general circulation model simulations of the temperature changes induced by a doubling of the $CO_2$ concentration.

### *Radiative-Convective Models*

Radiative-convective model simulations of a $2xCO_2$-$1xCO_2$ induced warming give values of $\Delta T_s$ which range from 0.48 to 4.2°C. Thus by Eq. (6) and $(\Delta T_s)_o$ = 1.2°C, f = -1.5 to 0.7. Several feedback mechanisms are likely to be the cause of this range in f, including as $T_s$ increases: 1) the increase in the amount of water vapor in the atmosphere as a consequence of the constancy of the relative humidity, 2) the change in the temperature lapse rate, 3) the increase in the cloud altitude as the clouds maintain their temperature, 4) the change in cloud amount, 5) the change in the cloud optical depth, and 6) the decrease in surface albedo.

The values of these feedbacks have been determined by Schlesinger [1985, 1988a] using the method given by Eq. (20), the results of which are presented in Table 1. This table shows that: 1) the water vapor, cloud altitude and surface albedo feedbacks are positive, with values that decrease in that order; 2) the cloud optical depth feedback is negative; 3) the temperature lapse rate feedback is either positive or negative, depending on whether the lapse rate is controlled by baroclinic adjustment (BADJ) or convective (MALR or PC) processes; and 4) the cloud cover feedback is unknown.

### *General Circulation Models*

*Analysis by use of an RCM.* For the particular analysis performed by Hansen et al. [1984], the

TABLE 1. Summary of the feedbacks f in RCM and GCM simulations of $CO_2$-induced surface air temperature change

| Feedback Mechanism | RCM [a] | GCM | |
|---|---|---|---|
| | | GISS [b] | GFDL [c] |
| Water Vapor | 0.3 to 0.4 | 0.66 | 0.41 |
| Lapse Rate [d] | | | |
| BADJ | 0.1 | | |
| MALR | -0.25 to -0.4 | | |
| PC | -0.65 | | |
| Total | | -0.26 | 0.05 |
| Cloud | | | |
| Altitude | 0.15 to 0.30 | 0.12 | |
| Cover | Unknown | 0.10 | |
| Altitude & Cover | | 0.22 | 0.09 |
| Optical Depth | 0 to -1.32 | | |
| Surface Albedo | 0.14 to 0.19 | 0.09 | 0.13 |
| Total | -1.5 to 0.71 | 0.71 | 0.68 |

a. Based on the analysis of Schlesinger [1985, 1988a].
b. Based on the results of Hansen et al. [1984]
c. Based on the results of Wetherald and Manabe [1988] with modifications as described in the text.
d. BADJ, MALR and PC denote baroclinic adjustment, moist adiabatic adjustment and penetrative convection, respectively.

effect of the 33% increase in total water vapor simulated by the GCM was estimated by increasing the water vapor at each level of the RCM by 33%, thereby giving the result for $\delta_1(\Delta T_s)$. To determine the effect of the change in the vertical distribution of water vapor simulated by the GCM, the latter was inserted into the RCM and the resulting temperature change was decreased by $\delta_1(\Delta T_s)$ to obtain $\delta_2(\Delta T_s)$. To determine the effect of the change in lapse rate simulated by the GCM, the latter was inserted into the RCM and gave $\delta_3(\Delta T_s)$. Similarly, $\delta_4(\Delta T_s)$ for the GCM-simulated change in surface albedo was obtained. The total cloud effect on temperature was obtained by changing the cloud amounts at all levels in the RCM in proportion to the changes obtained in the GCM. The effect of changing only cloud cover, $\delta_6(\Delta T_s)$, was obtained by inserting a uniform cloud change in the RCM equal to the total change in the GCM. Lastly, the effect of the cloud altitude change, $\delta_5(\Delta T_s)$, was obtained by subtracting $\delta_6(\Delta T_s)$ from the total cloud effect.

The results of the feedback analysis using the method given by Eq. (22) together with the Hansen et al. [1984] values of $\delta_j(\Delta T_s)$ are presented in Table 1. The feedback due to the changes in water vapor amount and vertical distribution is $f_W = 0.66$. This is considerably larger than the $f_W = 0.3$ to 0.4 given by the RCMs. The much larger $f_W$ estimated for the GISS GCM indicates that the relative humidity increased with doubled $CO_2$ in that model, unlike the constant relative humidity assumed by the RCMs. Indeed, Hansen et al. [1984] state that the average relative humidity increased by 1.5% with a maximum of 6% at the 200 mb level. The estimated

**lapse rate feedback, $f_{LR}$ = -0.26, lies at the smaller limit given by the RCMs for the moist adiabatic lapse rate case,** perhaps because the change in lapse rate of -0.2°C/km in the GCM simulation is less than the change in the moist adiabatic value of -0.5°C/km. The cloud altitude feedback, $f_{CA}$ = 0.12, also lies at the lower limit given by the RCMs. The cloud cover feedback estimated for the GCM is positive. This indicates that the global-mean low and middle clouds, whose albedo effect dominates their longwave effect giving $\partial N/\partial I_j < 0$ in Eq. (14), decreased and/or the global-mean high clouds, whose longwave effect dominates their albedo effect giving $\partial N/\partial I_j > 0$, increased. Finally, the surface albedo feedback, due largely to the reduced sea ice, is estimated as $f_{SA}$ = 0.09 which is somewhat smaller than the estimates given by the RCMs. The total feedback estimated for the GCM is f = 0.71, of which water vapor feedback, $f_W$ = 0.66, is the single most important positive contributor, followed by cloud feedback, $f_C$ = 0.22, and surface albedo feedback, $f_{CA}$ = 0.09, with the lapse rate feedback, $f_{LR}$ = -0.26, making a negative contribution.

*Analysis by use of a radiative transfer model.* Wetherald and Manabe [1988] present results for: 1) the change in the net radiation at the top of the atmosphere (TOA) due to the doubling of the $CO_2$ concentration, $[\Delta Q_{TOA}]_{ave}$ = 2.28 $Wm^{-2}$; 2) the total $CO_2$-induced temperature change $[\Delta T_s]_{ave}$ = 4.0°C, and 3) the $[\delta_j N_{TOA}]_{ave}$ values obtained for the $2xCO_2$-$1xCO_2$ change in surface albedo and the changes in the vertical profiles of temperature, water vapor and fractional cloud cover. From Eqs. (3) and (23) the total feedback can be expressed by

$$f_{TOA} = \begin{cases} 1 - \dfrac{G_o\,[\Delta Q_{TOA}]_{ave}}{[\Delta T_s]_{ave}} & \text{(24a)} \\[2ex] \dfrac{G_o}{[\Delta T_s]_{ave}} \displaystyle\sum_{j=1}^{J} [\delta_j N_{TOA}]_{ave} & \text{(24b)} \end{cases}$$

Using Eq. (24a) together with $G_o$ = 0.3°C/($Wm^{-2}$) and the above values of $[\Delta Q_{TOA}]_{ave}$ and $[\Delta T_s]_{ave}$ gives $f_{TOA}$ = 0.829. As can be seen from Fig. 2 and Table 1, this is a very large value of the feedback, much larger than the maximum value for the RCMs, f = 0.71, and the same value for the GISS GCM. The reason for this large value is the use above of $[\Delta Q_{TOA}]_{ave}$, that is, the change in the net radiation at the top of the atmosphere. Consequently, for the feedback analysis of the Wetherald and Manabe [1988] results to be consistent with the two other feedback analyses, it is necessary to convert the top-of-the-atmosphere results, $[\Delta Q_{TOA}]$ and $[\delta_j N_{TOA}]_{ave}$, to the corresponding tropopause results, $[\Delta Q_{TROP}]$ and $[\delta_j N_{TROP}]_{ave}$. (Wetherald and Manabe [1988] presented values of $[\Delta Q_{TOA}]$ and $[\delta_j N_{TOA}]_{ave}$ instead of $[\Delta Q_{TROP}]$ and $[\delta_j N_{TROP}]_{ave}$ because the profiles of the changed quantities extended throughout both the troposphere and stratosphere, and because it is difficult to define the tropopause in a GCM simulation. However, because: 1) the change in the net radiation flux is a maximum at the tropopause, 2) the sign of the $CO_2$-induced change in the stratospheric temperature is opposite to that of the tropospheric and surface temperature changes, and 3) the changes in the stratospheric quantities have little effect on the troposphere and the surface [Schlesinger 1985, 1988a], it is more correct to use $[\Delta Q_{TROP}]$ and $[\delta_j N_{TROP}]_{ave}$ than $[\Delta Q_{TOA}]$ and $[\delta_j N_{TOA}]_{ave}$.) We do this as described below.

For the forcing we take $[\Delta Q_{TROP}]_{ave}$ = $[\Delta Q_{TOA}]_{ave}$ + C, where C is a conversion factor. The value of C was obtained from the Intercomparison of Radiation Codes used in Climate Models (ICRCCM) study in which the radiation model of the GCM was used to compute the longwave fluxes for five atmospheric profiles [Ellis, 1987, personal communication]. From the average over the five profiles of these ICRCCM results for the difference between the changes in the longwave fluxes at the tropopause and at the top of the atmosphere, C = 2.03 $Wm^{-2}$. (The contribution to $[\Delta Q_{TROP}]_{ave}$ and $[\Delta Q_{TOA}]_{ave}$ by the change in the absorbed solar radiation due to the $CO_2$ doubling is negligible.) Thus for $[\Delta Q_{TOA}]_{ave}$ = 2.28 $Wm^{-2}$, the above conversion gives $[\Delta Q_{TROP}]_{ave}$ = 4.31 $Wm^{-2}$. Using this value in

$$f_{TROP} = 1 - \frac{G_o [\Delta Q_{TROP}]_{ave}}{[\Delta T_s]_{ave}} \tag{25}$$

gives $f_{TROP} = 0.677$ which is comparable to the maximum value for the RCMs and the same value for the GISS GCM.

Unfortunately, we do not have the analog of the ICRCCM results with which to convert the $[\delta_j N_{TOA}]_{ave}$ values to $[\delta_j N_{TROP}]_{ave}$ values. Consequently, we compute the components of $f_{TOA}$ from Eq. (24b) and convert them into the components of $f_{TROP}$ by

$$f_{TROP,\,j} = \left( \frac{f_{TROP}}{f_{TOA}} \right) f_{TOA,\,j} = 0.817\, f_{TOA,\,j} \tag{26}$$

Although this conversion is by no means unique, at least the resultant values of $f_{TROP,\,j}$ and $f_{TROP}$ satisfy Eq. (15).

The results of this feedback analysis of the Wetherald and Manabe [1988] study are presented in Table 1. It can be seen that the water vapor feedback for the GFDL GCM is comparable to that of the RCMs, and is about two-thirds that of the GISS GCM. Interestingly, the total lapse rate feedback is positive in the GFDL GCM in contrast to the negative lapse rate feedback of the GISS GCM. The combined cloud altitude and cloud cover feedback is positive for the GFDL GCM but is only about 40% of the corresponding positive feedback of the GISS GCM. Finally, the surface albedo feedback of the GFDL GCM is comparable to that of the GISS GCM and is on the low side of the RCM results.

A rigorous intercomparison of the quantitative values of the feedbacks for the GFDL and GISS GCMs is made difficult by the fact that the feedbacks of these models have been analyzed by two different methods, namely, the radiative-convective-model method for the GISS model and the radiative-transfer-model method for the GFDL model. This notwithstanding, the intercomparison indicates that, although these models simulate similar values of $[\Delta T_s]_{ave}$, they do so with values of their feedbacks which differ in both magnitude and sign, this despite the approximate agreement of their simulated values of $T_s$ for the $1xCO_2$ climate.

## Conclusion

In the Introduction we recommended that the quantitative analysis of the feedbacks in climatic-change simulations should be the first step of a systematic approach to answering the question: Why do models, even within the same category of the climate model hierarchy, simulate different climatic changes for the same forcing? To support this recommendation we have presented in this paper: 1) the concept and terminology of classical feedback analysis; 2) three different feedback analysis methods, one for radiative-convective models and two others for general circulation models; and 3) the application of these feedback analysis methods to the simulations of $CO_2$-induced climatic change.

The intercomparison of the results of these feedback analysis shows that the contribution of the individual feedback processes to the simulated climatic change is not the same for different models, even when these models are within the same category of the climate model hierarchy. This is a clear demonstration of the utility of the analysis of the feedbacks in climatic-change simulations.

To improve our understanding of the behavior of feedbacks, the feedbacks in the simulations of $CO_2$-induced climatic change obtained by the GISS, NCAR, OSU and UKMO (NCAR is the National Center for Atmospheric Research, Boulder, CO; OSU is Oregon State University, Corvallis, OR; and UKMO is the United Kingdom Meteorological Office, Bracknell, Berkshire) GCMs should be determined using the radiative-transfer-model method and compared with the feedbacks of the GFDL model which have been determined by this method. However, the correct application of this feedback-analysis method requires use of the changes in the net radiative fluxes at the tropopause, not at the top of the atmosphere.

After completion of these additional feedback analyses, the feedback processes can be ranked in terms of their contributions to the simulated climatic changes. Then it will be time to take the second step in the systematic approach, namely, the intercomparison of the parameterizations of the highest-ranked processes among themselves, with more-detailed models, and with observations.

*Acknowledgments.* I thank John Mitchell for his constructive review of this paper. This study was supported by the U.S. National Science Foundation and the Carbon Dioxide Research Division, Office of Basic Energy Sciences of the U.S. Department of Energy under grant ATM 87-12033.

## References

Bode, H. W., Network Analysis and Feedback Amplifier Design, Krieger, New York, 577 pp., 1975.

Cess, R. D., and G. L. Potter, A commentary on the recent $CO_2$-climate controversy, *Climatic Change, 6*, 365-376, 1984.

Dickinson, R. E., Convergence rate and stability of ocean-atmosphere coupling schemes with a zero-dimensional climate model, *J. Atmos. Sci., 38*, 2112-2120, 1981.

Hansen, J., A. Lacis, D. Rind, G. Russell, P. Stone, I. Fung, R. Ruedy and J. Lerner, Climate sensitivity: Analysis of feedback mechanisms, in *Climate Processes and Climate Sensitivity, Maurice Ewing Series, 5*, J. E. Hansen and T. Takahashi (Eds.), American Geophysical Union, Washington, D.C., pp. 130-163, 1984.

Manabe, S., and R. T. Wetherald, On the distribution of climate change resulting from an increase in $CO_2$ content of the atmosphere, *J. Atmos. Sci., 37*, 99-118, 1980.

Schlesinger, M. E., Feedback analysis of results from energy balance and radiative-convective models, in *The Potential Climatic Effects of Increasing Carbon Dioxide*, M.C. MacCracken and F.M. Luther (Eds.), U.S. Department of Energy, DOE/ER-0237, pp. 280-319, 1985. (Available from NTIS, Springfield, Virginia.)

Schlesinger, M. E., Equilibrium and transient warming induced by increased atmospheric $CO_2$, *Climate Dynamics, 1*, 35-51, 1986.

Schlesinger, M. E., Quantitative analysis of feedbacks in climate model simulations of $CO_2$-induced warming, in *Physically-Based Modelling and Simulation of Climate and Climatic Change*, M.E. Schlesinger (Ed.), NATO Advanced Study Institute Series, Reidel, Dordrecht, pp. 653-736, 1988a.

Schlesinger, M. E., Model Projections of the Climatic Changes Induced by Increased Atmospheric $CO_2$, in *NATO Climate/Geo-Sciences Symposium*, A. Berger and S.H. Schneider (Eds.), Kluwer Academic (in press), 1988b.

Schlesinger, M. E., and J.F.B. Mitchell, Model Projections of the Equilibrium Climatic Response to Increased Carbon Dioxide, in *The Potential Climatic Effects of Increasing Carbon Dioxide*, M.C. MacCracken and F.M. Luther (Eds.), U.S. Department of Energy, DOE/ER-0237, pp. 81-147, 1985. (Available from NTIS, Springfield, Virginia.)

Wetherald, R. T., and S. Manabe, Cloud feedback processes in a general circulation model. *J. Atmos. Sci., 45*, 1397-1415, 1988.